CSR, Sustainability, Ethics & Governance

Series Editors

In recent years the discussion concerning the relation between business and society has made immense strides. This has in turn led to a broad academic and practical discussion on innovative management concepts, such as Corporate Social Responsibility, Corporate Governance and Sustainability Management. This series offers a comprehensive overview of the latest theoretical and empirical research and provides sound concepts for sustainable business strategies. In order to do so, it combines the insights of leading researchers and thinkers in the fields of management theory and the social sciences – and from all over the world, thus contributing to the interdisciplinary and intercultural discussion on the role of business in society. The underlying intention of this series is to help solve the world's most challenging problems by developing new management concepts that create value for business and society alike. In order to support those managers, researchers and students who are pursuing sustainable business approaches for our common future, the series offers them access to cutting-edge management approaches.

CSR, Sustainability, Ethics & Governance is accepted by the Norwegian Register for Scientific Journals, Series and Publishers, maintained and operated by the Norwegian Social Science Data Services (NSD)

João Bento • Miguel M. Torres
Editors

Global Perspectives on Climate Change, Inequality, and Multinational Corporations

Springer

Editors
João Bento
Lisbon Accounting and Business School (ISCAL)
Polytechnic Institute of Lisbon (IPL)
Lisbon, Portugal

Miguel M. Torres
Kent Business School
University of Kent
Canterbury, UK

ISSN 2196-7075 ISSN 2196-7083 (electronic)
CSR, Sustainability, Ethics & Governance
ISBN 978-3-031-80796-1 ISBN 978-3-031-80797-8 (eBook)
https://doi.org/10.1007/978-3-031-80797-8

This Springer imprint is published by the registered company Springer Nature Switzerland AG
The registered company address is: Gewerbestrasse 11, 6330 Cham, Switzerland

Dedicated to our families

Foreword

Without doubt, the time for talking is over. Now is the time for action. But what action? In this volume João Bento and Miguel Torres assemble a convincing agenda, across a remarkably broad canvas, nested within a book series initiative to deliver on "the underlying intention [...] to help solve the world's most challenging problems" through "employing academic and practical discussion on innovative management concepts". In this collection, therefore, we find contributions that relate not only to the "big picture" aspirations such as are encapsulated within the United Nations' Sustainable Development Goals (SDGs) but also to micro-level activities that, if one takes the time to stop and think, are more likely within the grasp of most decision takers.

In reviewing this work, I find great appeal in its world view. It places inequality at the centre, not only of its title, but also at the centre of its ethos. Inequality is the bane of our world, both within nations and between nations. It incites fear in the hearts of those with the greatest wealth, with the most to lose, and despondency in those with the least. It promotes war and motivates the leaders of nations to pursue political manifestos of self-interest, just at the time when the world needs a manifesto for collective action.

The antidote to our seemingly global "death wish", and our best hope, may well be, as the editors and authors suggest, the multinational enterprise (MNE). Quite apart from the value-creating economic effects of foreign direct investment (FDI) poured over by researchers for decades and which, in principle, are pro-economic development, the MNE today has a new reason to merit our attention—its attraction in the context of the SDGs. For instance, the MNE has no vested interest in gender inequality. Well, from the advanced economies one would expect that. And one would expect that when investing in economies with weak commitment to gender equality (and all other forms of equality) the MNE would stand morally bankrupt if it were not to apply the same standards abroad as at home. Here, we put our faith in advanced economy investors but also in current innovations in investment quality indicators such as those for Environmental, Social and Governance (ESG). These are part of a movement to measure and manage sustainability risks and the opportunities presented to firms to outcompete rivals in terms of actually promoting sustainability, in all its guises. This is, to use a somewhat already hackneyed term, a potential "game changer". That the MNE should be the natural agent for the spread of game-changing radical innovation is hardly a surprise—the theory of the MNE emphasises the role of international business in accelerating the rate of technology diffusion. In principle, a good thing, especially when the technology has huge pro-welfare potential. Artificial Intelligence has this potential. While some might fear it could give with one hand and take away with the other (as with all MNE activity), it is worthy of debate—a debate kickstarted in this volume that usefully could be extended to all of the chapters, beyond those in which it already appears.

This volume is really all about transparency in, and knowledge of, practices relating to sustainability. We often change the way we think "if we only knew". Well, now we can glimpse a future in which we can see the consequences of our, and others', actions stripped of the rhetoric. MNEs that originate in the advanced economies almost invariably have operations in the developing economies. This, using the increasingly demanding yardsticks applied to investment, naturally produces a pressure to raise standards. The other way round, MNEs from developing economies, even if their investors are not international and subject to the same scrutiny (they may, indeed, be state-owned), nevertheless almost invariably wish to market their output in the advanced economies. If they do wish to do so, there are now more demanding and legitimate standards that can reasonably be expected of them, if they wish to benefit from a level playing field. Each of the contributions in this volume, therefore, aids transparency and the pro-welfare use of knowledge. And regarding the mission to "serve as a valuable guide for policymakers", the contributions collectively set out in plain sight the levers that legislators, and policy officials the world over, must heed. Taken together this volume gives us a language with which to better understand the key developments to achieving its overarching goal: "to help solve the world's most challenging problems".

Leeds University Business School
Leeds, UK

Jeremy Clegg

Preface

This book aims to inspire discussion, research, and the development of frameworks on the role of multinational enterprises (MNEs) in advancing the Sustainable Development Goals (SDGs) while offering insights into sustainable business practices. It provides valuable perspectives on the connections between climate change, economic inequality, and MNEs, featuring contributions from an international group of leading practitioners, researchers, and academics. Each chapter seeks to enhance our understanding of how MNEs can help reduce global inequality through ethical and sustainable business approaches.

We have selected some of the best papers for this edited volume. This publication collects 12 peer-reviewed papers, with the aim of sharing ideas and visions for a sustainable future and providing useful tips for academics, practitioners, and policymakers in the context of the 2030 Agenda for Sustainable Development of the United Nations.

The editors hope readers will find the book engaging and that its established networks will enable ongoing contributions towards fostering responsible business practices in challenging times.

Lisbon, Portugal — João Bento
Canterbury, UK — Miguel M. Torres

Global Perspectives on Climate Change, Inequality, and Multinational Corporations: An Introduction

The Sustainable Development Goals (SDGs) call for reducing inequalities within and among countries to create a fair society for all (Ghauri, 2022). There is a growing awareness of the need to consider that the inequalities to be addressed include poverty, climate change, environmental quality, and climate justice. Further research is needed on how multinational corporations (MNEs) can play a positive role in achieving these goals (Ghauri et al., 2018; Van Tulder et al., 2021; Kolk & Van Tulder, 2010; Van Zanten & Van Tulder, 2018).

We have sought manuscripts for this edited book as part of the Springer series CSR, Sustainability, Ethics, and Governance. We have accepted submissions that examine other SDGs whenever MNEs can play an essential role in implementing the 2030 Sustainable Development Agenda (United Nations, 2015). We have welcomed manuscripts that are conceptual or theoretical papers, literature reviews, qualitative or quantitative studies, and case studies. We have encouraged a comprehensive international representation that illustrates different sustainable business perspectives covering developing, emerging, and developed country contexts and multiple directions of international business flows. The potential non-exhaustive list of topics has included themes like ethics, philanthropy, corruption, transparency in economic transactions, advancement of human rights, innovation and other favourable spillover effects of MNEs; MNEs and local business interactions, promotion of educational opportunities, gender equality and access to technology for the disadvantages; MNEs and governance practices (code of good governance processes, quality reporting requirements, anti-corruption measures, and ethical business practices); challenges faced by MNEs to achieve SDGs and the role of international business in sustainable development; MNEs' modularity; business action on climate justice and climate change adaptation and mitigation.

This book serves as a valuable guide for policymakers. It deepens our understanding of how MNEs and their foreign direct investment activities can contribute to a more sustainable future and mitigate the climate crisis. This book aims to stimulate discussion, research, and frameworks on the role of MNEs in achieving SDGs and to provide insight into sustainable business practices. It adds valuable insights into the link between climate change, economic inequality, and MNEs by bringing together an international range of prominent practitioners, researchers, and academics. Each chapter aims to deepen our understanding of how MNEs can contribute to reducing inequalities through sustainable and ethical business practices. We hope you will enjoy and find the chapters useful and understandable.

The first chapter by Laura Emilia Hellqvist and Harald Heubaum, two scholars from the UK, is entitled "MNEs and the Advancement of SDGs in Developing Countries: A Gender Perspective on Energy Technologies". They argue that MNEs are players in advancing SDGs in emerging markets and developing economies, particularly in technological innovation and gender equity in accessing modern energy technologies. In Africa, MNEs often dominate the solar energy sector, raising questions about their overall impact on the continent's economic and social progress. While their support for innovative payment models and technological solutions such as pay-as-you-go (PAYGO) and M-PESA has helped close energy access gaps, a continued focus on market expansion and profit maximisation creates challenges for achieving equitable and sustainable energy for all. This underscores the urgent need for MNEs to adopt inclusive and equitable practices that drive technological and economic growth and ensure that benefits are widely distributed, particularly among women and underserved communities. Through an analysis of Kenya using the "innovation systems theory", this chapter critically examines the role of MNEs in advancing SDG 7 (Affordable and Clean Energy) to women in rural settings. The analysis underscores the importance of integrating local and gender perspectives into MNEs' strategies for advancing energy technologies, emphasising the need for inclusivity and social responsibility. A comprehensive incorporation of gender perspectives across all aspects of technological innovation and energy access considerations, from public policy development to the deployment of financial instruments and beyond, is needed if SDG7 is to be achieved.

The second chapter by İlayda İpek from Turkey, entitled "Green Business Strategies in MNE Research: Thematic Considerations and Methodological Features", deals with the burgeoning interest in environmental issues within the context of international business and academic research channelled into understanding the green business strategies of MNEs. The author systematically reviews and critically assesses the current knowledge of MNEs' green business strategies. For this purpose, this scholar employs a sample of 65 empirical articles extracted from 40 academic journals and published between 1995 and 2023. By doing this, we've gained valuable thematic and methodological insights into the specific characteristics of the related literature. In the chapter entitled "The Impact of MNEs on Ocean Sustainability", Arijit Goswani, a seasoned professional in the technology services and consulting sector in India, focuses on MNEs' Corporate Social Responsibility (CSR) towards producing goods and services but also on impacting local and distant ecosystems. Industrial activities often release chemical effluents into marine ecosystems via direct discharge and runoff. Fast fashion contributes 190 thousand tonnes of oceanic primary microplastics annually, and the fishing sector has overexploited 35% of fish stocks. Additional sources of marine pollution include oil spills, pharmaceutical discharges, semiconductor effluents, and shipping ballast water. Given their technical and financial capabilities, MNEs are responsible for conserving oceans and marine ecosystems. The environmental footprint of MNEs is crucial in achieving SDG 14. Various sectors, including automotive, fast fashion, fishing, manufacturing, oil and gas, pharmaceuticals, and shipping, have implemented technologies to treat wastewater, minimise water usage, and restrict the flow of toxic effluents into water streams that eventually reach the oceans. By deploying policy and technological interventions, MNEs can contribute significantly to the tenets of SDG 14. The role of MNEs in this domain depends on meaningful collaboration with public authorities, non-governmental organisations, international bodies, and civil society, along with accountability and transparency in reporting their environmental impact.

Lane Hakan, Charles Oakley, Tanzeem Hossain, Arijit Goswami, Rajan Kumar Sah, and Mohit Mishra, academics from Germany, the United States, India and Nepal, present their chapter on "MNEs, Climate Change, and Heart Disease: AI for Global Health Equity". Climate change is considered one of the most threatening factors to the health of the human population. Many impacts are attributed to climate change through direct calamities in extreme weather and natural disasters and prolonged exposure to heat, bacteria, vector-borne diseases, and multiple other pathways. The exposure to and the consequences of these health risks are often unequally distributed among people from different regions and socio-economic backgrounds. Using Artificial Intelligence (AI) constitutes a solution to address health inequities globally. It makes it possible to deliver quality health care virtually to communities where physical facilities are lacking, along with analysing disease trends and patterns to predict and prepare for future healthcare needs in certain populations. This approach can contribute to tailored treatment plans and preventive strategies through electronic health records, understanding the diversity in health needs and mitigating disparities in healthcare outcomes. Personalised medicine, a potentially transformative application, leverages AI to scrutinise individual health data, tailoring treatment plans and preventive strategies. This approach ensures a new understanding of diverse health needs, mitigating disparities in healthcare outcomes. AI's transformative impact extends to healthcare access and delivery, especially in regions with limited infrastructure. AI-driven chatbots and virtual assistants are accessible sources of basic healthcare information, offering critical support where traditional medical services are scarce. Their financial and technological resources give MNEs a fundamental role in equitably developing and introducing these services. The advancement of CSR and the largely untapped personal health market creates a strong business narrative, whether through direct business-to-customer sales or partnerships with health authorities and clinics. However, the widespread integration of AI in global health is not without challenges. Concerns regarding data privacy, limited technology access, and potential algorithmic biases pose significant hurdles. Despite these obstacles, noteworthy initiatives worldwide showcase AI's positive impact, ranging from AI-powered diagnostic tools for remote areas to predictive analytics optimising resource allocation. In navigating the delicate balance between innovation and ethical considerations, AI is a potent tool to drive health equity globally. The examples of successful AI applications underscore its transformative potential, compelling the case for continued exploration and integration in global healthcare frameworks. This study presents case studies from multiple continents and socio-cultural contexts. The authors examine how local communities may implement machine learning and other AI tools to ensure better access to healthcare and enable informed decisions for healthy lifestyle behaviours, specifically for combating non-communicable diseases (NCDs) like cardiovascular disease (CVD). Analysing large geographic datasets can help identify patterns and trends of CVDs in communities vulnerable to climate change and understand the risks of exposure while providing tools for prevention and treatment. The authors argue that these tools could be employed by policymakers, private citizens, or MNEs to identify measures to reduce the effect of the exposome on heart diseases.

In the fifth chapter, Suzanne Kieviet and Henriette Elsabe Scholtz from South Africa look at "Inequality in South African Banking: Executive Directors' Remuneration and Company Performance". Amidst growing income inequality in South Africa and recent scandals involving multinationals like Steinhoff and Tongaat Hulett, shareholders are increasingly scrutinising the size and structure of executive remuneration packages, particularly given the high levels of executive pay and excessive risk-taking by bank executives. Shareholders also demand greater transparency regarding the criteria and processes for awarding performance incentives. These questions are pertinent in the South African context where inequality, as measured by Gini coefficient, is amongst the highest in the world. Socio-economic factors affecting South Africa can significantly influence how MNEs contribute to reducing inequalities through sustainable and ethical business practices.

Moreover, research into the pay-performance relationship of executive directors in the banking industry is used to gain insight into how sustainable development goals can be achieved to reduce inequalities in South Africa. The scholars analyse panel data from listed banks' annual reports over 12 years to determine if there is a short- and long-term relationship between banks' executive directors' remuneration (EDR) and company performance in the South African banking sector. The detailed analysis considers market and accounting performance to measure company performance. Agency theory, optimal contract theory, and managerial power theory are used as the theoretical underpinning to consider the historical and current tendency in banks' executive directors' remuneration, including King IV and its "say-on-pay" provision. To address escalating EDR and to discourage short-termism focused solely on profit metrics, it is proposed that business educators instil a mindset of long-term, sustainable value creation among graduates. This study details some criticisms of EDR and associated disclosure thereof, followed by a discussion of previously proposed EDR reforms. The findings show a long-term relationship between EDR and company performance, but not between short-term and long-term EDR and company performance. The authors recommend that MNEs, the South African government, and the Institute of Directors South Africa should be cognisant of the socio-economic and political boundaries within which they operate when considering implementing international remuneration best practices locally. Furthermore, national development priorities should connect with international education agendas, reiterating education as the long-term solution to unemployment and reducing wage inequalities. The results of this study provide a clearer understanding of the socio-economic landscape and relationship between EDR and company performance in the South African banking sector. These insights can help develop sustainable business strategies to reduce inequalities.

The sixth chapter by Mesiet W. Kamihanda and Miguel M. Torres, scholars from Morocco and the UK, is entitled "Towards Internationalisation of African Banks: Strategy, Legitimacy, and Sustainability". In this chapter, they question how African MNE banks strategically address the United Nations' SDGs and what is the role of legitimacy and internationalisation performance in achieving those goals? They built a model to test the moderating role of legitimisation and internationalisation performance in adopting SDGs for a set of Moroccan Banks expanding into the West African Economic and Monetary Union. These scholars follow a case study methodology to three Moroccan banks. Data collected from the banks' published reports, coded data from interviews, and survey questionnaires point to the importance of legitimisation but do not support typical internationalisation strategies as significantly as those used during the banking expansion process.

Rifat Kamasak, Catherine Cai, and Aslihan Bozcuk, scholars from the UK and Turkey, present their chapter on "Red Tape Roadblocks on M&As: Emerging Economies' Regulators as Source of Inefficiency". The regulatory agencies are critical in addressing the competition and ethical concerns in M&A transactions. Thus, they should not prolong the duration of an M&A deal while performing their duties since adverse rulings by courts and regulatory agencies can deter deal completion and result in operational and financial consequences. Given the unique characteristics of emerging economies, the authors propose that regulators and antitrust authorities may operate differently than in developed economies, leading to extended deal durations. Drawing on Thomson SDC Mergers and Zephyr databases, they empirically show how regulatory agencies prolong the M&A deal duration of Brazilian acquirers. Their study finds a positive association between the involvement of regulatory agencies and prolonged M&A deal durations in emerging economies. However, the regulatory agencies of developed economies did not correlate with M&A deal durations. Accordingly, the results illuminate the inefficiencies of regulatory agencies, complex or sluggish bureaucratic structures, politics, and rent-seeking practices in emerging economies that may explain their findings.

In the eighth chapter, Shabnum Soomro, Jam Ghulam Murtaza Sahito, and Farahnaz Gilal take a look at "The Link Between Climate Change and Human Health". This is a relevant aspect for MNEs because it impacts the safeguarding of their operations and contributes to global efforts for sustainable development, ultimately ensuring their relevance and success in a rapidly changing world. These authors from Pakistan emphasise climate change's significant role in spreading infectious diseases, mainly through animals, mosquitoes, and ticks, impacting most human pathogenic diseases. Factors such as land use, warmer temperatures, and severe weather conditions contribute to this proliferation. Vector-borne diseases like malaria, dengue fever, and Zika virus are notably affected, resulting in approximately 700,000 deaths annually and constituting 17% of all infectious diseases. Diseases such as Zika, Lyme, malaria, and dengue fever are particularly susceptible to climate change, with mosquitoes and ticks serving as vectors for transmission. Extreme weather events, including heat waves and flooding, significantly influence disease transmission rates, contributing to a substantial increase in malaria cases and an estimated 249 million cases of cholera in 2022. Additionally, heat-related illnesses, exacerbated by human-induced climate change, have elevated mortality rates in individuals over 65, with symptoms such as nausea, muscle cramping, fatigue, and profuse sweating. The authors argue that the interconnectedness of air pollution and climate change poses severe environmental and health risks, with rising temperatures contributing to heat waves and wildfires worsening air pollution through increased greenhouse gas emissions.

The ninth chapter, entitled "Towards Understanding the Impact of a Refugee Crisis on MNEs: A Theoretical Essay in the Light of Hannah Arendt's Political Theory", is authored by Jenerton Arlan Schütz, Marcos de Moraes Sousa and Miguel Matos Torres, scholars from Brazil and the UK. In this chapter, they explore the issue of statelessness through the lens of Hannah Arendt's political theories, mainly focusing on her concept of the "right to have rights". Arendt's experiences as a stateless refugee inform her profound analysis of the loss of rights and identity experienced by stateless individuals. They examine the conditions under which statelessness arises, traditional solutions like repatriation and naturalisation inadequacies, and the broader implications for human rights in contemporary society. The chapter explores the diminishing role of public action and political engagement in addressing these issues, exacerbated by modernity's impact on these fundamental human activities. By linking Arendt's critique to the CSR practices of MNEs, the chapter highlights the potential of these entities to mitigate some of the adverse effects of statelessness. MNEs, leveraging their global influence and resources, can foster inclusive practices that provide stateless individuals with opportunities for social and economic integration. Through a critical analysis of Arendt's work and its relevance to today's globalised world, the chapter advocates for rethinking human rights frameworks and the role of corporate entities in promoting social justice and equity.

The tenth chapter by Hicham Nachit, Rachid Hachad, Saad Laraqui, and Miguel M. Torres, scholars from Morocco and the UK, entitled "Transforming Africa Byte by Byte: AI's Role in Achieving SDG7", examines the transformative potential of AI in advancing SDG 7 to ensure access to affordable, reliable, sustainable, and modern energy for all in Africa. It commences by delineating the evolving capabilities of AI and its critical importance within the contemporary sustainability-driven business context. The analysis investigates the confluence of AI and sustainability in Africa, particularly emphasising the renewable energy sector. A review of extant literature establishes a foundation for comprehending the global progression of AI technologies and their specific applications in this industry. The chapter elucidates Africa's distinctive challenges and opportunities and contrasts the continent's AI adoption rates in renewable energy with those of other regions. Central to this discourse is the Moroccan Agency for Sustainable Energy (MASEN) case study, which exemplifies the integration of AI into renewable energy strategies. This case study highlights AI applications in solar and wind energy optimisation, predictive maintenance, and grid management, demonstrating quantifiable benefits such as enhanced efficiency, cost reductions, and improved energy access. The chapter further proposes a strategic roadmap for AI-driven sustainability, underscoring the necessity for comprehensive AI strategies, policy recommendations, and international collaboration. It culminates with a call to action for stakeholders to engage in ethical AI development and a prospective outlook on integrating AI into sustainable practices, revolutionising Africa's energy landscape, byte by byte.

Ovidiu Bordean, Ioana Natalia Beleiu, Irina-Iulia Salanta, and Emil Lucian Crisan, scholars from Romania, authored the eleventh chapter entitled "The Role Played by Blockchain Technologies of Supply Chains in Fighting Inequality". Economic inequality between countries has been exacerbated to extreme levels in the last few years. The COVID-19 crisis has led to even greater differences in inequality that have precipitated long-term social and economic development. Reducing both within- and between-country inequality requires a joint effort on the part of multiple stakeholders. MNEs, as key actors, play a significant role in reducing inequality through various mechanisms and strategies, including the supply chains that they create to support the activities involved in producing and delivering goods and services to consumers. Supply chains have been involved recently in the digital transformation phenomenon, with blockchain technology being one of the most promising technologies in terms of economic benefits for all supply chain participants. Recent findings point to numerous benefits that blockchain technology may present to actors involved in SCs, such as transparency, trust, and security. However, there are a number of insights regarding how the use of blockchain technology in supply chains generates other outcomes, such as the ones related to sustainable goals. Based on a structured, systematic review that uses the Context-Intervention-Mechanism-Outcome (CIMO) analysis of 14 papers containing 17 cases, the authors have identified 5 primary mechanisms through which blockchain technology influences inequality in supply chains: access, transparency, safety, traceability, and regulation. Based on a critical evaluation, which considers both the functional and dysfunctional outcomes of such mechanisms, this study also sought to provide a promising research agenda for scholars that is likely to be highly important for policymakers, societies, and individuals.

In the last chapter, João Bento and Miguel M. Torres, scholars from Portugal and the UK, look at "Exploring the Relationship Between Democracy and the Environment: The Moderating Role of FDI". This chapter aims to examine the impact of five democracy varieties proxied by a composite democracy indicator on the relationship between FDI and the quality of the environment and to investigate the moderating role of democracy on the FDI-environment relationship. The authors present an empirical paper investigating the environmental effects of FDI and democracy using a panel dataset of 160 countries from 1980 to 2021 and instrumental variables within the framework of the GMM estimation technique. They address endogeneity and unobserved heterogeneity issues not considered in previous studies and use time-fixed effects. This study also examines the marginal effects of FDI at the mean and various percentile values of the democracy variables. The findings demonstrate that FDI significantly reduces environmental emissions as countries increase their level of democracy. Evidence shows that the interaction between FDI and democracy significantly improves environmental quality, supporting the pollution halo hypothesis. The mitigation effect is larger in countries that present higher levels of democracy and attract more FDI. While previous studies have focused mainly on how democracy facilitates the choice of FDI host-country locations, FDI, and democracy, resulting in environmental improvements, this study explores the interaction between FDI, democracy, and the natural environment. The authors argue that as long as a nation moves up the democratic ladder and attracts more FDI, the interplay between FDI and democracy will likely lead to environmental improvements in host countries.

The 12 chapters of the book from across the globe, in our view, have made immense contributions to improving our understanding of climate change, inequality, and MNEs and have set up ideas for improving our world and achieving the SDGs.

Lisbon Accounting and Business School (ISCAL) João Bento
Lisbon Polytechnical Institute (IPL)
Lisbon, Portugal

Kent Business School Miguel M. Torres
University of Kent
Canterbury, UK

References

Ghauri, P. (2022). The role of multinational enterprises in achieving sustainable development goals. *AIB Insights, 22*(1), 1–5.

Kanagaretnam, K., Lobo, G., & Zhang, L. (2022). Relationship between climate risk and physical and organizational capital. *Management International Review, 62*(2), 245–283.

Kolk, A., & Tulder, R. (2010). International business, corporate social responsibility and sustainable development. *International Business Review, 19*(2), 119–125.

Sachs, J. D., & Sachs, L. E. (2021). Business alignment for the "Decade of action". *Journal of International Business Policy, 4*(1), 22–27.

Stern, N. (2008). The economics of climate change. *American Economic Review, 98*(2), 1–37.

Tol, R. S. J. (2009). The economic effects of climate change. *Journal of Economic Perspectives, 23*(2), 29–51.

United Nations. (2015). *Transforming our world: The 2030 Agenda for Sustainable Development* (UN Doc. A/RES/70/1 (September 25, 2015)). United Nations Department of Economic and Social Affairs – Sustainable Development.

Van Tulder, R., Rodrigues, S. B., Mirza, H., & Sexsmith, K. (2021). The UN's sustainable development goals: Can multinational enterprises lead the decade of action? *Journal of International Business Policy, 4*(1), 1–21.

Wright, C., & Nyberg, D. (2017). An inconvenient truth: How organizations translate climate change into business as usual. *Academy of Management Journal, 60*(5), 1633–1661

Acknowledgements

The authors would like to extend their gratitude for the valuable comments received at the International Conference on Economics, Finance and Business, organised by the International Institute of Social and Economic Sciences (IISES) in Prague, held from 11 to 13 September 2023, as well as at the European Trade Study Group Annual Conference (ETSG 2024) at the University of Athens, held from 12 to 14 September 2024. Valuable insights were also gained during the Summer School at the Lisbon School of Economics and Management (ISEG) from 22 to 25 July 2024, in the course titled "The Economics of Inequality and Public Policy," led by Professor Niels Johannesen from Oxford University. This work has been financially supported by the Research Unit on Governance, Competitiveness and Public Policy (GOVCOPP) and national funds through The Foundation for Science and Technology and the Ministry of Science, Technology and Higher Education in Portugal (UIDB/04058/2020) + (UIDP/04058/2020).

Contents

MNEs and the Advancement of SDGs in Developing Countries: A Gender Perspective on Energy Technologies 1
Laura Hellqvist and Harald Heubaum

Green Business Strategies in MNE Research: Thematic Considerations and Methodological Features 29
İlayda İpek

The Impact of MNEs on Ocean Sustainability 45
Arijit Goswami

MNEs, Climate Change, and Heart Diseases: AI for Global Health Equity 77
Hakan Lane, Charles J. Oakley, Tanzeem Hossain, Arijit Goswami, Rajan Kumar Sah, and Mohit Mishra

Inequality in South African Banking: Executive Directors' Remuneration and Company Performance 109
Suzanne Kieviet and Henriette Elsabe Scholtz

Towards Internationalisation of African Banks: Strategy, Legitimacy and Sustainability 129
Mesiet W. Kamihanda and Miguel M. Torres

Red Tape Roadblocks on M&As: Emerging Economies' Regulators as a Source of Inefficiency 155
Catherine Cai, Rifat Kamasak, and Aslihan Bozcuk

The Link Between Climate Change and Human Health 183
Shabnum Soomro, Jam Ghulam Murtaza Sahito, and Farahnaz Gilal

Towards Understanding the Impact of a Refugee Crisis on MNEs: A Theoretical Essay in the Light of Hannah Arendt's Political Theory 209
Jenerton Arlan Schütz, Marcos de Moraes Sousa, and Miguel M. Torres

Transforming Africa Byte by Byte: AI's Role in Achieving SDG7 227
Hicham Nachit, Rachid Hachad, Saad Laraqui, and Miguel M. Torres

The Role Played by Blockchain Technologies in Supply Chains in Fighting Inequality 261
Ovidiu Niculae Bordean, Ioana Natalia Beleiu, Irina-Iulia Salanţă, and Emil Lucian Crişan

Exploring the Relationship Between Democracy and the Environment: The Moderating Role of FDI 283
João Bento and Miguel M. Torres

About the Editors and Contributors

About the Editors

João Bento, PhD in Economics, is a senior lecturer in economics at Lisbon Accounting and Business School (ISCAL), Polytechnic Institute of Lisbon (IPL). He is also an external affiliate at the Competitiveness, Innovation and Sustainability (CIS) research group within the research unit on Governance, Competitiveness and Public Policies (GOVCOPP) at the University of Aveiro (UA), Portugal. He has more than 20 years of experience in empirical research, academic knowledge, and expertise in economics. He holds a PhD in Economics from the University of Reading, UK. He works and has published in the areas of international business, tourism and environmental economics.

Miguel M. Torres, PhD in Economics, is a Lecturer in International Business at Kent Business School (KBS). His career includes consulting services to the European Commission and World Bank, applied research and teaching roles at Leeds, Milano, Aveiro, KU Leuven, and Uppsala universities, with appointments at the University Federal of Goiás, IF Goiano, and the University Catolica of Portugal. He has held multiple visiting professorships and is an INSOL International and Marie Skłodowska-Curie Fellow. He completed executive programs at Harvard, Carnegie Mellon University (USA), and University College London (UK). His research focuses on the intersection of international business, public policy, and strategic management, focusing on multinational management models, corporate political risk, and negotiation strategies, particularly emphasizing the complexities of cross-border restructuring and foreign direct investment drivers using cross-sectional, time-series, and panel data approaches. His work has been published in the *International Business Review, Journal of Business Ethics, Multinational Business*

Review, Journal of Economic Methodology, Arbitration International, International Insolvency Review, PLOS One, and AIB Insights. He co-edited the books *Global Perspectives on Climate Change, Inequality, and Multinational Corporations (Springer, 2025) and Dead Firms: Causes and Effects of Cross-Border Corporate Insolvency* (Emerald, 2017).

Contributors

Ioana Natalia Beleiu Babeş-Bolyai University, Faculty of Economics and Business Administration, Cluj-Napoca, Romania

João Bento Lisbon Accounting and Business School, Polytechnical Institute of Lisbon, Lisbon, Portugal

Ovidiu Niculae Bordean Babeş-Bolyai University, Faculty of Economics and Business Administration, Cluj-Napoca, Romania

Aslihan Bozcuk Akdeniz University, Antalya, Turkey

Catherine Cai University of the West of England, Bristol, UK

Emil Lucian Crişan Babeş-Bolyai University, Faculty of Economics and Business Administration, Cluj-Napoca, Romania

Marcos de Moraes Sousa Federal University of Goiás, Goiânia, Brazil

Farahnaz Gilal Department of Agricultural Economics, Singh Agriculture University, Tando Jam, Sindh, Pakistan

Arijit Goswami Markets and Markets, Bengaluru, Karnataka, India

Rachid Hachad Hassan II University of Casablanca, Casablanca, Morocco

Laura Hellqvist School of Oriental and African Studies, University of London, London, UK

Harald Heubaum School of Oriental and African Studies, University of London, London, UK

Tanzeem Hossain University of Freiburg, Freiburg im Breisgau, Germany

İlayda İpek Faculty of Business, Dokuz Eylül University, İzmir, Turkey

Rifat Kamasak University of Reading, Reading, UK

Mesiet W. Kamihanda UEMF, Business School, Fès, Morocco

Suzanne Kieviet Stellenbosch University, Stellenbosch, South Africa

Hakan Lane Johannes Gutenberg University Mainz, Mainz, Germany

Saad Laraqui ESCA Ecole de Management, Casablanca, Morocco

Mohit Mishra Outpayce from Amadeus, Bengaluru, Karnataka, India

Hicham Nachit ESCA Ecole de Management, Casablanca, Morocco

Charles J. Oakley Liberty University, Lynchburg, VA, USA

Jam Ghulam Murtaza Sahito Department of Agricultural Economics, Singh Agriculture University, Tando Jam, Sindh, Pakistan

Rajan Kumar Sah Tribhuvan University, Kathmandu, Nepal

Irina-Iulia Salanță Babeș-Bolyai University, Faculty of Economics and Business Administration, Cluj-Napoca, Romania

Henriette Elsabe Scholtz Stellenbosch University, Stellenbosch, South Africa

Jenerton Arlan Schütz Catholic University of Brasília, Brasília, Brazil

Shabnum Soomro Department of Agricultural Economics, Singh Agriculture University, Tando Jam, Sindh, Pakistan

Miguel M. Torres Kent Business School, University of Kent, Canterbury, UK

MNEs and the Advancement of SDGs in Developing Countries: A Gender Perspective on Energy Technologies

Laura Hellqvist and Harald Heubaum

1 Introduction

Taken together, the Paris Agreement on Climate Change and the United Nations (UN) Sustainable Development Goals (SDGs) are key initiatives guiding the international development agenda, setting out an ambitious and comprehensive plan for achieving sustainable, low-carbon, and climate-resilient development (UNFCCC, 2024). An integrated, strategic approach is essential for addressing climate change's extensive and complex effects on global socio-economic progress, with energy production and consumption responsible for about 90% of carbon dioxide (CO_2) emissions (Xu et al., 2023; Wei et al., 2023). The global energy transition sits at the heart of such an approach.

SDG 7 aims to ensure access to affordable, reliable, sustainable and modern energy for all and serves as a crucial objective to guide the transition, complete with indicators to track progress toward the Paris Agreement's objective of limiting the rise in global average surface temperatures to below 2 °C above pre-industrial levels (Wei et al., 2023). However, the achievement of SDG7 is challenged by inadequate energy access in the Global South, restricting both human and economic development (Paraschiv & Paraschiv, 2023; Woollacott et al., 2023; Pearl-Martinez, 2020).

Currently, 759 million people globally lack electricity access, while 2.8 billion experience unreliable energy services (Alda-Vidal et al., 2023). Off-grid solar (OGS) systems are often the most practical solution for delivering energy access to rural communities where extending the grid is excessively costly and inefficient (Wagner et al., 2021). However, a substantial funding and adoption gap persists in renewable energy sectors, and investment commitments continue to favour

L. Hellqvist (✉) · H. Heubaum
School of Oriental and African Studies, University of London, London, UK
e-mail: 311473@soas.ac.uk; hh15@soas.ac.uk

J. Bento, M. M. Torres (eds.), *Global Perspectives on Climate Change, Inequality, and Multinational Corporations*, CSR, Sustainability, Ethics & Governance,
https://doi.org/10.1007/978-3-031-80797-8_1

large-scale projects at the expense of smaller-scale renewable solutions that are vital for rural areas (Hermanus & Scholtz, 2018). It is perhaps unsurprising that progress toward achieving SDG 7 has plateaued across the Global South (WHO, 2022), and the anticipated energy transition remains tenuous (Wei et al., 2023).

Decarbonising energy systems offers significant socio-economic advantages to local communities by driving economic growth with reduced environmental impact (Gielen et al., 2019; Meirun et al., 2021; Kim & Wilson, 2019; Feng & Zheng, 2022). However, such a transition will not occur in isolation; the success of the SDGs crucially depends on the proactive engagement of multinational enterprises (MNEs) and their investment in these transformative technologies (van Tulder et al., 2021). MNEs are recognised as drivers of economic growth, technology transfer, and job creation in developing economies, playing a critical role in integrating them into the global market and supporting SDGs through inclusive growth and sustainable practices. At the same time, they are central to global innovation, significantly influencing research and development by facilitating the international exchange and integration of technological knowledge and expertise across borders (Kolk & van Tulder, 2010).

MNEs play an important role in enhancing energy access and promoting sustainable development, particularly in Kenya and the broader Global South. Many MNEs now address global challenges such as poverty, environmental degradation, and social inequality, extending sustainable development beyond environmental concerns to include vital social and economic dimensions necessary for holistic economic growth (Kolk, 2016; Montiel et al., 2021). This engagement is often driven by the need to align operations with global sustainability standards and to maintain legitimacy in diverse and dynamic environments. Drawing on their ability to foster innovation and create scalable solutions, MNEs can leverage their resources and capabilities to address energy poverty and promote sustainable economic development while achieving competitive advantage (Kolk & van Tulder, 2010; Scherer et al., 2013).

Establishing an MNE presence in the solar energy sector is crucial for advancing SDG 7 in developing countries, especially in rural areas. Through strategic investments in infrastructure, technology transfer, and capacity building, MNEs contribute to reducing the reliance on fossil fuels and enhancing social outcomes (Scherer et al., 2013). By providing reliable and affordable energy access, MNEs help improve living standards and economic opportunities in underserved communities (Montiel et al., 2021). Integrating sustainability concerns into innovation systems and transitioning to sustainable development are essential steps in this process (Scherer et al., 2013).

Innovation systems (IS) theory (IST) offers a robust framework for understanding the societal impacts of MNE-driven innovations and the environments that foster socio-economic development and achievement of the SDGs (Weber & Rohracher, 2012; Edquist, 1997). IS thinking sheds light on the interplay between institutions, policies, and socio-economic dynamics, highlighting systemic interactions among stakeholders—including governments, businesses, academia, and civil society—in fostering a conducive environment for technological advancements. It is particularly insightful for understanding the development, dissemination, and adoption of innovative technologies (such as renewables) and practices that drive economic growth while minimising environmental impact.

In Fig. 1, the cyclical relationship between the actors represents the interactions and influences between the different components in the IS, with *Sustainability Concerns* placed at the centre, highlighting the interconnected nature of these systems and their collective focus on sustainability. The interlinkages among MNEs, local small- and medium-sized enterprises (SMEs), the government, academia, civil society, and financial institutions create a dynamic and integrated system that fosters innovation and sustainability.

Yet while MNEs have begun to shift from a sole focus on profit maximisation to integrating broader societal, economic, and environmental impacts into their business operations, reflecting a broader trend towards adopting sustainable business

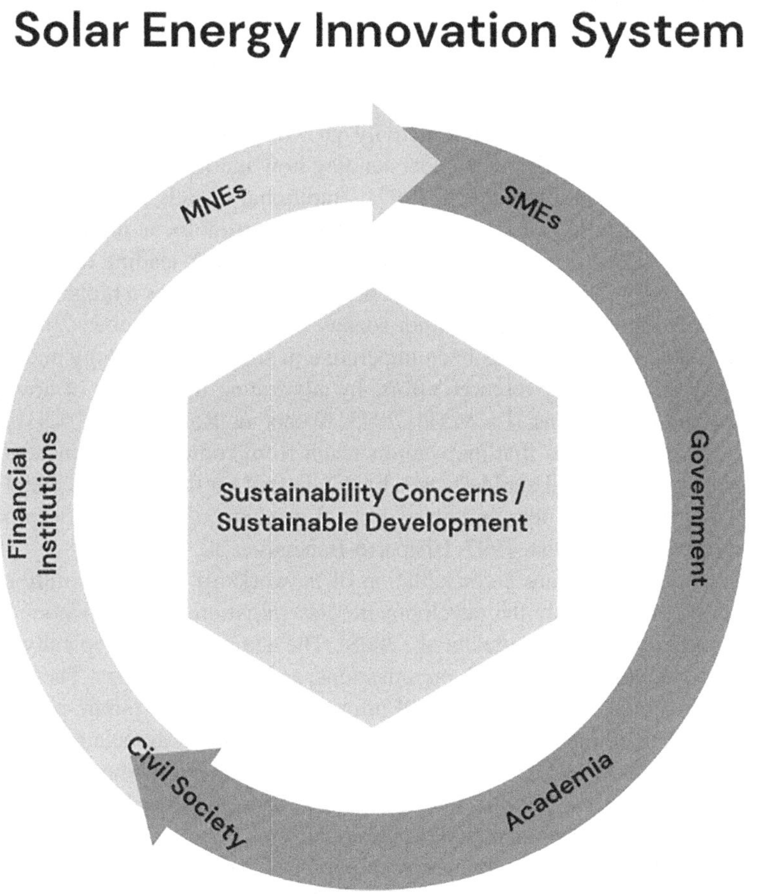

Fig. 1 Integration of sustainability concerns and sustainable development into solar energy innovation system. (Source: Authors' elaboration)

practices aligned with the SDGs, a number of significant challenges remain. First, cultural norms and gender biases hinder women's participation in the energy sector (Montiel et al., 2021). Second, MNEs also often focus on profitable urban markets, neglecting rural areas (Johnstone, 2020). Finally, the significant gap in after-sales services for solar products undermines the further diffusion of solar energy access (Gebreslassie et al., 2022). Financial models like pay-as-you-go (PAYGO) still leave many rural households unable to afford solar home systems (SHS) (Barry & Creti, 2020), and the reliance on external funding risks undermining the long-term sustainability of the approach (Sanyal et al., 2016). Additionally, digital infrastructure, literacy issues, and privacy concerns further complicate adopting these technologies (Baker, 2023).

2 Leveraging IST to Drive MNEs' Strategies for Advancing SDG 7 in Kenya

There is almost universal agreement that technological advancements and innovations are key contributors to productivity growth and greater economic wealth, hence the critical importance of understanding how innovations and technological change occur in societies (Edquist, 1997). Traditional models in evolutionary neoclassical economics have been criticised for their narrow focus on market failures and pursuing economic growth targets (Smith et al., 2010), leading to the development of IS theory. This alternative theoretical perspective offers a broader, systemic view of how innovations occur within society (Weber & Rohracher, 2012). Since the late 1980s, IS thinking has been imperative in shaping technology and innovation policy and directing research efforts by advocating the scaling of novel technologies and innovations (Li & Ge, 2023; Weber & Rohracher, 2012; Edquist, 2009). The theory posits that innovations result from complex, non-linear interactions among various stakeholders over time, centring on the emergence and diffusion of knowledge and learning around novel products and production processes (Fagerberg, 2009; Edquist, 1997; Chaparro-Banegas et al., 2023).

Innovation systems are a constellation of networks of actors, institutions, and interactions that facilitate the development, dissemination, and application of new products and processes (Bergek et al., 2008). The actors involved typically include businesses, universities, research organisations, and industry groups. These entities are interconnected through formal and informal networks like public-private partnerships (PPP), commercial relationships, and professional associations (Edquist, 1997; Breschi & Malerba, 1997). The institutional framework supporting innovation comprises cultural norms, laws, regulations, and established routines essential for technological development and progress.

Since its inception, IS thinking has evolved from models primarily focused on economic growth to addressing broader societal challenges. At first, IS theory aimed to enhance the institutional environment for innovation and improve methods to bolster business innovation capabilities to drive economic growth (Edquist, 1997). However, it has adapted to emerging challenges such as climate change and the urgent need for sustainable development through technological leapfrogging. This adaptation involves integrating niche approaches and broader concepts, revealing diverse mechanisms and processes that foster innovation across various contexts (Weber & Truffer, 2017; Markard et al., 2012).

Initially conceptualised on a national scale, IS thinking has expanded to include technological, regional, sectoral, and global perspectives. While initially centred on the industrialised economies of the Global North, such as the European Union and the Organisation for Economic Co-operation and Development (OECD) (Edquist, 1997), recent developments in IS theory have broadened its applicability to the Global South, underscoring its importance as a tool for addressing global challenges, in developed as much as developing and emerging markets (Hayashi, 2020; Szogs, 2008). New interpretations have further broadened the theory's scope, emphasising sustainability transitions and energy sector transformation while exploring alternative pathways for innovation (Szogs, 2008; Kim & Wilson, 2019; Sovacool et al., 2023). This shift is reflected in various strands of IS research, shaping policy and research in innovation globally (Larsen, 2019). The role of IS thinking in adapting to environmental imperatives has become increasingly significant in guiding the transition to renewable energy technologies and emphasises the importance of a multidisciplinary approach to address climate and energy challenges, particularly in the Global South (Bergek et al., 2015; Negro et al., 2008), amid the urgent need for sustainability transitions towards a net-zero, climate-resilient future.

By leveraging insights from IS thinking, MNEs can align their strategies with broader sustainable development goals, especially SDG 7, which targets affordable and clean energy—an urgent need in developing countries where energy access is a barrier to economic development and social equity (Montiel et al., 2021). The IS approach provides a valuable understanding of how innovations impact societies and identifies environments that foster innovation essential for socio-economic development and achieving the SDGs (Hoang et al., 2021; Danilda & Thorslund, 2011; Li & Ge, 2023; Niaz, 2022).

3 IST for Energy Access Among Kenyan Women

The systems of innovation approach is invaluable for examining the energy access landscape in Kenya as it extends beyond a sole focus on economic growth. It facilitates the inclusion of economic, institutional, organisational, social, and political factors, making it an interdisciplinary tool for evaluating some of the barriers to reaching universal energy access (Edquist, 1997)—a perspective enabling a more nuanced understanding of how innovations diffuse and impact various aspects of society (Hudson et al., 2011; Urmetzer & Pyka, 2019), such as gender, which is a crucial yet often overlooked element in energy innovation (Anditi et al., 2022). Decarbonisation efforts and sustainability transitions will have a greater chance of success if they account for local social, cultural, and organisational factors and characteristics (Stern et al., 2016; Borozan, 2022), especially when addressing challenges like expanding energy access to energy-deficient, rural off-grid areas or for particular groups, such as women.

Climate change impacts men and women differently (Reggers, 2019; Johnson et al., 2020); women are especially vulnerable to both short-term extreme weather events, such as floods and cyclones, and the long-term consequences of environmental change, such as land erosion, droughts, sea level rise, and salinity intrusion into soil and water (Md et al., 2022; Eastin, 2018). They further entrench persistent gender inequalities and negatively impact women's food and water security, health, and socio-economic development (Chowdhury et al., 2022).

Without sustainable, reliable, and accessible energy, women depend on natural resources for heating, lighting, and cooking (Md et al., 2022). By instead relying on larger OGS systems, such as SHSs, the time previously spent gathering wood, dung, and crop residue can be used for income generation, education, childcare, and social activities (Chowdhury et al., 2022). They enable women to keep businesses open longer, increasing income-generating opportunities and further improving their socio-economic status (Md et al., 2022; Khandker et al., 2014). Additionally, laws across a number of developing countries preventing women from playing a bigger role in the workforce and restricting them to traditional roles at home expose them to greater health risks stemming from the burning of kerosene, diesel and other high-polluting energy sources (Muzammil & Ahmed, 2019; Khandker et al., 2014). Nearly 2.4 billion women are denied equal economic opportunity; such a disparity in economic rights confines women to hazardous domestic environments (World Bank, 2022); thus, replacing polluting forms of energy with OGS reduces air pollutants, offering significant health benefits for women and children (Opiyo, 2019).

Integrating sustainability concerns into innovation systems and transitioning to sustainable development is essential. IS thinking supports the knowledge-based, evolutionary, and complex nature of innovation processes, helping MNEs navigate the multifaceted challenges associated with sustainable energy transitions. By integrating cutting-edge technological advancements and innovative practices into their operations, MNEs can enhance their capacity for clean energy innovation and sustainability. Engaging with local communities and stakeholders through this framework ensures that clean energy solutions are tailored to meet local needs and conditions, fostering social acceptance and long-term sustainability (Lindner et al., 2016), as illustrated in Fig. 2.

As a key stakeholder group, women benefit significantly from solutions tailored to local characteristics and socio-economic conditions, addressing immediate social needs and enhancing community resilience (Adaptation Fund, 2022). MNEs should ensure that the design and implementation of OGS projects and the associated technological solutions consider local women's specific needs and characteristics. MNEs can advance SDG 7 by promoting gender equality through policies that provide equal pay, career development opportunities, and supportive work environments for women. Additionally, MNEs can empower women by investing in education, supporting women entrepreneurs, and implementing inclusive business

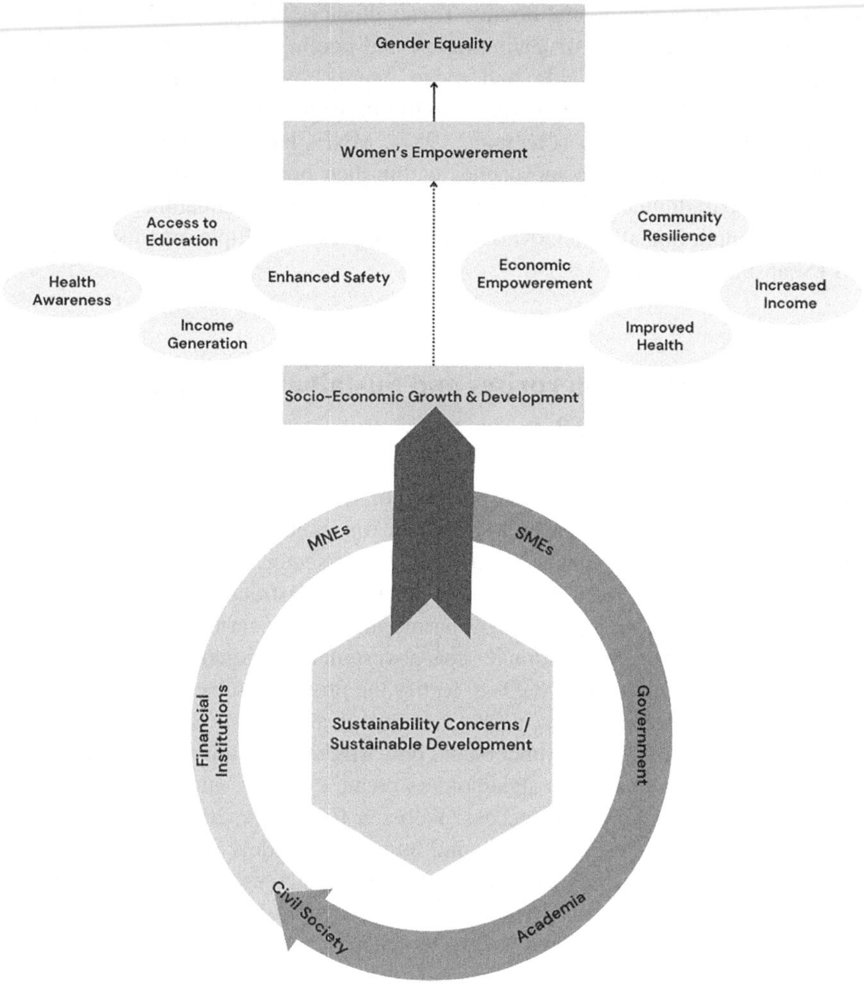

Fig. 2 Innovation system interlinkages driving sustainable development and gender equality. (Source: Authors' elaboration)

practices (Montiel et al., 2021). The integration of gender perspectives in innovation is crucial as women in rural Kenya are disproportionately affected by energy poverty. Adopting such an approach enables the creation of inclusive and sustainable energy solutions that provide electricity and empower women economically and socially, driving broader socio-economic development in rural areas (Danilda & Thorslund, 2011).

Moreover, IS frameworks encourage the inclusion of a broad range of stakeholders, from policymakers to local communities, ensuring that innovations are both technologically sound and socially equitable. Through collaborative networks, MNEs can form strategic alliances with other companies, including SMEs, research institutions, and governmental bodies. These alliances facilitate the sharing of best practices, co-development of technologies, and pooling of resources, accelerating the deployment of affordable and clean energy solutions. This collaborative approach helps MNEs overcome barriers to innovation and scale their sustainable initiatives more effectively (Freeman, 1987). MNEs have systematically adopted comprehensive and holistic approaches within their operational frameworks, illustrating a paradigmatic shift towards integrating sustainability and ESG (Environmental, Social, and Governance) principles at multiple organisational levels (Shapsugova, 2023).

4 Multinational Enterprises and Sustainable Energy Access Initiatives in Kenya

The IS approach emphasises the importance of systemic interactions among stakeholders in creating a conducive environment for technological advancements and sustainable practices (Edquist, 1997). In the solar sector, MNEs can drive energy access by investing in local projects that develop infrastructure to deliver reliable, clean energy to remote areas and by partnering with governments, multinational development banks, international non-governmental organisations and non-governmental organisations (NGO) to fortify the provision of services and technology in the local solar energy market (Arthur et al., 2024). By applying the IS theoretical lens, we can better understand how MNEs' investments in local projects and partnerships with various stakeholders create systemic interactions that drive innovation and enhance energy access (Weber & Rohracher, 2012).

In Kenya, MNEs have had a profound impact on the development of the energy sector. Since 2000, Kenya's electrification rate has increased dramatically, from a national access rate of 15.1% to the current 85%, with rural areas showing remarkable progress—79% now have access to energy, mainly through OGS systems (Pueyo & DeMartino, 2018; Winther et al., 2018; Roche & Blanchard, 2018). The rapid increase in Kenya's electrification rate can be attributed to the effective innovation systems established by MNEs, which facilitate the development and dissemination of renewable energy technologies (Bergek et al., 2008). The energy sector in Kenya has grown substantially through an increased focus on renewable resources like solar, wind, and geothermal, accounting for over 46% of the country's energy mix (Heinemann, 2022). Investments in these energy technologies, particularly OGS systems in rural regions, have significantly bolstered efforts to achieve universal energy access (Winther et al., 2018; Roche & Blanchard, 2018).

MNEs have been instrumental in advancing Kenya's solar sector through innovative mechanisms such as M-PESA and PAYGO solar systems, making energy more affordable and accessible to rural populations (Carr-Wilson & Pai, 2018). They have contributed to energy access dissemination by importing advanced technologies and investing in community development projects, such as large-scale solar power plants in northern parts of the country, which enhance regional power supplies and reduce reliance on non-renewable sources. Such projects also ensure continuous access to essential services by installing solar panels in schools and healthcare centres (Arthur et al., 2024). Additionally, MNEs' substantial investments have strengthened infrastructure development and established robust supply chains for solar products, attracting international funding and fostering cooperation through PPPs, contributing to the sector's overall growth (Sergi et al., 2018). The deployment of OGS systems by MNEs has facilitated income generation and poverty reduction by enhancing business performance and creating new economic opportunities for micro and small enterprises in rural areas (Pueyo & DeMartino, 2018). The substantial presence of MNEs, combined with donor funding and financing in the Kenyan solar market, has propelled Kenya to become the largest market for OGS systems globally.

While these large companies have helped drive Kenya to become the largest market for OGS systems globally (Ketelaars & Wheeldon, 2021; Volkert & Klagge, 2022), SHSs creating thousands of jobs along the entire value chain, questions remain about where the overall value and profit are retained (Johnstone, 2020). This suggests that while MNEs may contribute to economic activities and job creation, the overall value and profit generated often do not remain within the local economy, thereby limiting the long-term potential developmental benefits for the host country (Brandl et al., 2021). However, the strong presence of MNEs in the Kenyan solar sector has led to the introduction and certification of good-quality OGS systems and enhancing the overall quality of solar systems by primarily importing systems and components and setting standards for appliance quality, overseen by Kenya Bureau of Standards, thereby improving the reliability of OGS systems, fostering competition and sectoral growth, propelling sales (Galan, 2022; IRENA, 2018). The widespread adoption and effectiveness of OGS in rural areas are the result of systemic interactions and technological advancements, with high standards and competitive market growth promoted by functioning innovation systems leading to the successful introduction and certification of high-quality solar systems by MNEs (Edquist, 1997; Bergek et al., 2008).

Most OGS renewable energy systems sold on the market primarily include SHSs and smaller systems, including pico-lights (Blimpo & Cosgrove-Davies, 2019), ranging from 0.5 to 100 watts (W). Pico lanterns produce around 0.5–3 watt-peak (Wp) (Khan, 2020), while SHSs range between 11 and 100 Wp (Blimpo & Cosgrove-Davies, 2019; Khan, 2020). For instance, a 20 Wp solar system can power one light bulb, whereas a 75 Wp system can support up to five light bulbs (Khandker et al., 2014). The prices of SHSs vary depending on the system size and energy output level (Boliko & Ialnazov, 2019). In 2019 alone, sales included nearly 5 million Pico PV systems and 704,000 SHS (Galan, 2022). Despite earlier challenges faced in the OGS sector, such as financial missteps from a significant 1998 International Finance Corporation project, the participation of new multilateral investors like the World Bank has continued to strengthen the solar energy sector (Boliko & Ialnazov, 2019).

4.1 The Role of PPPs in Electrifying Rural Areas

Global interconnectivity has further heightened the interconnected role of states and the private sector in forming partnerships that leverage private expertise, resources, and efficiencies to extend energy access to financially challenging rural areas with government support (Azarian et al., 2023; Tijani, 2015; Baskaran & Coste, 2024). This synergy between international linkages and localised action underlines the effectiveness of PPPs. Entities like the World Bank enable international collaborations that deliver public services and infrastructure while also navigating challenges such as high costs and variable success in terms of private investment and social impact (Bayliss & Van Waeyenberge, 2017; Uzuegbunam, 2024; Fabre & Straub, 2021). PPPs can facilitate collaborative learning and innovation by fostering knowledge exchange, mobilising resources, and creating supportive institutional frameworks, leading to sustainable energy solutions in rural areas (Weber & Rohracher, 2012).

In the decade to 2022, the average debt ratio across sub-Saharan Africa doubled to 60% of GDP (IMF, 2023), with Kenya ranking above this average at 68.4% (Focus-Economics, 2023). An unsustainable debt burden coupled with liquidity problems has presented a challenge to Kenya's energy sector buildout and has increasingly shown the spotlight on PPPs as a way to mobilise greater infrastructure investment. Building on previous legislation, the 2021 Public-Private Partnership Act facilitates greater participation by the private sector in the financing, construction and operation of critical infrastructure and development projects such as power stations and transmission lines (Kenya Gazette, 2021).

The PPP model has enabled a risk transfer between public and private stakeholders, including independent power producers (IPPs) in geothermal and solar PV, the state-owned Kenya Power and Lighting Company (KPLC) and Kenya Electricity Transmission Company (KETRACO). In the case of geothermal energy, such as the 105-MW Menengai project or the 155-MW Olkaria III project, the public sector invested in the riskier upstream elements, while private MNEs took on the capital costs associated with plant construction, operation and maintenance (ThinkGeoEnergy, 2024). The private investment was further facilitated by risk insurance and a power purchasing agreement (PPA) between the private developers and the Kenyan government to ensure a return on investment (Africa Energy Portal, 2020). Such systemic interactions and collaborative efforts between public and private sector parties foster innovation, leading to effective risk management and enhanced energy infrastructure (Weber & Rohracher, 2012).

In the solar sector, the World Bank has been instrumental in shaping the OGS market through initiatives like the Lighting Kenya programmes, which have combined public and private investment and dramatically expanded access to solar energy across the country. Commencing with Lighting Kenya I project, part of the Lighting in Africa initiative, operational from 2009 to 2013, which accelerated the distribution of solar lanterns from 29,000 to 680,000 units, achieving an economic return rate as high as 2000%, the programme transitioned to Lighting Kenya II, from 2013 to 2018 further fostering the PAYGO business model as a delivery method for increased solar access dissemination (World Bank, 2011).

These efforts have been bolstered by substantial financial injections, such as USD 150 million for the Kenya Off-grid Solar Access Project (KOSAP) for Underserved Counties and USD 630 million including the Last Mile Connectivity Program, aiming for universal energy access by 2020 (Keller, 2018). Projects such as KOSAP have increased rural electrification rates by deploying SHSs in 14 rural counties with high poverty rates and minimal grid coverage (Harrington & Wambugu, 2021). Funded with a considerable budget, KOSAP underscores a PPP commitment to improving energy access among economically disadvantaged populations, addressing broader social and economic challenges by creating networks that foster innovation and enhance energy access by facilitating knowledge exchange and resource mobilisation (Inclusive Infrastructure, 2020). Despite regulatory challenges and the absence of subsidies for additional electrical appliances, these large-scale projects have boosted market penetration to 69% (Harrington & Wambugu, 2021; Bauer, 2019; Boliko & Ialnazov, 2019).

4.2 *Rural Electrification: M-PESA and the PAYGO Model*

Traditional energy solutions often fall short in rural areas due to infrastructural weaknesses, limited reach of the national grid, and high upfront costs associated with conventional energy projects. Against this backdrop, technological innovations, notably fintech mobile banking and solar technology solutions, have revolutionised rural electrification in the Global South, including Kenya. These innovations result from systemic interactions among various actors, including technology developers, financial institutions, and end-users, creating a conducive environment for innovation and adoption (Palm, 2022). MNEs, through foreign direct investment, loans, and grants, have catalysed the expansion of Kenya's solar sector. Significant investments from international donors, such as a USD 700 million investment from UN-led development agencies, have fueled technological advancements and innovations in the sector (IEA, 2021). Key investors like UK Aid, Azuri Technologies, Shell Foundation, and Safaricom have directly supported the development of PAYGO mobile solutions, which have been pivotal in rural solar energy dissemination (Busara, 2022).

PAYGO and M-PESA are some of Kenya's most noteworthy native innovations that have fostered financial inclusion and electricity access even in remote areas. The introduction of these solutions has revolutionised traditional banking by facilitating microfinance solutions and alternative financial transactions for the rural, cash-poor populations and has also empowered rural women by enhancing financial independence and access to essential energy services, promoting a shift towards gender-inclusive financial services (Ndung'u, 2017; Mazzoni, 2019; Kim, 2022; Ndiaye, 2019). These digital solutions support rural households in securing loans and credit for energy solutions have enabled customers to use mobile money to make small, regular payments for OGS systems, matching households' regular kerosene expenses (Winther et al., 2018), underscoring the transformative potential of digital finance in promoting sustainable energy development (Hossain et al., 2023; Khan et al., 2019; Aslam et al., 2018; Sergi et al., 2018).

M-PESA, launched in 2007 by a leading mobile network operator Safaricom, serves as a mobile money platform and an access point to formalised financial services for those without access to traditional banking services (Van Hove & Dubus, 2018; Pueyo & DeMartino, 2018; Costa & Ehrbeck, 2015). Safaricom is a mobile network giant, 40% owned by UK-based Vodafone and 30% owned by the government of Kenya (Baker, 2023). The wide adoption and spread of M-PESA is due to Safaricom's collaboration with 25 banks and over 700 businesses since 2009 (Ndung'u, 2018; Baker, 2023). The PAYGO model, similar to the M-PESA, allows payments to be made via mobile phones (Boliko & Ialnazov, 2019), having enabled automated control and plug-and-play systems in the OGS market, which has been crucial for unlocking the solar PV market in Kenya.

The utilisation of PAYGO towards SHSs began in 2012, driven by high mobile-money penetration; by 2016, 80% of Kenya's population had a mobile telephone (Baker, 2023; Chengo et al., 2019). To set up a PAYGO SHS contract, customers need a mobile banking account, like M-PESA, an identification card, and an upfront payment of 10–20% of the total equipment price. After installation, users make small, regular top-up payments in local currency (Barry & Creti, 2020). By spreading an SHS's otherwise prohibitively high upfront costs, PAYGO enables low-income energy users, especially those unable to access credit from the retail banking sector, to afford an SHS they could not previously pay for in cash. Over 90% of SHSs operate using a lease-to-own model, meaning that once the system cost is repaid, generally over one to three years, depending on the system size, the device is permanently unlocked, granting ownership to the consumer (IRENA, 2020). However, most SHSs are monitored and managed remotely by providers using hardware that allows activation upon receipt of payment or remote locking or disconnection in case of payment default, and the system is disconnected if payments are not made (Barry & Creti, 2020; Baker, 2023).

Solutions such as PAYGO and M-PESA offer a flexible and accessible platform for companies managing a large pool of dispersed energy customers (Adwek et al., 2020). They also provide significant value for mobile network operators through their established partnerships with SHS companies, which have been instrumental in introducing new off-grid electricity solutions while simultaneously expanding their customer base (REN21, 2020), as most Kenyan households use these digital services for periodic payments toward utilities and OGS systems (Boliko & Ialnazov, 2019). By 2015, these services had reached 84% of Kenyans living below USD 2 per day (Costa & Ehrbeck, 2015), highlighting their importance for the rural poor. This trajectory has persisted; by 2018, PAYGO accounted for 40% of all verified sales in the OGS market. Furthermore, the introduction of PAYGO has almost wholly replaced previous financing mechanisms, such as hire purchase schemes and cash payment models, which had limited success in extending rural electricity access (Lighting Global et al., 2020; Rolffs et al., 2015).

The synergy between PAYGO solar systems and M-PESA exemplifies the intersection of digital finance and sustainable energy, highlighting the adaptability of fintech-driven business models. These systems are expected to shape the future of the OGS sector, with PAYGO models driving current and future SHS sales (Mazzoni, 2019; Winther et al., 2018; Mawdsley, 2018; Chamarande et al., 2024). The innovations result directly from MNEs driving this development through substantial investments and strategic partnerships.

4.3 Challenges and Limitations of the PAYGO Model

The PAYGO model has shown potential for positive economic impacts by making sustainable energy accessible to low-income households and promoting economic activities by enabling local entrepreneurs to enter the renewable energy market, creating jobs, and fostering local business growth. However, while the PAYGO model offers numerous benefits, it also presents significant challenges. These systems often target consumers who can maintain regular payments, potentially excluding the poorest individuals who struggle even with small payments. Muchunku et al. (2018, p. 6) noted that PAYGO SHS in Kenya primarily serves households with incomes between USD 6 and USD 40 per day, not the poorest of the society that requires the first access to energy the most. Moreover, existing research indicates that many PAYGO SHS companies have shifted from providing small-scale systems for the energy-poor to larger systems for wealthier customers and productive users. This shift focuses more on profit than achieving SDG7, with many systems now out of reach for low-income households, especially in rural areas (Barry & Creti, 2020; Groenewoudt et al., 2020). Potentially perpetuating a new form of energy poverty, where the most economically vulnerable remain disconnected from essential energy services (Winther et al., 2018).

In Kenya, the impact of PAYGO SHS on energy access has been limited. Most sales are to customers who meet credit thresholds and live in urban or already electrified areas, using SHSs as a backup to an unreliable grid (Boamah, 2020; Ulsrud, 2020). The success of PAYGO systems depends on robust digital infrastructure, which is not uniformly available across all rural areas. Additionally, issues of digital literacy further impede the adoption and effectiveness of these systems (Baker, 2023). While the introduction of fintech solutions and artificial intelligence algorithms by MNEs has enhanced the deployment of SHSs in underserved regions by optimising installation locations and tailoring services to rural needs (Heinemann, 2022), these advancements also bring challenges. According to IS theory, the effectiveness of these technological solutions depends on the systemic interactions among various stakeholders, including technology providers, financial institutions, regulators, and local communities (Edquist, 1997).

There are concerns related to equity and ethics, such as the potential oversight of remote or impoverished communities and significant privacy risks from handling sensitive consumption data. The need for continuous data flow for operational efficiency raises severe privacy and security concerns, with potential dangers of exploitative practices such as data mining and surveillance without adequate consumer protection (Baker, 2023). As MNEs drive this type of energy access development and construct associated technological environments, implementing robust safeguards to ensure ethical practices and protect consumer rights is crucial. By focusing on the innovation system's systemic interactions, networks and feedback mechanisms, IS thinking can inform strategies to create more inclusive and resilient energy solutions (Weber & Rohracher, 2012). There is limited research specifically on the effects of PAYGO on women, leaving the positive or negative impacts currently unknown. More comprehensive studies are needed to fully understand these dynamics and develop strategies that mitigate the risks while enhancing the benefits for all community members, particularly women. These issues could reinforce existing inequalities and create dependency on advanced digital infrastructures (Ockwell et al., 2019).

5 Navigating Sustainable Energy Solutions: Balancing Market-Driven Approaches and Socio-economic Inclusivity

While market-driven digital approaches, such as the PAYGO model, have rapidly expanded energy access, they raise questions about sustainability and long-term impacts. Critics argue that these models prioritise profit over comprehensive energy access, potentially sidelining the poorest or those in more remote areas. This creates a paradox where greater energy access is provided but not necessarily for all, especially those suffering from extreme poverty. The introduction of the Rural Electrification Authority (REA) by the Ministry of Energy (MoE) in 2004 marked targeted efforts to enhance rural electrification through market-driven strategies and privatisation, aligning with global trends supported by international donors (Chamarande et al., 2024; Sergi et al., 2018). Despite efforts, achieving universal energy access by 2030 remains challenging, prompting project timeline extensions and strategy adjustments (Capital Business, 2022). The World Bank's Gender Action Plan and the African Development Bank's strategies emphasise economic empowerment, focusing on generating job opportunities and enhancing energy access for women and marginalised groups (World Bank, 2017). The KOSAP initiative aims to improve women's access to solar products and their involvement in the renewable energy sector, highlighting a concerted effort to address gender disparities. However, implementation faces substantial challenges, such as high costs and economic burdens in impoverished areas, necessitating a balance between enhancing accessibility and managing financial impacts on vulnerable communities (Phillips et al., 2020; GOGLA, 2021). For women in Kenya, these financial burdens can exacerbate existing inequalities, making it difficult for them to afford the initial costs of solar products despite the long-term savings on energy costs.

Additionally, the reliance on external finance and the financialization of energy access introduce long-term sustainability concerns, particularly if external economic conditions deteriorate. To create a competitive market for SHS provision, financing OGS systems should be leveraged from local sources from the outset, allowing sectoral knowledge to develop. This approach helps prevent the market from collapsing once donor support is withdrawn and emphasises the importance of not only focusing on profit-making. There can be significant consequences if subsidy regimes are not controlled, as uncontrolled subsidies can destabilise the market (Sanyal et al., 2016).

The reliance on PAYGO systems and M-PESA has facilitated innovative energy access but must be carefully managed to prevent exacerbating socio-economic disparities, especially among rural women. While of great value to financial transactions, M-PESA requires vigilant oversight to prevent financial exclusion or over-indebtedness among vulnerable populations (Bateman et al., 2019; Mas & Morawczynski, 2009). Kenya's advancements in OGS systems have been substantial. However, achieving universal energy access and equitable economic development requires addressing the risks associated with high transactional frequency and small debt accumulations through M-PESA, which pose challenges for rural populations and women (Mbiti & Weil, 2013). Critically, MNEs deepening the dependence and market penetration of systems such as PAYGO risks creating new cycles of debt for consumers, linking energy payments to financial vulnerabilities, which can impact creditworthiness and economic stability if consumers fail to keep up with payments, cutting off their energy access and further entrenching socio-economic challenges (Bateman et al., 2019; Baker, 2023).

To address these challenges and ensure sustainable transitions in energy access, it is essential to consider the role of digitalisation and its impacts on the innovation system. Digital solutions, when managed correctly, can drive innovation while balancing profit motivations with broader socio-economic inclusivity and SDGs (Satalkina & Steiner, 2020). This is particularly important for Kenyan women, as initiatives like KOSAP aim to improve women's access to solar products and their involvement in the renewable energy sector. However, high costs and economic burdens in impoverished areas necessitate a balance between enhancing accessibility and managing financial impacts on vulnerable communities. By fostering energy solutions that are both economically viable and socially equitable, digital innovations can ensure that these initiatives effectively empower women and marginalised groups in Kenya.

6 Small- and Medium-Sized Enterprises to the Rescue?

The technological advancements in solar power—such as improved photovoltaic cells and more efficient batteries—have complemented fintech innovations. Companies can now offer more reliable and durable SHSs, suitable for the varying demands of rural households. This synergy between mobile banking and solar technology has not only facilitated energy access but also spurred a plethora of new business opportunities. MNEs and SMEs alike are exploring new service models and market segments, ranging from basic lighting to powering small appliances, significantly enhancing daily living and economic activities in rural areas. These technological innovations have fundamentally altered the landscape of rural electrification by shifting from traditional, centralised energy systems to decentralised, consumer-driven models.

In their development of OGS solutions, however, larger MNEs often neglect rural areas due to a lack of bankable customers, focusing instead on more profitable urban markets. Local SMEs are better positioned to address the specific needs of rural communities, enhancing energy access for those who need it most. Empowering these smaller enterprises can ensure that the benefits of energy access from renewable sources such as solar reach the entire population, fostering sustained economic growth and improving living standards in underserved areas (Johnstone, 2020).

Due to their crucial role in providing comprehensive after-sales services and being vital to the supply chain, SMEs are pivotal in the OGS sector. After-sales services are essential for the dissemination of solar products as they address the significant barrier of maintaining continuous customer support post-purchase, which is often overlooked by companies focused solely on the technical aspects of their products (Anditi et al., 2022; Shuy, 2023). This service gap highlights the need for a robust local presence to handle issues like system repairs and customer education, which are critical for solar initiatives' long-term success and user satisfaction. SMEs facilitate the entire lifecycle of solar products—from distribution and installation to ongoing maintenance—ensuring that rural consumers have reliable support when technical issues arise.

This involvement is crucial for innovation diffusion and knowledge sharing. Importantly, without these services, customers might revert to traditional energy sources like kerosene or biomass, undermining the climate and environmental benefits of renewable energy solutions (Fingleton-Smith, 2018; Biswass et al., 2022). Therefore, the comprehensive involvement of SMEs helps ensure that solar energy remains a viable and reliable resource for rural populations, directly contributing to the broader goals of energy access and sustainability (Heinemann, 2022; Cross & Murray, 2018). However, while the cooperation between MNEs and SMEs can support the expansion of OGS in Kenya's rural communities, the lack of inclusion of women remains a particular challenge that is holding back further progress. Addressing these barriers is crucial for achieving broader socio-economic benefits and sustainable development goals (Women's World Banking, 2020).

7 Impact of Gender-Focused Strategies on Energy Access: Empowering Change

The innovation systems approach determines 'the system' in which innovation activities take place, such as the development, diffusion, and use, including all principal, political, social, economic, organisational, institutional, and other factors that might not be recognised a priori (Edquist, 1997). This approach is particularly relevant for addressing MNEs' challenges in spreading energy access to rural areas and the inclusion of women. Empowering SMEs can help ensure the benefits of energy access from renewable sources such as solar reach the entire population, including women, fostering sustained economic growth and improving living standards in underserved areas (Johnstone, 2020).

The challenges of achieving gender inclusivity within the operations and internally within MNEs are complex and deeply rooted in global and local socio-economic conditions. MNEs, while uniquely positioned to promote gender inclusivity across borders, often face barriers such as cultural norms that undervalue women's work, lack of access to education and vocational training for women, and insufficient legal protections against gender discrimination. Gender bias in recruitment, promotion, and remuneration practices persists, undermining efforts to achieve gender parity (Montiel et al., 2021). These challenges are further exacerbated in the energy sector, where women's access to education, employment, and entrepreneurship has been historically limited (Clark, 2021; Pearl-Martinez, 2020).

Particular attention should be given to women as a consumer group because they are most affected by energy scarcity and environmental degradation. In low-income countries, significant gender gaps and inadequate access to clean energy hinder women's access to labour-saving technologies, limiting girls' education as they spend most of their time on household chores. Energy poverty denies women a basic standard of living and an escape from drudgery. Besides poverty, factors such as lack of credit, land ownership rights, decision-making power, and prevailing social and cultural contexts also hinder women's access to energy services.

Policies and economic instruments for sustainable energy projects have primarily focused on financial terms, market forces, and technological advances, often neglecting consumers' socio-cultural and socio-economic profiles. This oversight has resulted in poor adoption in rural areas of developing nations, particularly impacting women. Most rural electrification projects use gender-neutral approaches, risking the neglect of the specific energy needs of rural women. Energy access programmes in developing countries must adapt to the unique energy poverty situations, emphasising the inclusion of rural consumer characteristics and addressing the specific needs of women.

To support gender-inclusive energy access, policy recommendations for MNEs and governments include prioritising investments in renewable energy projects that consider women's specific needs and roles in local communities. Policies should encourage the participation of women in the renewable energy sector, from decision-making processes to employment and entrepreneurship opportunities. Given that only 13% of women hold well-paid positions in STEM fields and 16% in non-STEM technical roles, increasing women's technical skills through educational programmes in renewable energy technologies is vital (IFC, 2022; Moss-Racusin et al., 2021). In collaboration with MNEs, governments should implement regulatory frameworks that ensure women's equal access to land and credit, facilitating their involvement in renewable energy initiatives (Montiel et al., 2021).

MNEs must adopt more comprehensive and culturally sensitive strategies to promote gender inclusivity within their global operations. Incorporating gender perspectives into MNE operations requires a strategic and holistic approach. Best practices include conducting gender impact assessments for all new projects to understand their effects on gender equality, establishing gender-inclusive recruitment and career advancement policies, and implementing mentorship and training programmes specifically designed to empower women. Furthermore, fostering a workplace culture that actively challenges gender stereotypes and supports work-life balance is crucial. MNEs can also leverage their influence to advocate for gender-responsive policies and practices within their supply chains and in the countries where they operate. Alternative motivations beyond profit-making are essential for the successful spread of energy access projects. Ensuring these projects and programmes are tailored to local characteristics, particularly those affecting women, is critical for their adoption and sustainability. By aligning gender-focused strategies with the broader goals of the innovation system, stakeholders can ensure that energy access solutions are not only technologically sound but also socially equitable, driving sustainable development and economic growth in rural areas. This alignment integrates sustainability concerns into all innovation system components, including MNEs, SMEs, public policies, academic institutions, civil society, and financial institutions—fostering systemic interactions and collective efforts towards inclusive and sustainable energy access.

8 Conclusion

Multinational enterprises are critical in advancing SDGs in the Global South, particularly through their substantial investments in innovative, renewable energy technologies. In Kenya, their support for OGS systems has significantly contributed to the solar sector's growth. PAYGO models and mobile banking solutions such as M-PESA have facilitated the widespread adoption of SHSs offered through KOSAP, enhancing financial inclusion and providing sustainable energy access to remote areas. Kenya's national electrification rate has surged from 15% in 2000 to 85% today, with rural areas achieving a remarkable 79% access primarily through OGS systems.

However, while enhancing energy access through innovative payment models and technological solutions, MNEs also face the challenge of balancing market expansion and profit maximisation with the achievement of equitable and sustainable energy for all. This is especially true in Kenya's rural areas, which are often neglected due to a lack of bankable customers. Here, the relationship between MNEs and local SMEs and their ability to facilitate the entire lifecycle of solar products, from distribution and installation to ongoing maintenance, holds the key to addressing the specific needs of rural communities.

Additionally, MNEs need to pay closer attention to the needs of rural women if equitable energy access is to be achieved. The introduction of PAYGO models and mobile banking has revolutionised energy access, enabling women, in particular, to afford SHSs through microfinance solutions. Yet key findings from the Kenyan context highlight the challenges MNEs continue to face, such as cultural barriers and gender bias, underscoring the need for policies that encourage women's participation in the renewable energy sector. Here, too, empowering local SMEs is crucial for addressing the specific needs of rural women.

The study acknowledges several limitations. Firstly, while highlighting the significant role of MNEs in advancing SDGs, it does not fully address the potential negative impacts of market-driven approaches on local economies and the environment. The focus on profit maximisation by MNEs may sometimes conflict with achieving equitable and sustainable energy access. Furthermore, the study primarily centres on the Kenyan context, which may limit the generalisability of the findings to other regions with different socio-economic and cultural conditions. Future research should explore these dynamics in a broader range of settings to develop a more comprehensive understanding of the interplay between sustainability, innovation systems, and gender inclusivity.

Some practical applications of IS theory include identifying and leveraging local strengths to enhance regional development. For instance, fostering collaboration among regional actors—such as local SMEs, academic institutions, and civil society organisations—can lead to more tailored and effective energy solutions that address specific local needs. In Kenya, regional strategies integrating local SMEs into the energy supply chain can improve the sustainability and effectiveness of renewable energy projects. Moreover, promoting environmentally and socially sustainable innovations requires incorporating SDGs into national and regional innovation policies. This can be achieved by establishing regulatory frameworks that encourage MNEs to adopt sustainable business practices and invest in clean energy technologies. In Kenya, policies that mandate gender impact assessments and support women's participation and technical skills development in the renewable energy sector can ensure that energy innovations are both inclusive and sustainable. A symbiotic relationship between MNEs and emerging and developing markets can be achieved by strategically aligning MNE operations with SDG targets, emphasising inclusivity and social responsibility.

Policymakers and stakeholders must prioritise inclusive policies that ensure equitable distribution of benefits from renewable energy projects, particularly for women and underserved communities. MNEs should leverage their resources to foster innovation, support local SMEs, and advocate for gender-responsive policies. By championing these initiatives, MNEs can enhance their competitiveness and sustainability while significantly contributing to achieving global sustainable development goals.

References

Adaptation Fund. (2022). *A study on intersectional approaches to gender mainstreaming in adaptation-relevant interventions*. Retrieved from https://www.adaptation-fund.org/wp-content/uploads/2022/02/AF-Final-Version_clean16Feb2022.pdf

Adwek, G., et al. (2020). The solar energy access in Kenya: A review focusing on pay-as-you-go solar home systems. *Environment, Development and Sustainability, 22*, 3897–3938.

Africa Energy Portal. (2020). *Kenya: Construction of the Menengai geothermal power plant is completed.* Retrieved from https://www.africa-energy-portal.org/news/kenya-construction-menengai-geothermal-power-plant-completed

Alda-Vidal, C., Khalid, R., Foulds, C., Royston, S., & Greene, M. (2023). Gender imaginaries in energy transitions: How professionals construct and envision gender equity in energy access in the Global South. *World Development, 168*, 106258. https://doi.org/10.1016/j.worlddev.2023.106258

Anditi, C., Musango, K. J., Smit, S., & Ceschin, F. (2022). Addressing gender dimensions in energy innovations: A gender analysis framework for informal urban settlements in Africa. *Energy Research & Social Science, 88*, 102476. https://doi.org/10.1016/j.erss.2021.102476

Arthur, B., Saha, M., Sarpong, F. A., & Dutta, K. D. (2024). Unlocking Africa's potential: The transformative power of foreign direct investment for sustainable development. *Heliyon, 10*(5), e26507. https://doi.org/10.1016/j.heliyon.2024.e26507

Aslam, A., Eugster, J., Ho, G., Jaumotte, F., & Piazza, R. (2018, April). Globalization helps spread knowledge and technology across borders. *IMF Blog*. Retrieved from https://www.imf.org/en/Blogs/Articles/2018/04/09/globalization-helps-spread-knowledge-and-technology-across-borders. Accessed 14 Feb 2024.

Azarian, M., Shiferaw, A. T., Stevik, T. K., Lædre, O., & Wondimu, P. A. (2023). Public-private partnership: A bibliometric analysis and historical evolution. *Buildings, 13*(8), 2035. https://doi.org/10.3390/buildings13082035

Baker, L. (2023). New frontiers of electricity capital: Energy access in sub-Saharan Africa. *New Political Economy, 28*(2), 206–222. https://doi.org/10.1080/13563467.2022.2084524

Barry, M., & Saliou, C. A. (2020). Pay-as-you-go contracts for electricity access: Bridging the "last mile" gap? A case study in Benin. *Energy Economics, 90*, 104843. https://doi.org/10.1016/j.eneco.2020.104843

Baskaran, G., & Coste, S. (2024). *Achieving universal energy access in Africa amid global decarbonization. CSIS | Center for Strategic & International Studies.* https://www.csis.org/analysis/achieving-universal-energy-access-africa-amid-global-decarbonization

Bateman, M., Duvendack, M., & Loubere, N. (2019). Is fin-tech the new panacea for poverty alleviation and local development? Contesting Suri and Jack's M-Pesa findings published in science. *Review of African Political Economy, 46*(161), 480–495. https://doi.org/10.1080/03056244.2019.1614552

Bauer, L. (2019). *Power isn't water: Learnings from Kenya's rural electrification efforts.* Blum Centre for Developing Economies. Retrieved from. https://blumcenter.berkeley.edu/power-isnt-water-learnings-from-kenyas-rural-electrification-efforts/

Bayliss, K., & Van Waeyenberge, E. (2017). Unpacking the public-private partnership revival. *Journal of Development Studies, 54*, 577. https://doi.org/10.1080/00220388.2017.1303671

Bergek, A., Jacobsson, S., Carlsson, B., Lindmark, S., & Rickne, A. (2008). Analyzing the functional dynamics of technological innovation systems: A scheme of analysis. *Research Policy, 37*(3), 407. https://doi.org/10.1016/j.respol.2007.12.003

Bergek, A., Hekkert, M., Jacobsson, S., Markard, J., Sandén, B., & Truffer, B. (2015). Technological innovation systems in contexts: Conceptualizing contextual structures and interaction dynamics. *Environmental Innovation and Societal Transitions, 16*, 51–64. https://doi.org/10.1016/j.eist.2015.07.003

Biswass, R., Sharmin, A., Islam, M. A., & Ashaduzzaman, M. (2022). Consumption of biomass fuels by rural households and estimation of their organic carbon in the Khulna region of Bangladesh. *Research Square*. https://doi.org/10.21203/rs.3.rs-1964354/v1

Blimpo, M. P., & Cosgrove-Davies, M. (2019). *Electricity access in sub-Saharan Africa: Uptake, reliability, and complementary factors for economic impact* (Africa Development Forum series). World Bank. https://doi.org/10.1596/978-1-4648-1361-0

Boamah, F. (2020). Emerging low-carbon energy landscapes and energy innovation dilemmas in the Kenyan periphery. *Annals of the American Association of Geographers, 110*(1), 145–165. https://doi.org/10.1080/24694452.2019.1644990

Boliko, C. M., & Ialnazov, D. S. (2019). An assessment of rural electrification projects in Kenya using a sustainability framework. *Energy Policy, 133*, 110928. https://doi.org/10.1016/j.enpol.2019.110928

Borozan, D. (2022). Asymmetric effects of policy uncertainty on renewable energy consumption in G7 countries. *Renewable Energy, 189*, 412–420. https://doi.org/10.1016/j.renene.2022.02.055

Brandl, K., Moore, E., Meyer, C., & Doh, J. (2021). The impact of multinational enterprises on community informal institutions and rural poverty. *Journal of International Business Studies, 53*, 1133–1152. https://doi.org/10.1057/s41267-020-00400-3

Breschi, S., & Malerba, F. (1997). Sectoral innovation systems: Technological regimes, Schumpeterian dynamics, and spatial boundaries. In C. Edquist (Ed.), *Systems of innovation: Technologies, institutions and organisations* (Science Technology and the International Political Economy Series) (pp. 130–156). Pinter.

Busara. (2022). *Evaluating and maximizing the impact of PayGo on lives and livelihoods*, Final Report. Shell Foundation. Retrieved from https://shellfoundation.org/app/uploads/2022/08/PayGo-Final-Internal-Report-V3-1.pdf

Capital Business. (2022). Firms receive Sh390 million to sell solar units in 14 counties. *Article*, published June 28th. Retrieved from https://www.capitalfm.co.ke/business/2022/06/firms-receive-sh390-million-to-sell-solar-units-in-14-counties/. Accessed 28 Feb 2024.

Carr-Wilson, S., & Pai, S. (2018). Pay-as-you-go: How a business model is helping light millions of rural Kenyan homes with solar. *CSE, 2*, 1–10. https://doi.org/10.1525/CSE.2017.000737

Chamarande, T., Etienne, E., & Mathy, S. (2024). Sizing isolated mini-grids in Kenya: Risk transfer to deal with multidimensional uncertainties and constraints. *Renewable and Sustainable Energy Transition, 5*, 100078. https://doi.org/10.1016/j.rset.2024.100078

Chaparro-Banegas, N., Mas-Tur, A., & Roig-Tierno, N. (2023). Driving research on eco-innovation systems: Crossing the boundaries of innovation systems. *International Journal of Innovation Studies, 7*, 218. https://doi.org/10.1016/j.ijis.2023.04.004

Chengo, V., et al. (2019). *Understanding the sustainable development prospects of mobile-enabled solar home systems in Kenya*. Climate Resilient Economies Working Paper no. 006/2019. Available from: https://www.ash-net.org/wpcontent/uploads/2019/06/Working-Paper-_Understanding-the-sustainable-development-prospects-of-mobile-enabled-solar-home.pdf

Chowdhury, M. A., Hasan, M. K., & Islam, S. L. U. (2022). Climate change adaptation in Bangladesh: Current practices, challenges and the way forward. *The Journal of Climate Change and Health, 6*, 100108. https://doi.org/10.1016/j.joclim.2021.100108

Clark, L. (2021, May). Powering households and empowering women: The gendered effects of electrification in sub-Saharan Africa. *Journal of Public & International Affairs*. Retrieved from https://jpia.princeton.edu/news/powering-households-and-empowering-women-gendered-effects-electrification-sub-saharan-africa

Costa, A., & Ehrbeck, T. (2015). A market-building approach to financial inclusion. *Innovations, 10*(1–2), 53.

Cross, J., & Murray, D. (2018). The afterlives of solar power: Waste and repair off the grid in Kenya. *Energy Research & Social Science, 44*, 100. https://doi.org/10.1016/j.erss.2018.04.034

Danilda, I., & Thorslund, G. J. (Eds.). (2011). *Innovation & gender*. Vinnova. Edita Västra Aros AB.

Eastin, J. (2018). Climate change and gender equality in developing states. *World Development, 107*, 289–305. https://doi.org/10.1016/j.worlddev.2018.02.021

Edquist, C. (1997). Systems of innovation approaches—Their emergence and characteristics. In C. Edquist (Ed.), *Systems of innovation: Technologies, institutions and organisations* (Science Technology and the International Political Economy Series). Pinter.

Edquist, C. (2009). Systems of innovation: Perspectives and challenges. In J. Fagerberg & C. D. Mowery (Eds.), *The Oxford handbook of innovation*. Oxford University Press.

Fabre, A., & Straub, S. (2021). *The impact of public-private partnerships (PPPs) in infrastructure, health and education*. Toulouse School of Economics, No. 986. https://www.tse-fr.eu/sites/default/files/TSE/documents/doc/wp/2019/wp_tse_986.pdf

Fagerberg, J. (2009). Innovation: A guide to the literature. In J. Fagerberg & C. D. Mowery (Eds.), *The Oxford handbook of innovation*. Oxford University Press.

Feng, G.-F., & Zheng, M. (2022). Economic policy uncertainty and renewable energy innovation: International evidence. *Innovation and Green Development, 1*, 100010. https://doi.org/10.1016/j.igd.2022.100010

Fingleton-Smith, E. (2018). The lights are on but no (men) are home: The effect of traditional gender roles on perceptions of energy in Kenya. *Energy Research & Social Science, 40*, 211. https://doi.org/10.1016/j.erss.2018.01.006

Focus-Economics. (2023). *Public debt (% of GDP)*. Retrieved from https://www.focus-economics.com/economic-indicator/public-debt/

Freeman, C. (1987). *Technology policy and economic performance: Lessons from Japan*. Pinter.

Galan, M. (2022). Governing the scalar politics of solar energy: Global production and national regulation in Kenyan and Indian off-grid solar markets. *Energy Research & Social Science, 90*, 102607. https://doi.org/10.1016/j.erss.2022.102607

Gebreslassie, M. G., Cuvilas, C., Zalengera, C., et al. (2022). Delivering an off-grid transition to sustainable energy in Ethiopia and Mozambique. *Energy Sustainability Society, 12*, 23. https://doi.org/10.1186/s13705-022-00348-2

Gielen, D., Boshell, F., Saygin, D., Bazilian, M. D., Wagner, N., & Gorini, R. (2019). The role of renewable energy in the global energy transformation. *Energy Strategy Reviews, 24*, 38. https://doi.org/10.1016/j.esr.2019.01.006

GOGLA. (2021). *Conditional cash transfers in Kenya for off-grid solar energy: Cash Plus / Mwangaza Mashinani Programme*. Retrieved from https://www.gogla.org/sites/default/files/resource_docs/case_study_-_kenya.pdf

Groenewoudt, A., Romijn, H., & Alkemade, F. (2020). From fake solar to full service: An empirical analysis of the solar home systems market in Uganda. *Energy for Sustainable Development, 58*, 100–111.

Harrington, E., & Wambugu, A. W. (2021). Beyond technical standards: Creating an ecosystem for quality and repair in Kenya's off-grid solar sector. *Energy Research & Social Science, 77*, 102101. https://doi.org/10.1016/j.erss.2021.102101

Hayashi, D. (2020). Harnessing innovation policy for industrial decarbonization: Capabilities and manufacturing in the wind and solar power sectors of China and India. *Energy Research & Social Science, 70*, 101644. https://doi.org/10.1016/j.erss.2020.101644

Heinemann, G. (2022). Innovating in the off-grid sector: Sustainable supply chains and business models for solar home system provision in Bangladesh and Kenya. *Energy Research & Social Science, 94*, 102853. https://doi.org/10.1016/j.erss.2022.102853

Hermanus, L., & Scholtz, L. (2018). *Small-scale renewable energy technologies in East Africa: An overview*. WWF. Retrieved from. https://wwfint.awsassets.panda.org/downloads/report___small_scale_renewable_energy_in_east_africa.pdf

Hoang, N., Nahm, D., & Dobbie, M. (2021). Innovation, gender, and labour productivity: Small and medium enterprises in Vietnam. *World Development, 146*, 105619. https://doi.org/10.1016/j.worlddev.2021.105619

Hossain, M., Park, S., & Shahid, S. (2023). Frugal innovation for sustainable rural development. *Technological Forecasting and Social Change, 193*, 122662. https://doi.org/10.1016/j.techfore.2023.122662

Hudson, L., Winskel, M., & Allen, S. (2011). The hesitant emergence of low carbon technologies in the UK: The micro-CHP innovation system. *Technology Analysis & Strategic Management, 23*(3), 297–312.

Inclusive Infrastructure. (2020). *Inclusive Infrastructure and Social Equity*. Retrieved from https://inclusiveinfra.gihub.org/

International Energy Agency (IEA). (2021). *Financing clean energy transitions in emerging and developing economies*. IEA.

International Finance Corporation (IFC). (2022). *Women's participation in the renewable energy workforce in sub-Saharan Africa: Identifying barriers and opportunities for women as leaders & employees*. Retrieved from https://www.ifc.org/en/insights-reports/2022/womens-participation-in-the-renewable-energy-workforce-in-sub-saharan-africa

International Monetary Fund (IMF). (2023). *How to avoid a debt crisis in sub-Saharan Africa*. Authors: F. Comelli, A. David, L. Eyraud, P. Kovacs, J. Montoya, and A. Sode. Retrieved from https://www.imf.org/en/News/Articles/2023/09/26/cf-how-to-avoid-a-debt-crisis-in-sub-saharan-africa

International Renewable Energy Agency (IRENA). (2018). *Off-grid renewable energy solutions*. Abu Dhabi. Retrieved from https://www.irena.org//media/Files/IRENA/Agency/Publication/2018/Jul/IRENA_Off-grid_RE_Solutions_2018.pdf

International Renewable Energy Agency (IRENA). (2020). Innovation landscape brief: Pay-as-you-go models. Abu Dhabi.

Johnson, O.W., Han, J.Y-C., Knight, A-L., Mortensen, S., Aung, M.T., Boyland, M., & Resureccion, B. P. (2020). Intersectionality and energy transitions: A review of gender, social equity and low-carbon energy. Energy Research & Social Science, 7. https://doi.org/10.1016/j.erss.2020.101774, 101774.

Johnstone, K. (2020). *Where's the capital for homegrown companies? Why energy access investments in Kenya need to go local*. IIED. Retrieved from https://www.iied.org/wheres-capital-for-homegrown-companies-why-energy-access-investments-kenya-need-go-local.

Keller, A. (2018). *Last mile connectivity program*. Global Infrastructure Hub. Retrieved from. https://inclusiveinfra.gihub.org/case-studies/last-mile-connectivity-program-kenya/

Kenya Gazette. (2021). *Kenya Gazette Supplement Acts 2021, The Public Private Partnerships Act, 2021*. Republic of Kenya. Retrieved from https://kenyalaw.org/kl/fileadmin/pdfdownloads/Acts/2021/ThePublicPrivatePartnershipsAct_No.14of2021.pdf

Ketelaars, S., & Wheeldon, S. (2021). Discussion paper: How end-user subsidies can help achieve universal energy access. *Gogla*. Retrieved from https://gogla.org/sites/default/files/resource_docs/gogla_discussion-paper-subsidies_def_2.pdf

Khan, I. (2020). Impacts of energy decentralization viewed through the lens of the energy cultures framework: Solar home systems in the developing economies. *Renewable and Sustainable Energy Reviews, 119*, 109576. https://doi.org/10.1016/j.rser.2019.109576

Khan, T., Khanam, S. N., Rahman, M. H., & Rahman, S. M. (2019). Determinants of microfinance facility for installing solar home system (SHS) in rural Bangladesh. *Energy Policy, 132*, 299–308. https://doi.org/10.1016/j.enpol.2019.05.047

Khandker, S. R., Samad, H. A., et al. (2014). Surge in solar-powered homes: Experience in off-grid rural Bangladesh. *Directions in Development. The World Bank*. https://doi.org/10.1596/978-1-4648-0374-1

Kim, K. (2022). Assessing the impact of mobile money on improving the financial inclusion of Nairobi women. *Journal of Gender Studies, 31*(3), 306. https://doi.org/10.1080/09589236.2021.1884536

Kim, Y. J., & Wilson, C. (2019). Analysing energy innovation portfolios from a systemic perspective. *Energy Policy, 134*, 110942. https://doi.org/10.1016/j.enpol.2019.110942

Kolk, A. (2016). The social responsibility of international business: From ethics and the environment to CSR and sustainable development. *Journal of World Business, 51*, 23. https://doi.org/10.1016/j.jwb.2015.08.010

Kolk, A., & van Tulder, R. (2010). International business, corporate social responsibility, and sustainable development. *International Business Review, 19*(2), 119–125.

Larsen, H. (2019). *Capabilities, networks, and directionality: Innovation policy for sustainable development goals*. (PhD Thesis). Imperial College, Centre for Environmental Policy, Faculty of Natural Sciences.

Li, D., & Ge, A. (2023). New energy technology innovation and sustainable economic development in the complex scientific environment. *Energy Reports, 9*, 4214–4223. https://doi.org/10.1016/j.egyr.2023.03.029

Lighting Global, GOGLA, & ESMAP. (2020). *Off-grid solar market trends report*. Retrieved from https://www.lightingglobal.org/wpcontent/uploads/2020/03/VIVID%20OCA_2020_Off_Grid_Solar_Market_Trends_Report_Full_High.pdf

Lindner, R., Daimer, S., Beckert, B., Heyen, N., Koehler, J., Teufel, B., Warnke, P., & Wydra, S. (2016). *Addressing directionality: Orientation failure and the systems of innovation heuristic*. Towards Reflexive Governance.

Markard, J., Raven, R., & Truffer, B. (2012). Sustainability transitions: An emerging field of research and its prospects. *Research Policy, 41*, 955–967.

Mas, I., & Morawczynski, O. (2009). Designing mobile money services: Lessons from M-PESA. *Innovations: Technology, Governance, Globalization, 4*, 77–91. https://doi.org/10.1162/itgg.2009.4.2.77

Mawdsley, E. (2018). Development geography II: Financialization. *Progress in Human Geography, 42*(2), 264. https://doi.org/10.1177/0309132516667874

Mazzoni, D. (2019). *The current status and prospective of PAYG off-grid solar solutions*. Fondazione Eni Enrico Mattei (FEEM). Retrieved from. http://www.jstor.com/stable/resrep21860.6

Mbiti, I., & Weil, D. (2013). The home economics of e-money: Velocity, cash management, and discount rates of M-Pesa users. *The American Economic Review, 103*(3), 369–374. https://doi.org/10.1257/AER.103.3.369

Md, A., Gomes, C., Dias, J. M., & Cerda, A. (2022). Exploring gender and climate change nexus, and empowering women in the south western coastal region of Bangladesh for adaptation and mitigation. *Climate, 10*, 172. https://doi.org/10.3390/cli10110172

Meirun, T., Mihardjo, W. L., Haseeb, M., Khan, R. A. S., & Jermsittiparset, K. (2021). The dynamics effect of green technology innovation on economic growth and CO2 emission in Singapore: New evidence from bootstrap ARDL approach. *Environmental Science and Pollution Research International, 28*(4), 4184–4194. https://doi.org/10.1007/s11356-020-10760-w

Montiel, I., Cuervo-Cazurra, A., Park, J., Antolin-Lopez, R., & Husted, B. W. (2021). Implementing the United Nations' Sustainable Development Goals in international business. *Journal of International Business Studies, 52*, 999. https://doi.org/10.1057/s41267-021-00445-y

Moss-Racusin, C. A., Pietri, E. S., Van Der Toorn, J., & Ashburn-Nardo, L. (2021). Boosting the sustainable representation of women in STEM with evidence-based policy initiatives. *Policy insights from the. Behavioral and Brain Sciences, 8*(1), 50. https://doi.org/10.1177/2372732220980092

Muchunku, C., et al. (2018). Diffusion of solar PV in East Africa: What can be learned from private sector delivery models? *WIREs Energy and Environment, 7*, 282.

Muzammil, M., & Ahmed, R. U. (2019). Chapter 12: Solar home systems in Bangladesh. In S. Huq, J. Chow, A. Fenton, C. Stott, J. Taub, & H. Wright (Eds.), *Confronting climate change in Bangladesh: Policy strategies for adaptation and resilience* (The Anthropocene: Politik--Economics--Society--Science Series) (Vol. 28, pp. 175–192). Springer International Publishing AG.

Ndiaye, O. K. (2019). Is the success of M-Pesa empowering Kenyan rural women? *Feminist Africa, 18*. Retrieved from https://feministafrica.net/wpcontent/uploads/2019/10/standpoints_is_the_success_of_m-pesa_empowering_kenyan_rural_women_.pdf.

Ndung'u, N. (2017). M-*Pesa—A success story of digital financial inclusion*. Practitioner's Insight, Blavatnik School of Government, University of Oxford. Retrieved from https://www.bsg.ox.ac.uk/sites/default/files/2018-06/2017-07-M-Pesa-Practitioners-Insight.pdf.

Ndung'u, N. (2018). The M-Pesa technological revolution for financial services in Kenya: A platform for financial inclusion. In D. L. K. Chuen & R. Deng (Eds.), *Handbook of blockchain, digital finance, and inclusion, volume 1: Cryptocurrency, FinTech, InsurTech, and regulation* (pp. 37–56). Academic Press.

Negro, S. O., Hekkert, M. P., & Smits, R. E. H. M. (2008). Stimulating renewable energy technologies by innovation policy. *Science and Public Policy, 35*(6), 403–416.

Niaz, U. M. (2022). Socio-economic development and sustainable development goals: A road-map from vulnerability to sustainability through financial inclusion. *Economic Research-Ekonomska Istraživanja, 35*(1), 3243–3275. https://doi.org/10.1080/1331677X.2021.1989319

Ockwell, D., Atela, J., Mbeva, K., Chengo, V., Byrne, R., Durrant, R., Kasprowikz, V., & Ely, A. (2019). Can pay-as-you-go, digitally enabled business models support sustainability transformations in developing countries? Outstanding questions and a theoretical basis for future research. *Sustainability, 11.*

Opiyo, N. N. (2019). Impacts of neighbourhood influence on social acceptance of small solar home systems in rural western Kenya. *Energy Research & Social Science, 52*, 91–98. https://doi.org/10.1016/j.erss.2019.01.013

Palm, A. (2022). Innovation systems for technology diffusion: An analytical framework and two case studies. *Technological Forecasting and Social Change, 182*, 121821. https://doi.org/10.1016/j.techfore.2022.121821

Paraschiv, L. S., & Paraschiv, S. (2023). Contribution of renewable energy (hydro, wind, solar and biomass) to decarbonization and transformation of the electricity generation sector for sustainable development. *Energy Reports, 9*(Supplement), 9. https://doi.org/10.1016/j.egyr.2023.07.024

Pearl-Martinez, R. (2020). Global trends impacting gender equality in energy access. In A. Pueyo & M. Maestre (Eds.), *Gender and energy: Opportunities for all* (IDS Bulletin Transforming Development Knowledge) (Vol. 51). No. 1. Institute of Development Studies.

Phillips, J., Davies, G., & Plutshack, V. (2020, May 4). Future development - An off-grid energy future requires learning from the past. *Brookings*. Retrieved from https://www.brookings.edu/blog/future-development/2020/05/04/an-off-grid-energy-future-requires-learning-from-the-past/

Pueyo, A., & DeMartino, S. (2018). The impact of solar mini-grids on Kenya's rural enterprises. *Energy for Sustainable Development, 45*, 28–37.

Reggers, A. (2019). Chapter 8 climate change is not gender neutral: Gender inequality, rights and vulnerabilities in Bangladesh. In S. Huq, J. Chow, A. Fenton, C. Stott, J. Taub, & H. Wright (Eds.), *Confronting climate change in Bangladesh: Policy strategies for adaptation and resilience* (The Anthropocene: Politik--Economics--Society--Science Series) (Vol. 28, pp. 103–118). Springer International Publishing AG.

REN21. (2020). *Renewables 2020 global status report*. Retrieved from https://www.ren21.net/wp-content/uploads/2019/05/gsr_2020_full_report_en.pdf

Roche, O. M., & Blanchard, R. E. (2018). Design of a solar energy Centre for providing lighting and income-generating activities for off-grid rural communities in Kenya. *Renewable Energy, 118*, 685–694. https://doi.org/10.1016/j.renene.2017.11.053

Rolffs, P., Ockwell, D., & Byrne, R. (2015). Beyond technology and finance: Pay-as-you-go sustainable energy access and theories of social change. *Environment and Planning A: Economy and Space, 47*, 2609–2627. https://doi.org/10.1177/0308518X15615368

Sanyal, S., Pinchot, A., Prins, J., & Visco, F. (2016). *Stimulating pay-as-you-go energy access in Kenya and Tanzania: The role of development finance*. World Resources Institute. Retrieved from https://www.wri.org/research/stimulating-pay-you-go-energy-access-kenya-and-tanzania-role-development-finance

Satalkina, L., & Steiner, G. (2020). Digital entrepreneurship and its role in innovation systems: A systematic literature review as a basis for future research avenues for sustainable transitions. *Sustainability, 12*(7). https://doi.org/10.3390/su12072764

Scherer, A. G., Palazzo, G., & Seidl, D. (2013). Managing legitimacy in complex and heterogeneous environments: Sustainable development in a globalized world. *Journal of Management Studies, 50*(2), 259–284.

Sergi, B., Babcock, M., Williams, N. J., Thornburg, J., Loew, A., & Ciez, R. E. (2018). Institutional influence on power sector investments: A case study of on- and off-grid energy in Kenya and Tanzania. *Energy Research & Social Science, 41*, 59. https://doi.org/10.1016/j.erss.2018.04.011

Shapsugova, M. (2023). ESG principles and social responsibility. *E3S Web of Conferences*. https://doi.org/10.1051/e3sconf/202342006040

Shuy, C.-W. (2023). Lessons from the World Bank's solar home system-based rural electrification projects (2000–2020): Policy implications for meeting Sustainable Development Goal 7 by 2030. *Energy Reports, 9*, 2820. https://doi.org/10.1016/j.egyr.2023.01.128

Smith, A., Voß, J.-P., & Grin, J. (2010). Innovation studies and sustainability transitions: The allure of the multi-level perspective and its challenges. *Research Policy, 39*, 435–448.

Sovacool, K. B., Iskandarova, M., & Hall, J. (2023). Industrializing theories: A thematic analysis of conceptual frameworks and typologies for industrial sociotechnical change in a low-carbon future. *Energy Research & Social Science, 97*, 102954. https://doi.org/10.1016/j.erss.2023.102954

Stern, P., Janda, K., Brown, M., Steg, L., Vine, E., & Lutzenhiser, L. (2016). Opportunities and insights for reducing fossil fuel consumption by households and organizations. *Nature Energy, 1*, 16043. Retrieved from https://www.nature.com/articles/nenergy201643

Szogs, A. (2008). *The role of mediator organisations in the making of innovation systems in least developed countries: Evidence from Tanzania* (CIRCLE, Paper No. 2008/19). Centre for Innovation, Research and Competence in the Learning Economy (CIRCLE), Lund University.

ThinkGeoEnergy. (2024). *PPPs as an enabler of project financing for geothermal in Africa*. Author: C. Cariaga. Retrieved from https://www.thinkgeoenergy.com/ppps-as-an-enabler-of-project-financing-for-geothermal-in-africa/

Tijani, H. O. (2015). Globalisation, institution and public-private partnership: The Nigerian experience. *Journal of Law, Policy and Globalization, 43*. Retrieved from https://core.ac.uk/download/pdf/234650399.pdf

Ulsrud, K. (2020). Access to electricity for all and the role of decentralized solar power in sub-Saharan Africa. *Norsk Geografisk Tidsskrift—Norwegian. Journal of Geography, 74*(1), 54–63.

United Nations Framework Convention on Climate Change (UNFCCC). (2024). *Kenya*. Retrieved from https://unfccc.int/node/61092

Urmetzer, S., & Pyka, A. (2019). Innovation systems for sustainability. In W. Filho Leal, M. A. Azul, L. Bandli, P. G. Özuyar, & T. Wall (Eds.), *Encyclopedia of the UN sustainable development goals: Decent work and economic growth*. Springer. https://link.springer.com/referenceworkentry/10.1007/978-3-319-71058-7_43-1

Uzuegbunam, I. (2024). Culture and investment in green public-private partnerships. *Journal of Cleaner Production, 443*, 141122. https://doi.org/10.1016/j.jclepro.2024.141122

Van Hove, L., & Dubus, A. (2018). M-PESA and financial inclusion in Kenya: Of paying comes saving? *Sustainability, 11*(3), 568. https://doi.org/10.3390/su11030568

van Tulder, R., Rodrigues, S. B., Hafiz, M., & Sexsmith, K. (2021). The UN's sustainable development goals: Can multinational enterprises lead the decade of action? *Journal of international Business. Policy, 4, 1.*

Volkert, M., & Klagge, B. (2022). Electrification and devolution in Kenya: Opportunities and challenges. *Energy for Sustainable Development, 71*, 541. https://doi.org/10.1016/j.esd.2022.10.022

Wagner, N., Rieger, M., Bedi, A. S., Vermeulen, J., & Demena, B. A. (2021). The impact of off-grid solar home systems in Kenya on energy consumption and expenditures. *Energy Economics, 99*, 105314. https://doi.org/10.1016/j.eneco.2021.105314

Weber, M., & Rohracher, H. (2012). Legitimising research, technology and innovation policies for transformative change - Combining insights from innovation systems and multi-level perspective in a comprehensive 'failures' framework. *Research Policy, 41*, 1037–1047.

Weber, M., & Truffer, B. (2017). Moving innovation systems research to the next level: Towards an integrative agenda. *Oxford Review of Economic Policy, 33*(1), 101–121.

Wei, J., Wen, J., Wang, X.-Y., Ma, J., & Chang, C.-P. (2023). Green innovation, natural extreme events, and energy transition: Evidence from Asia-Pacific economies. *Energy Economics, 121*, 1–17. https://doi.org/10.1016/j.eneco.2023.106638

Winther, T., Ulsrud, K., & Saini, A. (2018). Solar powered electricity access: Implications for women's empowerment in rural Kenya. *Energy Research & Social Science, 44*, 61–74. https://doi.org/10.1016/j.erss.2018.04.017
Women's World Banking. (2020). *Empowering MSMEs: Creating a better banking experience for women-led micro, small, and medium enterprises in Kenya*. Retrieved from https://www.womensworldbanking.org/wp-content/uploads/2020/11/MSME_Report_2020.pdf
Woollacott, J., Henry, C., Bean de Hernandez, A., DiVenanzo, L., Oliveira, H., Cai, Y., & Larson, J. (2023). Quantifying the local economic supply chain impacts of renewable energy investment in Kenya. *Energy Economics, 106810*, 106810. https://doi.org/10.1016/j.eneco.2023.106810
World Bank (World Bank). (2011, June 6). *Lighting Africa: Promoting the use of high quality, clean electricity*. Retrieved from https://www.worldbank.org/en/news/feature/2011/06/06/lighting-africa-promoting-the-use-of-high-quality-clean-electricity
World Bank (World Bank). (2017). *Kenya: Off-grid solar access project for underserved counties*, Report No: PAD2008. Retrieved from https://documents1.worldbank.org/curated/fr/212451501293669530/pdf/Kenya-off-grid-PAD-07072017.pdf
World Bank (World Bank). (2022). Nearly 2.4 billion women globally don't have same economic rights as men. *Press Release*, March 1st. Retrieved from https://worldbank.org/en/news/press-release/2022/03/01/nearly-2-4-billion-women-globally-don-t-have-same-economic-rights-as-men
World Health Organization (WHO). (2022). *Tracking SDG7: The Energy Progress Report 2022*. Retrieved from https://www.who.int/publications/m/item/tracking-sdg7%2D%2Dthe-energy-progress-report-2022
Xu, J., Moslehpour, M., Tran, T. K., Ngo, T. Q., & Huy, P. Q. (2023). The role of institutional quality, renewable energy development and trade openness in green finance: Empirical evidence from south Asian countries. *Renewable Energy, 207*, 687. https://doi.org/10.1016/j.renene.2023.03.015

Laura Hellqvist is a doctoral researcher at SOAS, University of London. Her research explores the transformative role of innovation and technology in driving sustainable development, with a focus on energy access, women's empowerment, and inclusive economic growth in the Global South. She has worked with various UN agencies, international organizations, and development banks, contributing her expertise to global policy initiatives and high-level dialogues.

Dr. Harald Heubaum is Chair of the Centre for Energy and Climate Policy and Deputy Director of the Centre for Sustainable Finance, both at SOAS University of London. He convenes the MSc Global Energy and Climate Policy and also serves as Deputy Academic Director of the University Network for Strengthening Macrofinancial Resilience to Climate and Environmental Change, which forms part of the Resilience and Adaptation Mainstreaming Programme (RAMP). RAMP is a major international effort to partner with universities and governments across the Global South, building capacity to manage climate change risks and access adaptation finance.

Green Business Strategies in MNE Research: Thematic Considerations and Methodological Features

İlayda İpek

1 Introduction

Grand social and environmental challenges surrounding the global context have urged businesses to consider sustainability an essential focus (Burritt et al., 2020; Ghauri et al., 2021). In this sense, businesses have leaned towards a transformational change in the pursuit of 'a Nature Positive approach' by touching on various environmental issues such as decarbonization, sustainable global food systems, and circular economy (UNEP, 2019). On the other side, remarkably, in the way of understanding the relevance of the spheres of sustainability to businesses, the importance of international trade activities has been notably underlined primarily due to accelerating consumer-related, competitive, and institutional burdens on a global basis (UNCTAD, 2021; Zeriti et al., 2014) and the role of international trade as 'an engine for inclusive economic growth and poverty reduction, [which] contributes to the promotion of sustainable development' (UNCTAD, 2020). However, the domain of international business has been particularly interested in how MNEs deal with societal issues, given that MNEs are highly instrumental in the dynamics of both home and host countries (Burritt et al., 2020; Kolk & van Tulder, 2010).

Drawing on these tenets, in the field of international business, valuable insights have been increasingly provided into the understanding of MNEs' practices for sustainability and corporate social responsibility (Buckley et al., 2017). Nevertheless, despite these efforts, the international business literature has been criticised as being limited in fully capturing whether MNEs are contributors to the United Nations Sustainable Development Goals (UN SDGs) (Buckley et al., 2017; Ghauri, 2022), which reflects an ongoing debate in the related literature (Aguilera-Caracuel et al.,

İ. İpek (✉)
Faculty of Business, Dokuz Eylül University, İzmir, Turkey
e-mail: ilayda.gungor@deu.edu.tr

J. Bento, M. M. Torres (eds.), *Global Perspectives on Climate Change, Inequality, and Multinational Corporations*, CSR, Sustainability, Ethics & Governance,
https://doi.org/10.1007/978-3-031-80797-8_2

2011; Wood et al., 2021). In that vein, special emphasis as a promising research path has been placed on the manner MNEs approach environmental sustainability (Ghauri et al., 2021). This also entails the requirement to elucidate the peculiar characteristics of the body of empirical research covering the green business strategies of MNEs.

Green business strategy has been defined as corporate activities surrounded by 'a meaningful and enduring reason to exist that profitably contributes to a sustainable economy while protecting and restoring the environmental dependencies the firm relies upon across the value chain' (UNEP, 2019: 22). Previously, significant research attempts have been centred on conducting review studies and conceptual/theoretical papers on the topic of interest (e.g., Kolk, 2016; Nielsen et al., 2023; Yu et al., 2023). On the other hand, no systematic review has yet been dedicated to the empirical literature on MNEs' green business strategies with particular stress on the thematic perspectives and methodological aspects. Building on the aforementioned issues, a need emerges to systematically analyse, consolidate, and aggregate the empirical line of research on MNEs' green business strategies from a holistic perspective, which is of great importance to unveil how environmental concerns shape competitive strategies in international markets (Porter & Linde, 1995). In this sense, this review study seeks to answer the research question: 'What are the thematic and methodological idiosyncrasies of the empirical literature on MNEs' green business strategies?' Specifically, three main objectives to be achieved are recognised as follows: (a) to delve into the thematic considerations ingrained in the related research stream; (b) to uncover the methodological features concerned in the eligible studies; and (c) to propose worthwhile future research directions grounded on the knowledge gaps delineated.

2 Background

Mainly endeavouring to create a better future for all, the UN SDGs adopted in 2015 have been acknowledged as offering both dark problems and favourable opportunities concerning MNEs (Kolk & van Tulder, 2010), which leads to two opposite arguments in the field of international business regarding whether MNEs can support the achievement of the UN SDGs (Aguilera-Caracuel et al., 2011; Wood et al., 2021). Despite the inadequate attention attracted to the interplay between international business, MNEs, and the UN SDGs, conflicting standpoints discussing the act of MNEs in the environmental pillar of sustainability occupy a prominent place in the pertinent literature (Ghauri, 2022). This further demonstrates why sustainable development and climate change have been firmly accentuated as one of the new realities of international business research notably in the context of MNEs (Ghauri et al., 2021).

From one perspective, it has been strongly defended that MNEs' involvement is indispensable to accomplishing the UN SDGs (Ghauri, 2022), mainly owing to MNEs' assistance to decrease the effects of institutional voids in economies where the level of institutional quality is relatively low (Khanna & Palepu, 2004; Rondinelli, 2007). Specifically, favourable circumstances differing from well-established public services to environmentally friendly applications provided by MNEs have been suggested to boost national development (Kolk, 2016; Wood et al., 2021). Besides, no matter how the regulatory institutions in the host country are developed, the inclination of MNEs to globally standardise their green business practices utilising sophisticated ecological policies and perspectives positively influences environmental sustainability in widespread areas (Aguilera-Caracuel et al., 2011; Christmann, 2004).

From another point of view, MNEs have been largely accepted as one of the primary roots of inequalities surrounding societies (Ghauri, 2022; Kourula et al., 2017), which are closely intertwined with environmentally, socially, and/or economically devastating behaviours (Burritt et al., 2020; Wood et al., 2021; Yu et al., 2023). In particular, it has been advocated that by capitalising on the gaps in the social and environmental institutions, MNEs take advantage of 'moral free space' and enjoy formulating strategies depending on their codes of conduct (Donaldson, 1996; Kolk & van Tulder, 2010). In this respect, MNEs have been considered the major players in environmental degradation, especially within emerging and least-developed economies where firms are more apt to design business activities producing high environmental impact (Aguilera-Caracuel et al., 2011; Villo et al., 2020). In a nutshell, these notions have indicated that MNEs bring about environmental troubles at the global level (Yu et al., 2023).

3 Methods

In this study, a systematic review methodology was employed to encapsulate and synthesise the empirical literature on MNEs' green business strategies. Accordingly, the search strategy primarily involved the screening of the electronic databases (i.e., Science Direct, EBSCO Business Source Complete, and Web of Science (ISI)) with the identified keywords 'environmental', 'ecological', 'sustainable', and 'green' in combination with 'MNE', 'MNEs', 'MNC', 'MNCs', 'international', 'multinational', 'transnational', 'global', 'FDI', 'foreign direct investment', and 'subsidiary', which are based on the previous review studies on the similar research subject (Christofi et al., 2021; Leonidou et al., 2011). The electronic databases were specifically queried in line with the following eligibility criteria: (a) to elaborate on green business strategies in the domain of MNEs; (b) to address MNEs' green business strategies from the sphere of environmental sustainability; (c) to appear in business-related publication outlets; and (d) to be of an empirical nature with the inclusion of primary and/or secondary data. As for the quality check of the search results, the Academic Journal Guide by the Association of Business Schools (ABS) was concentrated, being widely identified as a directory in the relevant streams of research (e.g., Adams et al., 2016; Paul et al., 2017).

Along with carefully evaluating the electronic search results, the relevant articles on the subject were determined for further investigation. Nevertheless, to avoid the likelihood of missing some eligible studies in the electronic search, the manual search strategy was also pursued by scanning the reference lists of the shortlisted studies. Herewith, the final set of articles was recognized as 65 at the end of the search process. All eligible articles were subjected to content analysis by two independent coders with sufficient knowledge of international business. In doing so, a coding framework was developed with a specific focus on thematic considerations (i.e., theme, subtheme, and research topics addressed) and methodological features (i.e., industrial sector, time emphasis, unit of analysis, data collection, and statistical method). Using open coding and axial coding, concepts were initially identified; thereafter, sub-themes and respective themes were defined to cluster thematic considerations and methodological features. The coding process demonstrated an inter-coder reliability of 95–100% (Kassarjian, 1977), obtained by comparing the analysis of the independent coders.

The final sample of this systematic review encompasses 65 empirical articles extracted from 40 academic journals and published between 1995 and 2023. The descriptive analysis indicates that green business strategies in the field of MNEs have been mostly handled in international business and area studies (26.2%), general management, ethics, and social responsibility (15.4%), and regional studies, planning, and environment (15.4%). Particularly, the top contributing publication outlets refer to *Business Strategy and the Environment* (12.3%), *International Business Review* (9.2%), *Journal of Business Research* (4.6%), *Journal of Business Ethics* (4.6%), and *Multinational Business Review* (4.6%).

4 Findings

This part presents the findings of this systematic review under two sub-groups, namely, thematic considerations and methodological features.

4.1 Thematic Considerations

In terms of thematic considerations (Fig. 1), the empirical literature on green business strategies of MNEs has been categorised into five broad research themes: *green reporting* (33.9%), *green innovation & new product development (NPD)* (24.6%), *green corporate governance* (23.1%), *green value chain* (12.3%), and *green international marketing* (6.2%).

Drawing the greatest scholarly attention in the relevant literature (Table 1), the first research stream covers the studies exploring mainly the way MNEs share environmental impact with public bodies, which have been specifically focused on two subthemes, including environmental transparency and disclosure and environmental

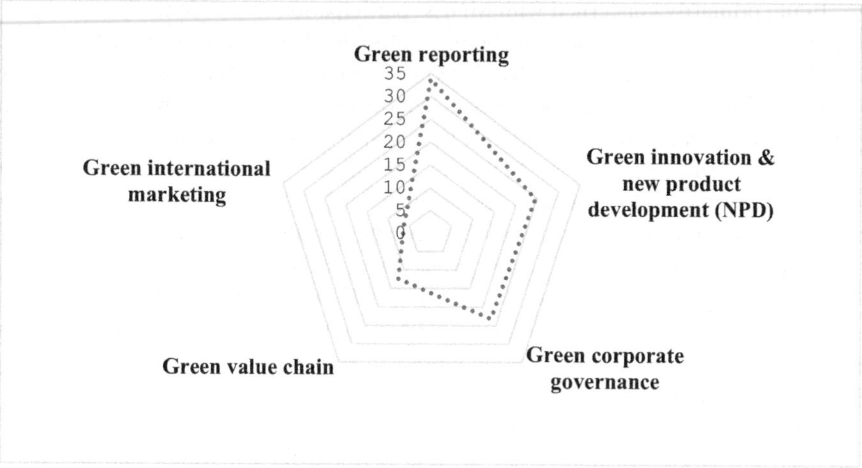

Fig. 1 Radar chart of thematic considerations. (Source: Authors' elaboration)

Table 1 Thematic considerations in MNE research (Source: Authors' elaboration)

Theme	Subtheme	Research topics addressed
Green reporting ($n_1 = 22$)	Environmental transparency and disclosure ($n_{1a} = 21$) Environmental reporting mechanisms ($n_{1b} = 1$)	Carbon transparency and disclosure; Water transparency and disclosure; Assurance of sustainability Environmental reporting policies
Green innovation & NPD ($n_2 = 16$)	Green product and process innovation ($n_{2a} = 14$) International green patent development ($n_{2b} = 2$)	Green product and process development; Eco-friendly technology adoption International green patent performance; Green innovation offshoring
Green corporate governance ($n_3 = 15$)	Strategic green orientation and structure ($n_{3a} = 14$) SDGs ($n_{3b} = 1$)	Green marketing orientation; Environmental proactivity; Structure of environmental strategy; Environmental certification SDG engagement
Green value chain ($n_4 = 8$)	Environmental collaboration ($n_{4a} = 5$) Green supply chain management mechanisms ($n_{4b} = 3$)	Environmental cooperation with suppliers; Sustainable sourcing; Green transportation Sustainable supply chain policies
Green international marketing ($n_5 = 4$)	Green cross-border investment ($n_{5a} = 3$) Green communication ($n_{5b} = 1$)	Green FDI; Internationalisation of green R&D Environmental communication standardisation

reporting mechanisms. For the first subtheme, scholars have largely examined the extent to which MNEs are transparent and publish publicly available environmental information (e.g., Ellimäki et al., 2023; Gallego-Álvarez & Ortas, 2017; Kolk & Fortanier, 2013).

Being primarily concerned with MNEs' participation in a wide array of environmental issues such as waste management, climate change, renewable energy etc., this line of research evaluates MNEs' environmental transparency and disclosure by either following a holistic approach or conducting special research inquiries. In terms of the former, empirical articles have endeavoured to illuminate the environmental disclosure activities of MNEs in general (e.g., Cormier et al., 2004; Cui et al., 2020; Delgado-Márquez & Pedauga, 2017). Concerning the latter, for instance, scholars have embraced MNEs' green reporting issues about carbon transparency and disclosure, water transparency and disclosure, and assurance of sustainability statement (e.g., Choi & Luo, 2021; Kolk & Perego, 2010; Peng et al., 2023). On the other side, the subtheme of environmental reporting mechanisms has received scant interest, with a single empirical research channelling into environmental reporting policies of MNEs (Gallego-Álvarez et al., 2017).

Another bulk of studies in the thrust of research on MNEs' green business strategies fall into the green innovation & NPD research theme, which fundamentally deals with green product and process innovation and international green patent development. Within this research theme, green product and process innovation have been heavily studied to shed light on MNEs' attempts to lessen environmental impacts by investing in new products, processes, and technologies (Gómez-Bolaños et al., 2022; Konara et al., 2021). Particularly, the development of environmentally friendly products mainly involves the inclusion of various green considerations into strategic product decisions such as sustainable design, ecological packaging, and eco-labelling, while green process innovations mostly aim to enhance the environmental performance of business operations (e.g., Ellimäki et al., 2023; Zhou et al., 2023). As for international green patent development, scholars have delved into MNEs' innovativeness in environmental subjects within the context of international trade activities (e.g., Leyva-de la Hiz et al., 2019; Marin & Zanfei, 2019).

In the third research theme, green business strategies of MNEs have been widely evaluated at the corporate level, with the vast majority of studies scrutinising strategic green orientation and structure. Specifically, several efforts exist associating the degree to which MNEs actively engage in anticipating and responding to environmental trends with strategic orientation and proactivity (e.g., Borah et al., 2022; Bueno-García et al., 2022). Besides, the level of centralization of environmental strategy, the existence of a formal environmental management system, and the ownership of environmental certifications in MNEs represent other important research topics for the subtheme of green orientation and structure (e.g., Epstein & Roy, 2007; Kawai et al., 2018; Margaret et al., 2024). Despite its significance, the relevance of the UN SDGs to MNEs has remained untouched in the extant literature such that only one article has been explicitly dedicated to SDG engagements in the setting of MNEs (DasGupta et al., 2022).

Although the cruciality of the integration between global value chains and environmental concerns has been strongly highlighted (Golgeci et al., 2021; Marchi et al., 2013), green business strategies of MNEs have not been intensified on the research theme of green value chain, encompassing environmental collaboration and green supply chain management mechanisms. In respect of environmental collaboration, cooperative relationships with value chain partners to enhance green value chain networks such as sustainable sourcing and green transportation have been ascribed to the empirical research on MNEs' green business strategies (e.g., Bouguerra et al., 2021; Hsu et al., 2014; Pazirandeh & Jafari, 2013). Apart from these, scholars have also been interested in green supply chain mechanisms, with special stress put on sustainable supply chain policies dealing primarily with whether MNEs adopt green protocols for the governance of global value chains (e.g., Das, 2024).

Turning to the last research theme identified in this systematic review, the studies in green international marketing have spotlighted either green cross-border investment or green communication decisions in the body of research on green business strategies of MNEs. Capturing mainly the role of green issues in internationalisation activities, green cross-border investment has been mainly channelled into green FDI projects and international distribution of green R&D locations (e.g., Amendolagine et al., 2023; De Beule et al., 2022; Noailly & Ryfisch, 2015). As for the subtheme of green communication, the degree to which environmental messages in advertising and public communication are standardised by MNEs has been investigated in the existing literature (Christmann, 2004).

4.2 Methodological Features

Pursuant to methodological features (Table 2), this review study assesses the empirical body of research on MNEs' green business strategies under five different characteristics: industrial sector, time emphasis, unit of analysis, data collection, and statistical method.

Concerning the industrial sector, the great majority of articles (73.9%) collected data from multiple industries mainly for the purpose of strengthening observed variance and cultivating the generalizability of the research results (e.g., Samiee & Chirapanda, 2019; Sousa & Lengler, 2009). On the other hand, about one-quarter of related studies (23.1%) followed a single-industry perspective, with a particular concentration on such industries as agri-food (e.g., García-Granero et al., 2020; García-Sánchez et al., 2021; Marco-Lajara et al., 2023), energy and energy-related utility (e.g., Amendolagine et al., 2023; Ellimäki et al., 2021; Gómez-Bolaños et al., 2022), and chemistry (e.g., Christmann, 2004; Ellimäki et al., 2023; Levy, 1995). The remaining portion of the articles (3.0%) provided no specific indication in terms of industrial aspects.

In relation to the time emphasis, the content analysis reveals approximately even distribution across longitudinal (47.7%) and cross-sectional research designs (52.3%). Longitudinal studies using panel data enable us to understand the act of green business strategies among MNEs from a long-term approach. While the general tendency in the related body of research as to time emphasis reflects the theme of green reporting, longitudinal research played a dominant role in the research areas of green innovation & NPD and green international marketing (e.g.,

Table 2 Methodological features in MNE research (Source: Authors' elaboration)

Methodological aspect	Total (n = 65) %	Theme 1: Green reporting (n_1 = 22) %	Theme 2: Green innovation & NPD (n_2 = 16) %	Theme 3: Green corporate governance (n_3 = 15) %	Theme 4: Green value chain (n_4 = 8) %	Theme 5: Green international marketing (n_5 = 4) %
Industrial sector						
Multiple	73.9	95.5	37.5	80	87.5	50
Single	23.1	4.5	56.3	13.3	12.5	50
Not available	3	–	6.2	6.7	–	–
Time emphasis						
Longitudinal	47.7	50	62.5	40	12.5	75
Cross-sectional	52.3	50	37.5	60	87.5	25
Unit of analysis						
Multi-level	46.2	68.2	37.5	26.7	37.5	50
Firm-level	43.1	31.8	62.5	53.3	25	25
Subsidiary-level	7.7	–	–	20	25	–
Other	3	–	–	–	12.5	25
Data collection						
Secondary data	67.7	95.5	75	46.7	12.5	75
Survey	33.8	4.5	31.3	53.3	87.5	25
Statistical method						
Regression-based	67.7	86.4	75	60	25	50
SEM	18.5	–	25	26.7	50	–
Econometric	6.2	9.1	–	–	–	50
Other	10.8	9.1	6.3	13.3	25	–

Amendolagine et al., 2023; Ha, 2021; Konara et al., 2021), which was considered in the rest of the literature on a less-frequent basis.

With regard to the unit of analysis, considerable numbers of the total studies rested either on multi-level (46.2%) or firm-level measurements (43.1%). The preponderance of multi-level investigation closely conforms to the suggestion identifying sustainability as a cross-level concept (Parboteeah et al., 2012) and that 'a micro or a macro lens alone yields incomplete understanding at either level' (Hitt et al., 2007: 1385). Aside from multi-level and firm-level analysis, subsidiary-level (7.7%) and other units of analysis (3.0%) (e.g., business unit, industry, and function) were also recorded to a lesser extent in the line of research on MNEs' green business strategies.

In terms of the data collection, the findings of this systematic review denote a small variety of data sources employed by scholars in the field, which solely include secondary data and survey. Given that panel data allows removing the estimates of the hidden biases with the inclusion of an adequate period length (Smith, 2008), the heavy reliance on secondary data (67.7%) in the extant stream of research is accountable. This methodological trend is largely evident in the green reporting research theme, which firmly corresponds to the nature of integrated reporting (Nwachukwu, 2022). Besides, due to its several advantages such as greater convenience and simpler usage (Li et al., 2023), survey data also accounted for an important percentage of the data collection methods (33.8%).

As for the statistical method, a notable tendency here is the vast adoption of advanced analytical perspectives, specifically regression analysis (67.7%), structural equation modelling (18.5%), and econometric modelling (6.2%). Among these, regression-based analytical methods were performed to a great extent, which can be explained by the frequent involvement in longitudinal research design within the line of research on green business strategies of MNEs. On the other hand, structural equation modelling was the most popular statistical method in the theme of green value chain, probably owing to its wide application to cross-sectional data. Furthermore, other statistical methods in this thrust of research can be exemplified as multi-level analysis (e.g., Bouguerra et al., 2021, 2023), univariate data analysis (e.g., Epstein & Roy, 2007), and event study analysis (e.g., Dam & Petkova, 2014).

5 Suggestions for Future Research

Thematically, some subthemes under green reporting, green innovation & NPD, and green corporate governance have remained relatively untouched. Specifically, in terms of the green reporting theme, future studies are strongly suggested to lean more towards environmental reporting mechanisms. In that vein, the detailed investigations of environmental reporting frameworks and standards in the landscape of MNEs would be fruitful. Moreover, the existing empirical evidence demonstrates a huge inclination to handle MNEs' green reporting from a descriptive perspective. Accordingly, in further research, emphasis should be placed on the difference between text (declarative) and quantitative information reporting, which have been deemed vital concerns in terms of deeply understanding the environmental reporting issues of firms (Dagiliene et al., 2020).

In addition, within the scope of green innovation & NPD, international green patent development merits further interest. This is mainly because cross-country inventions offer numerous advantages of technological diversity to MNEs (Kafouros et al., 2022), and the integration of international patents with green concerns has gained great importance at the global level (Noailly & Ryfisch, 2015). Apart from that, as a novel research inquiry, it would be worthwhile to explore MNEs' green innovation activities in the light of innovation ambidexterity (i.e., radical and incremental innovation) being closely attached to environmental sustainability (Vesal et al., 2022). In connection with the research theme of green corporate governance, it is indisputable that explicit reference to the UN SDGs is worth taking into account in the empirical literature on MNEs' green business strategies, given that the UN

SDGs have increasingly become critical for both businesses and scholarly research in the B2B context (Pizzi et al., 2020; Voola et al., 2022).

Referring to the limited attention attracted to the research themes of green value chain and green international marketing, in terms of the former, since relatively high stress has been put on linking ecological practices with upstream supply chain activities, notably, sustainable sourcing and partnerships with suppliers, it would be insightful to understand MNEs' green business strategies at the intersection of environmental sustainability and downstream supply chain activities (e.g., transportation, warehousing, and inventory management) or reverse logistics (Kumar, 2016). Respecting green international marketing, future studies should address the thematic holes associated with the ignored edges of the green marketing mix in the realm of MNEs, all of which have been widely acknowledged as key contributors to building a good corporate image (Mukonza & Swarts, 2020).

Methodologically, in terms of the industrial sector, the content analysis indicates the preeminence of the studies resting on multiple industries. This can be evaluated as a strength in the scope of research since a single industry focus might become restricted to generalising the study findings despite its benefit to handling extraneous inter-industry forces (Shaw et al., 2001). Besides, almost equal attention devoted to longitudinal and cross-sectional studies in the related stream of research signals another favourable characteristic of the research methodology due to the various advantages associated with each research design. However, future studies are strongly recommended to rest more on the longitudinal approach, which allows for performing tests of causality among MNEs (Aharoni et al., 2011) and inspecting the mosaic features and lagged structures of MNEs over time (e.g., Ozkan et al., 2022; Yang & Driffield, 2022).

Concerning the unit of analysis, the frequent employment of multi-level assessments in the pertinent body of research closely corresponds to the nature of sustainability (Parboteeah et al., 2012) and the proposals given to rely on multi-level analysis in an attempt to create a sufficient understanding of the phenomenon under investigation (e.g., Hitt et al., 2007). However, it should be noted that the 'correct level of analysis' is strictly based on the research purpose (Sousa et al., 2008). Furthermore, a narrow array of data collection tools signifies a critical shortcoming in research methodology, as the essence of environmental sustainability is persistent and complicated (Loorbach et al., 2010). In that vein, for future studies, the triangulation of quantitative and qualitative data and the adoption of new research methodologies (e.g., artificial intelligence and machine learning) are strongly suggested (Christofi et al., 2024). Finally, although the extensive usage of advanced analytical perspectives demonstrates a methodological quality in the landscape of MNEs' green business strategies (Leonidou & Katsikeas, 2010), a wider range of statistical methods and more specialised techniques for data analysis need a call for further research.

6 Conclusion

Spotlighting the mosaic characteristics of the empirical literature on green business strategies in the context of MNEs, this systematic review is expected to considerably contribute to the knowledge of how environmental sustainability is firmly relevant to MNEs in different ways. First, it unveils the thematic grounds of the thrust of research on MNEs' green business strategies. Second, it compiles the methodological aspects associated with the existing stream of research. Third, it recognizes the knowledge and research gaps in the pertinent literature and proposes further research directions that would push this domain forward meaningfully. The findings of this review study indicate that environmental issues are of great importance among MNEs, which echoes a wealth of research on the subject.

References

Adams, R., Jeanrenaud, S., Bessant, J., Denyer, D., & Overy, P. (2016). Sustainability-oriented innovation: A systematic review. *International Journal of Management Reviews, 18*(2), 180–205. https://doi.org/10.1111/ijmr.12068

Aguilera-Caracuel, J., Aragón-Correa, J. A., & Esther Hurtado-Torres, N. (2011). Extending the literature on the environmental strategy of MNEs. *Multinational Business Review, 19*(4), 299–310. https://doi.org/10.1108/15253831111190153

Aharoni, Y., Tihanyi, L., & Connelly, B. L. (2011). Managerial decision-making in international business: A forty-five-year retrospective. *Journal of World Business, 46*(2), 135–142. https://doi.org/10.1016/j.jwb.2010.05.001

Amendolagine, V., Hansen, U. E., Lema, R., Rabellotti, R., & Ribaudo, D. (2023). Do green foreign direct investments increase the innovative capability of MNE subsidiaries? *World Development, 170*, 106342. https://doi.org/10.1016/j.worlddev.2023.106342

Borah, P. S., Pomegbe, W. W. K., & Dogbe, C. S. K. (2022). Mediating role of green marketing orientation in stakeholder risk and new product success relationship among European multinational enterprises in Ghana. *Society and Business Review, 17*(4), 485–505. https://doi.org/10.1108/SBR-02-2021-0035

Bouguerra, A., Gölgeci, I., Gligor, D. M., & Tatoglu, E. (2021). How do agile organizations contribute to environmental collaboration? Evidence from MNEs in Turkey. *Journal of International Management, 27*(1), 100711. https://doi.org/10.1016/j.intman.2019.100711

Bouguerra, A., Hughes, M., Cakir, M. S., & Tatoglu, E. (2023). Linking entrepreneurial orientation to environmental collaboration: A stakeholder theory and evidence from multinational companies in an emerging market. *British Journal of Management, 34*(1), 487–511. https://doi.org/10.1111/1467-8551.12590

Buckley, P. J., Doh, J. P., & Benischke, M. H. (2017). Towards a renaissance in international business research? Big questions, grand challenges, and the future of IB scholarship. *Journal of International Business Studies, 48*, 1045–1064. https://doi.org/10.1057/s41267-017-0102-z

Bueno-García, M., Delgado-Márquez, B., Georgallis, P., & Aragón-Correa, J. A. (2022). How do shareholders influence international firms' environmental strategies? The differential impact of strategic and financial investors. *Long Range Planning, 55*(6), 102183. https://doi.org/10.1016/j.lrp.2022.102183

Burritt, R. L., Christ, K. L., Rammal, H. G., & Schaltegger, S. (2020). Multinational enterprise strategies for addressing sustainability: The need for consolidation. *Journal of Business Ethics, 164*, 389–410. https://doi.org/10.1007/s10551-018-4066-0

Choi, B., & Luo, L. (2021). Does the market value greenhouse gas emissions? Evidence from multi-country firm data. *British Accounting Review, 53*(1), 100909. https://doi.org/10.1016/j.bar.2020.100909

Christmann, P. (2004). Multinational companies and the natural environment: Determinants of global environmental policy. *Academy of Management Journal, 47*(5), 747–760. https://doi.org/10.5465/20159616

Christofi, M., Hadjielias, E., Hughes, M., & Plakoyiannaki, E. (2024). Advancing Research methodologies in management: Revisiting debates, setting new grounds for pluralism. *British Journal of Management, 35*(1), 24–35. https://doi.org/10.1111/1467-8551.12791

Christofi, M., Vrontis, D., & Cadogan, J. W. (2021). Micro-foundational ambidexterity and multinational enterprises: A systematic review and a conceptual framework. *International Business Review, 30*(1), 101625. https://doi.org/10.1016/j.ibusrev.2019.101625

Cormier, D., Gordon, I. M., & Magnan, M. (2004). Corporate environmental disclosure: Contrasting management's perceptions with reality. *Journal of Business Ethics, 49*, 143–165. https://doi.org/10.1023/B:BUSI.0000015844.86206.b9

Cui, X., Peng, X., Jia, J., & Wu, D. (2020). Does board independence affect environmental disclosures by multinational corporations? Moderating effects of national culture. *Applied Economics, 52*(52), 5687–5705. https://doi.org/10.1080/00036846.2020.1770681

Dagiliene, L., Frendzel, M., Sutiene, K., & Wnuk-Pel, T. (2020). Wise managers think about circular economy, wiser report and analyze it. Research of environmental reporting practices in EU manufacturing companies. *Journal of Cleaner Production, 274*, 121968. https://doi.org/10.1016/j.jclepro.2020.121968

Dam, L., & Petkova, B. N. (2014). The impact of environmental supply chain sustainability programs on shareholder wealth. *International Journal of Operations & Production Management, 34*(5), 586–609. https://doi.org/10.1108/IJOPM-10-2012-0482

Das, A. (2024). Predictive value of supply chain sustainability initiatives for ESG performance: A study of large multinationals. *Multinational Business Review, 32*(1), 20–40. https://doi.org/10.1108/MBR-09-2022-0149

DasGupta, R., Kumar, S., & Pathak, R. (2022). Multinational enterprises' internationalization and adoption of sustainable development goals. *International Journal of Managerial Finance, 18*(4), 617–638. https://doi.org/10.1108/IJMF-09-2021-0416

De Beule, F., Dewaelheyns, N., Schoubben, F., Struyfs, K., & Van Hulle, C. (2022). The influence of environmental regulation on the FDI location choice of EU ETS-covered MNEs. *Journal of Environmental Management, 321*, 115839. https://doi.org/10.1016/j.jenvman.2022.115839

Delgado-Márquez, B. L., & Pedauga, L. E. (2017). Environmental behavior and MNEs: A strategy pulled by stakeholder engagement. *Business Strategy and the Environment, 26*(7), 927–939. https://doi.org/10.1002/bse.1955

Donaldson, T. (1996). Values in tension: Ethics away from home. *Harvard Business Review, 74*(5), 48–62.

Ellimäki, P., Aguilera, R. V., Hurtado-Torres, N. E., & Aragón-Correa, J. A. (2023). The link between foreign institutional owners and multinational enterprises' environmental outcomes. *Journal of International Business Studies, 54*, 910–927. https://doi.org/10.1057/s41267-022-00580-0

Ellimäki, P., Gómez-Bolaños, E., Hurtado-Torres, N., & Aragón-Correa, J. A. (2021). Do global firms increase their environmental disclosure and performance? Symbolic versus effective operations and the moderating role of liability of origin. Legitimation implications. *Industrial Marketing Management, 92*, 354–363. https://doi.org/10.1016/j.indmarman.2019.10.015

Epstein, M. J., & Roy, M. J. (2007). Implementing a corporate environmental strategy: Establishing coordination and control within multinational companies. *Business Strategy and the Environment, 16*(6), 389–403. https://doi.org/10.1002/bse.545

Gallego-Álvarez, I., & Ortas, E. (2017). Corporate environmental sustainability reporting in the context of national cultures: A quantile regression approach. *International Business Review, 26*(2), 337–353. https://doi.org/10.1016/j.ibusrev.2016.09.003

Gallego-Álvarez, I., Ortas, E., Vicente-Villardón, J. L., & Alvarez Etxeberria, I. (2017). Institutional constraints, stakeholder pressure and corporate environmental reporting policies. *Business Strategy and the Environment, 26*(6), 807–825. https://doi.org/10.1002/bse.1952

García-Granero, E. M., Piedra-Munoz, L., & Galdeano-Gómez, E. (2020). Measuring eco-innovation dimensions: The role of environmental corporate culture and commercial orientation. *Research Policy, 49*(8), 104028. https://doi.org/10.1016/j.respol.2020.104028

García-Sánchez, I. M., Gallego-Álvarez, I., & Zafra-Gómez, J. L. (2021). Do independent, female and specialist directors promote eco-innovation and eco-design in agri-food firms? *Business Strategy and the Environment, 30*(2), 1136–1152. https://doi.org/10.1002/bse.2676

Ghauri, P. N. (2022). The role of multinational enterprises in achieving sustainable development goals. *AIB Insights, 22*(1), 1–5. https://doi.org/10.46697/001c.31077

Ghauri, P., Strange, R., & Cooke, F. L. (2021). Research on international business: The new realities. *International Business Review, 30*(2), 101794. https://doi.org/10.1016/j.ibusrev.2021.101794

Golgeci, I., Makhmadshoev, D., & Demirbag, M. (2021). Global value chains and the environmental sustainability of emerging market firms: A systematic review of literature and research agenda. *International Business Review, 30*(5), 101857. https://doi.org/10.1016/j.ibusrev.2021.101857

Gómez-Bolaños, E., Ellimäki, P., Hurtado-Torres, N. E., & Delgado-Márquez, B. L. (2022). Internationalization and environmental innovation in the energy sector: Exploring the differences between multinational enterprises from emerging and developed countries. *Energy Policy, 163*, 112867. https://doi.org/10.1016/j.enpol.2022.112867

Ha, Y. J. (2021). Attention green aliens? Activities of multinational enterprises in host countries and eco-innovation diffusion. *Journal of Business Research, 123*, 32–43. https://doi.org/10.1016/j.jbusres.2020.09.038

Hitt, M. A., Beamish, P. W., Jackson, S. E., & Mathieu, J. E. (2007). Building theoretical and empirical bridges across levels: Multilevel research in management. *Academy of Management Journal, 50*(6), 1385–1399. https://doi.org/10.5465/amj.2007.28166219

Hsu, P. F., Hu, P. J. H., Wei, C. P., & Huang, J. W. (2014). Green purchasing by MNC subsidiaries: The role of local tailoring in the presence of institutional duality. *Decision Sciences, 45*(4), 647–682. https://doi.org/10.1111/deci.12088

Kafouros, M., Hashai, N., Tardios, J. A., & Wang, E. Y. (2022). How do MNEs invent? An invention-based perspective of MNE profitability. *Journal of International Business Studies, 53*(7), 1420–1448. https://doi.org/10.1057/s41267-021-00499-y

Kassarjian, H. H. (1977). Content analysis in consumer research. *Journal of Consumer Research, 4*(1), 8–18. https://doi.org/10.1086/208674

Kawai, N., Strange, R., & Zucchella, A. (2018). Stakeholder pressures, EMS implementation, and green innovation in MNC overseas subsidiaries. *International Business Review, 27*(5), 933–946. https://doi.org/10.1016/j.ibusrev.2018.02.004

Khanna, T., & Palepu, K. G. (2004). Globalization and convergence in corporate governance: Evidence from Infosys and the Indian software industry. *Journal of International Business Studies, 35*, 484–507. https://doi.org/10.1057/palgrave.jibs.840010

Kolk, A. (2016). The social responsibility of international business: From ethics and the environment to CSR and sustainable development. *Journal of World Business, 51*(1), 23–34. https://doi.org/10.1016/j.jwb.2015.08.010

Kolk, A., & Fortanier, F. (2013). Internationalization and environmental disclosure: The role of home and host institutions. *Multinational Business Review, 21*(1), 87–114. https://doi.org/10.1108/15253831311309500

Kolk, A., & Perego, P. (2010). Determinants of the adoption of sustainability assurance statements: An international investigation. *Business Strategy and the Environment, 19*(3), 182–198. https://doi.org/10.1002/bse.643

Kolk, A., & Van Tulder, R. (2010). International business, corporate social responsibility and sustainable development. *International Business Review, 19*(2), 119–125. https://doi.org/10.1016/j.ibusrev.2009.12.003

Konara, P., Lopez, C., & Shirodkar, V. (2021). Environmental innovation in foreign subsidiaries: the role of home-ecological institutions, subsidiary establishment mode and post-establishment experience. *Journal of World Business, 56*(6), 101261. https://doi.org/10.1016/j.jwb.2021.101261
Kourula, A., Pisani, N., & Kolk, A. (2017). Corporate sustainability and inclusive development: Highlights from international business and management research. *Current Opinion in Environmental Sustainability, 24*, 14–18. https://doi.org/10.1016/j.cosust.2017.01.003
Kumar, P. (2016). State of green marketing research over 25 years (1990-2014) literature survey and classification. *Marketing Intelligence & Planning, 34*(1), 137–158. https://doi.org/10.1108/MIP-03-2015-0061
Leonidou, L. C., & Katsikeas, C. S. (2010). Integrative assessment of exporting research articles in business journals during the period 1960–2007. *Journal of Business Research, 63*(8), 879–887. https://doi.org/10.1016/j.jbusres.2010.01.005
Leonidou, L. C., Leonidou, C. N., Palihawadana, D., & Hultman, M. (2011). Evaluating the green advertising practices of international firms: A trend analysis. *International Marketing Review, 28*(1), 6–33. https://doi.org/10.1108/02651331111107080
Levy, D. L. (1995). The environmental practices and performance of transnational corporations. *Transnational Corporations, 4*(1), 44–67. https://doi.org/10.18356/d6842206-en
Leyva-de la Hiz, D. I., Hurtado-Torres, N., & Bermúdez-Edo, M. (2019). The heterogeneity of levels of green innovation by firms in international contexts: A study based on the home-country institutional profile. *Organization & Environment, 32*(4), 508–527. https://doi.org/10.1177/1086026618761623
Li, F., Larimo, J., & Leonidou, L. C. (2023). Social media in marketing research: Theoretical bases, methodological aspects, and thematic focus. *Psychology & Marketing, 40*(1), 124–145. https://doi.org/10.1002/mar.21746
Loorbach, D., van Bakel, J. C., Whiteman, G., & Rotmans, J. (2010). Business strategies for transitions towards sustainable systems. *Business Strategy and the Environment, 19*(2), 133–146. https://doi.org/10.1002/bse.645
Marchi, V. D., Maria, E. D., & Micelli, S. (2013). Environmental strategies, upgrading and competitive advantage in global value chains. *Business Strategy and the Environment, 22*(1), 62–72. https://doi.org/10.1002/bse.1738
Marco-Lajara, B., Úbeda-García, M., Zaragoza-Sáez, P., & Manresa-Marhuenda, E. (2023). The impact of international experience on firm economic performance. The double mediating effect of green knowledge acquisition & eco-innovation. *Journal of Business Research, 157*, 113602. https://doi.org/10.1016/j.jbusres.2022.113602
Margaret, I., Schoubben, F., & Verwaal, E. (2024). When do investors see value in international environmental management certification of multinational corporations? A study of ISO 14001 certification after the Paris Agreement. *Global Strategy Journal, 14*, 25–55. https://doi.org/10.1002/gsj.1490
Marin, G., & Zanfei, A. (2019). Does host market regulation induce cross-border environmental innovation? *The World Economy, 42*(7), 2089–2119. https://doi.org/10.1111/twec.12784
Mukonza, C., & Swarts, I. (2020). The influence of green marketing strategies on business performance and corporate image in the retail sector. *Business Strategy and the Environment, 29*(3), 838–845. https://doi.org/10.1002/bse.2401
Nielsen, B. B., Wechtler, H., & Zheng, L. G. (2023). Disasters and international business: Insights and recommendations from a systematic review. *Journal of World Business, 58*(4), 101458. https://doi.org/10.1016/j.jwb.2023.101458
Noailly, J., & Ryfisch, D. (2015). Multinational firms and the internationalization of green R&D: A review of the evidence and policy implications. *Energy Policy, 83*, 218–228. https://doi.org/10.1016/j.enpol.2015.03.002
Nwachukwu, C. (2022). Systematic review of integrated reporting: Recent trend and future research agenda. *Journal of Financial Reporting and Accounting, 20*(3/4), 580–598. https://doi.org/10.1108/JFRA-10-2020-0308

Ozkan, K. S., Khan, H., Deligonul, S., Yeniyurt, S., Gu, Q. C., Cavusgil, E., & Xu, S. (2022). Race for market share gains: How emerging market and advanced economy MNEs perform in each other's turf. *Journal of Business Research, 150*, 208–222. https://doi.org/10.1016/j.jbusres.2022.04.040

Parboteeah, K. P., Addae, H. M., & Cullen, J. B. (2012). Propensity to support sustainability initiatives: A cross-national model. *Journal of Business Ethics, 105*, 403–413. https://doi.org/10.1007/s10551-011-0979-6

Paul, J., Parthasarathy, S., & Gupta, P. (2017). Exporting challenges of SMEs: A review and future research agenda. *Journal of World Business, 52*(3), 327–342. https://doi.org/10.1016/j.jwb.2017.01.003

Pazirandeh, A., & Jafari, H. (2013). Making sense of green logistics. *International Journal of Productivity and Performance Management, 62*(8), 889–904. https://doi.org/10.1108/IJPPM-03-2013-0059

Peng, X., Lan, Y. C., Li, J., & Fan, H. (2023). Board gender diversity, national culture, and water disclosure of multinational corporations. *Applied Economics, 55*(14), 1581–1602. https://doi.org/10.1080/00036846.2022.2098240

Pizzi, S., Caputo, A., Corvino, A., & Venturelli, A. (2020). Management research and the UN sustainable development goals (SDGs): A bibliometric investigation and systematic review. *Journal of Cleaner Production, 276*, 124033. https://doi.org/10.1016/j.jclepro.2020.124033

Porter, M. E., & Linde, C. V. D. (1995). Toward a new conception of the environment-competitiveness relationship. *Journal of Economic Perspectives, 9*(4), 97–118. https://doi.org/10.1257/jep.9.4.97

Rondinelli, D. A. (2007). Globalization of sustainable development: Principles and practices in transnational corporations. *Multinational Business Review, 15*(1), 1–24. https://doi.org/10.1108/1525383X200700001

Samiee, S., & Chirapanda, S. (2019). International marketing strategy in emerging-market exporting firms. *Journal of International Marketing, 27*(1), 20–37. https://doi.org/10.1177/1069031X18812731

Shaw, J. D., Gupta, N., & Delery, J. E. (2001). Congruence between technology and compensation systems: Implications for strategy implementation. *Strategic Management Journal, 22*(4), 379–386. https://doi.org/10.1002/smj.165

Smith, J. (2008). That elusive elasticity and the ubiquitous bias: Is panel data a panacea? *Journal of Macroeconomics, 30*(2), 760–779. https://doi.org/10.1016/j.jmacro.2007.06.002

Sousa, C. M., & Lengler, J. (2009). Psychic distance, marketing strategy and performance in export ventures of Brazilian firms. *Journal of Marketing Management, 25*(5–6), 591–610. https://doi.org/10.1362/026725709X461876

Sousa, C. M., Martínez-López, F. J., & Coelho, F. (2008). The determinants of export performance: A review of the research in the literature between 1998 and 2005. *International Journal of Management Reviews, 10*(4), 343–374. https://doi.org/10.1111/j.1468-2370.2008.00232.x

UNCTAD. (2020). *Fostering green exports through voluntary sustainability standards*. Retrieved January 14, 2024, from: https://unctad.org/system/files/official-document/ditctab2020d1_en.pdf

UNCTAD. (2021). *Better trade for sustainable development: The role of voluntary sustainability standards*. Retrieved January 14, 2024, from: https://unctad.org/system/files/official-document/ditctab2021d2_en.pdf

UNEP. (2019). *Adapt to survive: Business transformation in a time of uncertainty.* Retrieved July 05, 2024, from: https://content.yudu.com/web/2y3n2/0A2y3n3/AdaptSurviveBusTrans/html/index.html?origin=reader

Vesal, M., Siahtiri, V., & O'Cass, A. (2022). Do senior managers hold the keys to unlock innovation and environmental sustainability? *Industrial Marketing Management, 103*, 83–96. https://doi.org/10.1016/j.indmarman.2022.03.010

Villo, S., Halme, M., & Ritvala, T. (2020). Theorizing MNE-NGO conflicts in state-capitalist contexts: Insights from the Greenpeace, Gazprom and the Russian state dispute in the Arctic. *Journal of World Business, 55*(3), 101068. https://doi.org/10.1016/j.jwb.2019.101068

Voola, R., Bandyopadhyay, C., Voola, A., Ray, S., & Carlson, J. (2022). B2B marketing scholarship and the UN sustainable development goals (SDGs): A systematic literature review. *Industrial Marketing Management, 101*, 12–32. https://doi.org/10.1016/j.indmarman.2021.11.013

Wood, G., Pereira, V., Temouri, Y., & Wilkinson, A. (2021). Exploring and investigating sustainable international business practices by MNEs in emerging markets. *International Business Review, 30*(5), 101899. https://doi.org/10.1016/j.ibusrev.2021.101899

Yang, Y., & Driffield, N. (2022). Leveraging the benefits of location decisions into performance: A global view from matched MNEs. *Journal of Business Research, 139*, 468–483. https://doi.org/10.1016/j.jbusres.2021.09.071

Yu, H., Bansal, P., & Arjaliès, D. L. (2023). International business is contributing to environmental crises. *Journal of International Business Studies, 54*, 1151–1169. https://doi.org/10.1057/s41267-022-00590-y

Zeriti, A., Robson, M. J., Spyropoulou, S., & Leonidou, C. N. (2014). Sustainable export marketing strategy fit and performance. *Journal of International Marketing, 22*(4), 44–66. https://doi.org/10.1509/jim.14.0063

Zhou, S., Tiruneh, W. A., & Legese, M. A. (2023). The effect of corporate social responsibility on environmental performance: The mediating role of green innovation and green human resource management. *International Journal of Emerging Markets, 1-21.* https://doi.org/10.1108/IJOEM-02-2022-0211

Ilayda Ipek PhD, Dokuz Eylül University, MBA, Middle East Technical University, is an Associate Professor in Marketing at the Faculty of Business, Dokuz Eylül University. From February to September 2024, she was a Visiting Scholar at the School of Business, University of Leicester. Her research interests fall mainly into international marketing, business-to-business marketing, and sustainability. Her articles have appeared in *Industrial Marketing Management*, *Business Strategy and the Environment*, *International Business Review*, *International Marketing Review*, and *the Journal of Marketing Management.*

The Impact of MNEs on Ocean Sustainability

Arijit Goswami

1 Introduction

Marine life worldwide has been immeasurably impacted by the toxicity and pollutants added by Multinational Enterprises (MNEs) as part of their business operations. MNEs, operating in diverse sectors such as manufacturing, energy, and agriculture, have historically contributed to water pollution through industrial discharges, chemical runoff, and improper waste disposal. This has resulted in devastating consequences for marine ecosystems, including habitat destruction, loss of biodiversity, and water quality degradation. The environmental footprint of these enterprises is a critical determinant in achieving Sustainable Development Goal (SDG) 14 (Ullah et al., 2023). From shipping activities and oil drilling to large-scale fishing and manufacturing, diverse types of industries impact marine ecosystems and marine life daily. Several other industrial activities by MNEs destroy coastal marine habitats, which have cascading effects on marine biodiversity that further impact the migration cycles of marine fauna, and reproductive cycles of aquatic animals and reduce the survivability of marine species. Even the business activities of MNEs that are distant from the oceans, seas, and rivers act as major contributors to the pollution of underground water reservoirs and erosion of the water table. Underwater life alone is not affected, but even the seabirds and amphibious species are adversely impacted by the collapse of marine ecosystems. Precisely, the economic growth and technological advancement that the MNEs offer come at the cost of environmental damage and loss of health of marine life.

The perpetual contamination of marine biospheres and available water sources adds massive pressure on the sustenance of the growing population. Disease

A. Goswami (✉)
Markets and Markets, Bengaluru, Karnataka, India
e-mail: arijitg21@iimklive.com

J. Bento, M. M. Torres (eds.), *Global Perspectives on Climate Change, Inequality, and Multinational Corporations*, CSR, Sustainability, Ethics & Governance, https://doi.org/10.1007/978-3-031-80797-8_3

outbreaks and the escape of non-native species into the wild further threaten marine biodiversity. Affecting life below water disrupts the food chain, affects the food supply, and impacts the livelihood of people, thereby amplifying the socioeconomic divide among the population of a region. This has a strong bearing on several SDGs—no poverty, zero hunger, good health & well-being, reduced inequalities, sustainable cities & communities, decent work & economic growth, and climate action. Therefore, the multifaceted impact of MNEs on life below water makes it critical for businesses, policymakers, academics, and governments to build laws and ethical standards to regulate the influence of MNEs on marine ecosystems and conserve life below water. A deeper comprehension of the financial and non-financial benefits of environmental, social, and governance (ESG) frameworks for the MNEs can elucidate the role MNEs must play in the conservation of life below water (Wilkinson et al., 2022).

The narrative shifts when considering the transformative potential of MNEs in protecting marine life. Growing awareness and proactive measures among multinational corporations to operate sustainably, using advanced technologies, are key. These technologies enable effective water management, pollution control, and waste reduction, enhancing environmental stewardship. Thus, MNEs play a crucial role in monitoring, measuring, and mitigating their impact on seas and oceans, contributing significantly to SDG 14 and the broader sustainability agenda (Tawari & Davies, 2020). The availability of significant capital and manpower positions MNEs to drive sustainability initiatives for life below water. Their extensive supply chains and global operations enable them to improve water quality and marine health across continents. This global presence allows MNEs to influence sustainable water policies locally and internationally. Ensuring water sustainability mitigates risks to their reputation and operations, as several firms have been penalized for regional water pollution. Moreover, protecting marine life aligns with corporate social responsibility (CSR) goals, empowering communities with safe aquatic ecosystems (Martin & O'Neill, 2021). Focusing on sustainable water bodies can enhance brand loyalty, competitive edge, and business reputation.

Considering the positive and negative impacts that MNEs can have on aquatic ecosystems, it can be inferred that MNEs have a fundamental role to play in the achievement of SDG 14 (Hart & Zingales, 2017). By implementing water conservation measures within their operations and supply chains, investing in innovative technologies, and partnering with stakeholders, MNEs can make significant contributions to safeguarding water resources for future generations (Schoenmaker, 2020). This chapter explores the role of MNEs in the health and sustenance of marine biospheres and establishes how MNEs can act as determinants of life below water. By citing real-world examples of MNEs affecting marine ecosystems, the chapter investigates the impact that MNEs have on marine ecosystems through effluent discharge, wastewater production, release of toxins, chemical contamination, and microbial pollution in different sectors. It further elucidates the strategies adopted by the various industries—such as automotive, pharmaceutical, oil and gas, semiconductors, shipping, and textiles—in wastewater treatment, effluent management, waste consumption optimization, and biological and chemical treatment.

Furthermore, the chapter reiterates the connection between MNEs and SDG 14 by establishing the economic rationale for MNEs to invest in marine sustainability and the partnerships they can leverage for such initiatives.

2 Impact of MNEs on Marine Life

Industrialization has had a significant impact on marine pollution, particularly in energy production, mining, manufacturing, and transportation (Ullah et al., 2023). Industrial setups often discharge chemical effluents into marine ecosystems through direct release into water bodies or runoff. These effluents contain toxic substances, heavy metals, pesticides, and non-biodegradable industrial wastes. This pollution collapses marine habitats, disrupts food chains, and destabilizes local ecosystems. Additionally, industrial plants emit airborne toxins that dissolve in oceans, causing global ocean acidification. Figure 1 depicts the falling pH levels in the oceans worldwide, implying increasing acidification.

2.1 Sectoral Impact on Life Below Water

Accidental oil spills caused by multinational oil and gas companies devastate aquatic organisms, kill entire habitats, affect migratory birds, and crumple marine ecosystems in the affected area. Routine oil and gas activities have detrimental effects on the marine environment during the phases of exploration, production, and

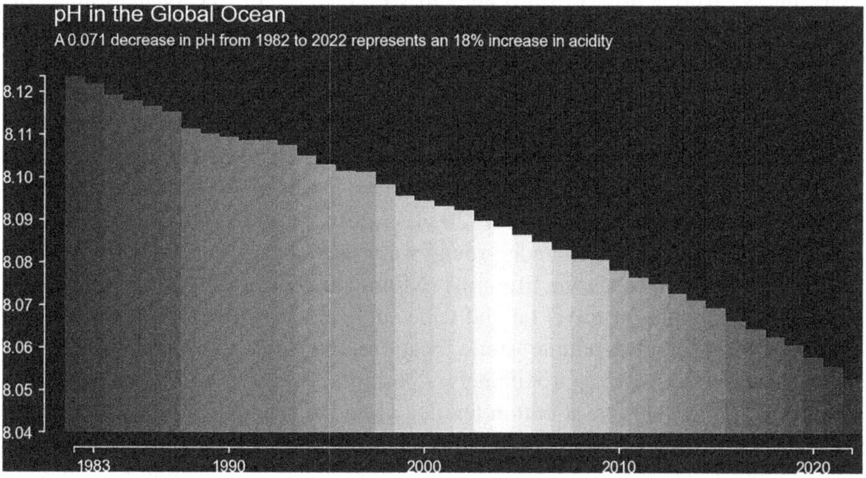

Fig. 1 Increase in acidification of global oceans. (Gruber and Gregor, 2024)

decommissioning (Cordes et al., 2016). During exploration, indirect disturbances like sound and traffic impact marine life as much as direct disturbances, such as drilling fluids and anchor chains. The production phase disturbs marine flora and fauna through contact with metal pipes. During decommissioning, contaminants can renter the marine environment. Acoustic disturbances from seismic surveys and deep-water oil and gas development affect underwater animals' feeding, breeding, resting, and migration behaviours. Moreover, considerable amounts of artificial light in oil fields can potentially affect the ecological processes in the upper ocean. Placement of structures on the seafloor causes a transient increase in local sedimentation, while anchors of the drilling rig damage benthic organisms and disturb fragile habitats of corals and sponges. Additionally, produced water is contaminated with dissolved organic salts, dispersed hydrocarbons, dissolved gasses, radioactive substances, minerals, and trace metals.

The automotive manufacturing process produces wastewater, which contains metals, grease, oils, and toxic chemicals from paint residue (Nixon & Saphores, 2007). The pollutants from automobiles and transportation infrastructure include sediments from the construction phase, oils and grease from leaks, heavy metals from car exhaust, engine parts and worn tires, and road salts. Millions of gallons of improperly discharged motor oil contaminate water bodies and coastal areas annually. MNEs in the pharmaceutical industry discharge active pharmaceutical ingredients (APIs) during their manufacture, use, and disposal. The APIs feminize fish, increase their susceptibility to predation, and increase the mortality rates of fish (Wilkinson et al., 2022). Pharmaceutical active compounds are one of the major water contaminants that have an inimical effect on aquatic organisms even at a very low concentration, as they don't degrade easily and persist in the aquatic environment for prolonged periods due to their stable structure (Samal et al., 2022).

The fast fashion industry sector plays a significant role in water pollution, as it consumes 79 trillion litres of water annually, which leads to 20 percent of industrial wastewater (Bailey et al., 2022). The industry also contributes to 35 percent of the oceanic primary microplastics, which amount to 190 thousand tonnes annually (Niinimäki et al., 2020). A Swedish study in 2014 found that out of 2450 chemicals related to textile manufacturing, 10 percent had high potential concern for human health, including fragrances and azo dyes, alongside reproductive toxins, such as fluoridated water, brominated flame retardants, phthalates, and stain repellents (KEMI Swedish Chemicals Agency, 2014). Antibacterial agents added to textiles travel with effluents to reach underground water and other water bodies to generate antibiotic resistance in marine lifeforms (Niinimäki et al., 2020). Such toxins from the fast fashion industry travel far and wide, and chemicals used to waterproof textiles have been found in remote Arctic locations and inside the bodies of seals and polar bears. On average, approximately 9.36 million litres of water is required for the production of one ton of cotton textile, while the production of a pair of jeans and one cotton T-shirt requires 10,850 litres and 2720 litres of water, respectively (Familglietti et al., 2022). Overfishing in the oceans is also leading to a massive decline of fish populations below water, resulting in a sharp decline of fish stocks within biologically sustainable levels. Figure 2 depicts the decline in the proportion

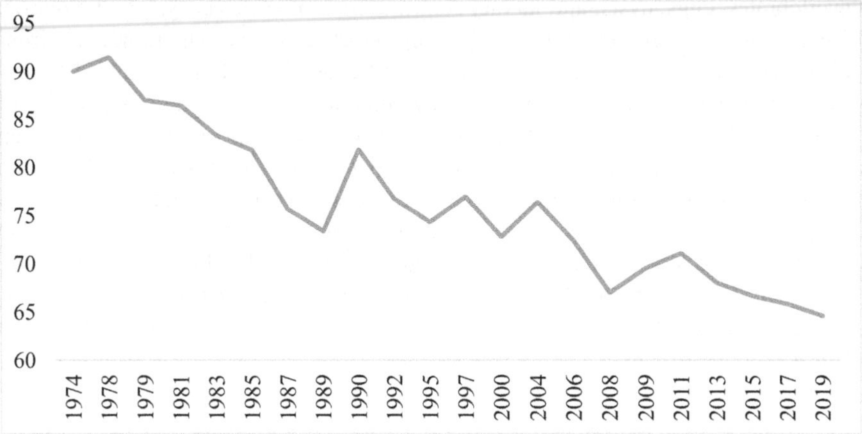

Fig. 2 The proportion of fish stocks within non-exploited biologically sustainable levels. (Food and Agriculture Organization, https://de-public-statsuite.fao.org/)

of fish stocks under sustainable levels, based on data sourced from the Food and Agriculture Organization (Food and Agriculture Organization of the United Nations, 2024).

Additionally, the semiconductor industry poses a major challenge to life below water. While water is essential for the fabrication of ultra-thin films, back grinding and chemical mechanical polishing produce wastewater (Sadafale & Gaikwad, 2023). The effluents coming from the semiconductor manufacturing plants have high turbidity and have acidic and basic ions, trace metals, volatile compounds, organic compounds, a high chemical oxygen demand concentration, and fine oxide particles that pollute water bodies into which they are released. The release of wastewater into streams and water bodies leads to the degradation of water quality, toxicity to marine life, eutrophication, acidification, and modification of composition (Villard et al., 2015). Moreover, various liquids used for lithography and etching are released into water bodies that threaten aquatic ecosystems. The use of ultra-pure water leads to intensive use of water, which results in increased water stress and threat to aquatic ecosystems.

The shipping industry contributes significantly to the pollution of marine habitats as the operation of ships leads to spills of grease, lubricating oil, fuel, and water into bilges (Iduk & Samson, 2015). Emulsified water and oil, if not separated, contribute to water pollution. Ballast water, with oil residues, pollutes oceans, as does cleaning process water containing detergent and solid matter. Without enforcement, ships dump garbage and waste, including glass, paper, cardboard, aluminium and steel cans, and plastics, which can be non-hazardous or hazardous after entering oceans. Operational errors, malfunctioning pumps and valves, and pipe ruptures can lead to oil spills during crude oil loading and discharge at offshore terminals. Similarly, loading bunker oil or lubricating oil for ship engines may result in operational spills. Ballast water, necessary for safe shipping, carries bacteria, microbes, invertebrates, eggs, cysts, and larvae, contributing to marine ecosystem disruption. These transferred species may establish reproductive populations in new environments, becoming invasive and disrupting native ecosystems. For example, 140 new exotic organisms have been introduced by shipping vessels to the Great Lakes of North America (Mills et al., 1993). Foreign species, such as European crab, Asian clam, and coast comb jellyfish, have been found to destroy fisheries on the American Atlantic coast (Council, 1996). Additionally, vessel-based sewage pollution has comparatively higher potency than domestic sewage as it has less water available per unit volume of sewage and wastes when discharged (Mobilik & Hassan, 2016). Figure 3 offers a snapshot of the contaminants that different industries add to water during upstream, midstream, and downstream operations, which eventually make their way to the oceans.

2.2 *Examples of MNEs Threatening Life Below Water*

Several incidents around the world have highlighted the severe consequences MNEs can have on the health of water bodies. From the food and beverages industry, Coca-Cola serves as a prime example. In Kerala, Coca-Cola's plant spans 34 acres, utilizing six borewells and two open wells for water. Dr. R.N. Athawale's 2022 report reveals it needs 0.64 M litters of water daily, producing 0.15–0.3 M litres of wastewater (Government of Kerala, 2010). Pfizer's 2018–19 water usage of 5.32 billion litres at its operated manufactured and owned sites threatened aquatic ecosystems, potentially depleting groundwater (Familglietti et al., 2022).

The semiconductor and materials sector has not been far behind. In 2011, Apple's key supplier in China—Meiko Electronics—was identified to have caused massive water pollution by its manufacturing activities, much to the agony of the residents.

Agents of ocean pollution and threat to marine ecosystems across the supply chain in major industries

	Upstream	Midstream	Downstream
	Metals and toxic substances in car paints, wastewater of car wash	Metals, grease, oils, and toxic chemicals from paint residue in wastewater	Grease from leaks, heavy metals, engine parts, worn tires, motor oil
	Acoustic disturbances, artificial light, drilling fluids, sedimentation	Organic salts, hydrocarbons, dissolved gases, radioactive substances	Decommissioning of plants, contaminant leakage
	Microplastics, wastewater, reproductive toxins, azo dyes, Antibacterial agents	Oil and contaminant leakage during shipping	Marine pollution through landfills, release of dyes by disposed clothes
	Chemical polishes, lithography liquids, basic ions, trace metals	Plant cooling facilities, storage cooling facilities intaking excess water	E-waste, heavy metals from disposed semiconductor equipment
	Active pharmaceutical ingredients, heavy metals, hormones, antibiotics	Chemical leakages from storages, sewage leakages	Wastewater, plastic coverings of drugs, expired or damaged medicines
	Hull paint, leaching of antifouling paints, wastewater from shipbuilding yards	Accidental discharge of oil in the sea, loss from deposit tankers and pipeline	Oil spills, ballast water, microorganisms, organic waste disposal

Fig. 3 Agents of ocean pollution across industries. (Source: Authors' elaboration)

The worst oil spill in South Korea occurred in December 2007 when a crane barge owned by Samsung Heavy Industries collided with the Hong Kong-registered oil tanker 'Habei Spirit', affecting the protected coastal area within Taean National

Table 1 Examples of a how MNEs threaten marine sustainability (Source: Authors' elaboration)

Sectors	Overconsumption	Contamination	Ecosystem destruction
Automotive	Producing a single vehicle requires over 40,000 gallons of water (Rakshit, 2022).	A 2020 report found that tyre dust was the biggest single source of microplastics, contributing 78% of the total mass (LaBrecque, 2023).	The New Jersey Department of Environmental Protection alleged Ford illegally disposed of paint sludge and other contaminants at a site in Bergen County, which led to contamination of groundwater, soils, sediments, surface waters, and freshwater wetlands (Melillo, 2022).
Fast fashion	In Levi Strauss & Co., a single pair of jeans has been known to consume up to 3800 litres of water (Naik, 2023).	Four of eleven H&M factories in Bangladesh discharged contaminated water, turning entire rivers biologically dead (Izmir, 2022).	The fashion industry accounts for approximately 20% of industrial water pollution and contributes 35% to oceanic microplastic pollution (Asia Europe Environment Forum, 2020).
Food & beverage	Coca-Cola consumed 0.64 M litres of water daily, producing 0.15–0.3 M litres of wastewater (Government of Kerala, 2010).	Cadmium, copper, chromium, and lead contamination exceed WHO limits by Coca-Cola.	Pepsi's packaging of beverages was found to introduce microplastics in Lake Erie, which supplies Buffalo's drinking water, as well as in fish species that are known to inhabit the lake (Khan, 2023).
Oil & gas	Between 2018 and 2021, oil and gas companies in California used nearly 3 billion gallons of fresh water for drilling—water that could have supplied domestic needs (Beaumont, 2023).	ExxonMobil's oil exploration impacted the primary fishing area bordering Namibia.	British Petroleum Deepwater Horizon oil disaster in the Gulf of Mexico led dolphins to suffer lung issues, anemia, and pregnancy problems (Holcomb, 2023).
Pharmaceutical	Pfizer's 2018–19 water usage of 5.32 billion litres depleted groundwater in several areas.	Arkema polluted the Rhône Valley with toxic PFAS.	Solvay in Salindres discharges up to 100 kilograms of TFA into the Neckar River daily.
Semiconductor	Google's data Centre withdraws 5.68 million litres daily from the strained Charleston aquifer.	Minera del Altiplano consumes millions of litres of water for mining lithium in the Salar del hombre Muerto salt flats.	Samsung owned crane barge caused an oil spill in the Taean National Park coastal area, impacting the marine life ecosystem.

Park (Global Atlas of Environmental Justice, 2024a). In 2021, Google's expansion plans for a data centre in South Carolina included withdrawing 5.68 million litres daily from the strained Charleston Aquifer (Familglietti et al., 2022). This plan faced opposition from the local population concerned about the health of aquifers and marine life (Table 1).

Clusters of water-polluting MNEs pose significant challenges in localized regions. For instance, the Rhône Valley south of Lyon is heavily contaminated with toxic per- and polyfluorinated substances (PFAS), largely due to companies like Arkema (Global Atlas of Environmental Justice, 2024b). Contaminated drinking water has resulted in over 200,000 people suffering from slow poisoning. Similarly, in Europe, the Solvay chemical plant in Salindres is one of France's five per- and polyfluoroalkyl substances (PFAS) production sites. The Solvay plant in Bad Wimpfen was found to be discharging up to 100 kilograms of TFA into the Neckar River per day, with TFA levels not closely monitored or included in environmental reviews (Global Atlas of Environmental Justice, 2024c). In 2021, the Angolan government approved an oil exploration and extraction project by ExxonMobil in southern Angola's Namibe region, a primary fishing area bordering Namibia. The Namibe fishing zone represents 65% of all fishing activity in Angola, offering a diverse range of fishing resources (Global Atlas of Environmental Justice, 2024d). Since 1997, Livent's subsidiary Minera del Altiplano has been mining lithium in the Salar del Hombre Muerto salt flats, using millions of litres of water in the extraction and processing of the mineral (Global Atlas of Environmental Justice, 2024e). This activity has led to an acceleration of groundwater flow, the scarcity or irreversible disappearance of water resources, and contamination of freshwater layers, endangering a fragile ecosystem and negatively impacting animal and human communities.

3 Strategies for MNEs to Mitigate Water Pollution

Across all sectors, MNEs are responsible and must be held accountable for the strategies they are implementing to ensure the safety of life below water. By investing in emerging technologies and identifying viable solutions to mitigate the negative implications of business, MNEs can reduce their environmental footprint. This section discusses a few strategies for MNEs to adopt across diverse sectors.

3.1 Strategies to Promote Sustainable Fishing

The application of advanced technologies and industrialized harvesting methods by multinational fishing enterprises has resulted in the collapse of fisheries (Wild, 2008). Traditional fisheries are losing commercial viability and fish harvesting is extending to formerly marginal areas, while one-third of the ocean fisheries have collapsed. MNEs must operate within their transferable quotas assigned by the government to set maximum catch limits on specific commercial species in certain locations, as has been implemented in New Zealand. MNEs also need to implement voluntary controls on bottom trawling activities in major seas and oceans and the harvest of endangered bycatch species, such as swordfish, rare turtles, marine birds, marine mammals, and petrels.

Multinational fishing enterprises collaborate with local fish farmers, aggregating harvests for global supply chains. To support small fish farmers, MNEs can host training sessions on modern techniques, covering gear use, fish seed multiplication, disease management, and weed control (Tawari & Davies, 2020). MNEs can establish a sustainable credit system for fishers' needs, supporting responsible fishing practices. Moreover, eco-labelling promotes sustainably caught fish, creating demand and linking producers to premium markets. Additionally, there is a growing trend towards socially responsible fisheries value chains that prioritize both environmental sustainability and the well-being of workers. MNEs can incentivize the inclusion of artisanal producers, empower women, and ensure fair working conditions. They can influence local policies, advocating for quotas to prevent overfishing and promoting equitable distribution. MNEs establish cooperatives for sustainable fishing, reducing by-catch and promoting equitable distribution. Cooperatives foster individual accountability, support scientific research, and encourage innovative fishing practices.

AI has become vital in combating illegal fishing. AI and automation aid in monitoring and managing marine resources, offering innovative solutions to research challenges. Machine learning techniques analyse underwater photos and acoustic data, using vision sensors, cameras, and AI drones, and assist in tracking endangered fish species and their habitats. Transmitters on marine species like humpback

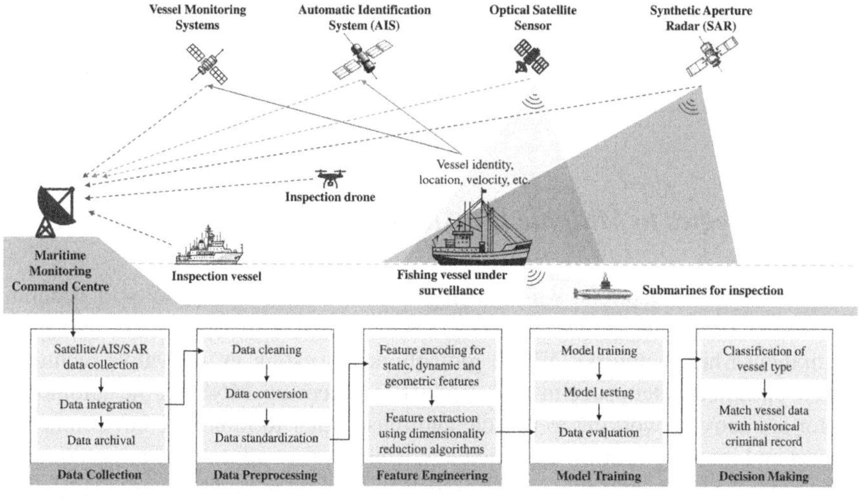

Fig. 4 Application of satellite monitoring, drones, and AI for tracking illegal fishing vessels. (Source: Authors' elaboration adapted from Arsenault & Le Billon, 2022)

whales enable real-time movement monitoring (Mohale et al., 2023). Figure 4 illustrates the use of satellites and AI in detecting illegal ships that are responsible for overfishing.

For instance, OceanMind, a marine enforcement organization, aids countries in adhering to the Port State Measures Agreement. It monitors around 5 million square kilometres of marine protected areas across all oceans, analysing the behaviour of tens of thousands of vessels for compliance with fisheries regulations. Using AI, OceanMind processes data from sources like transponders, radar images, satellite imagery, and cell phone signals. Its machine-learning algorithms predict fishing behaviour based on vessel location, spotting suspicious and potentially illegal activities such as fishing in prohibited areas (Thornton, 2019). Global Fishing Watch, a collaboration between Google, Oceana, and SkyTruth, uses AI and satellite imagery to monitor over 65,000 commercial fishing vessels globally. It combines satellite imagery analysis with AIS signals, radar, and optical imagery to detect vessels not broadcasting their locations because of illegal fishing activities (Woollacott, 2024).

To further enhance monitoring capabilities, efforts are underway to develop AI-powered submarine robots equipped with underwater sensors. Efforts are ongoing to develop AI-powered submarine monitoring robots with underwater sensors to analyse underwater sounds for identifying fishing activities and distinguishing between natural ocean sounds and those from fishing vessels (Mühling, 2023). AI technology aids researchers and conservationists in understanding and addressing overfishing challenges, promoting sustainable marine resource management. In aquaculture, AI plays a significant role in breeding, rearing, and harvesting aquatic organisms. Studies explore AI applications for disease detection, feeding optimization, and fish reproduction management (Rather et al., 2024). AI prediction models optimize feeding rates based on fish species, growth rates, and environmental factors, reducing antibiotic use and disease risks. These models forecast the expected yield and harvest date, enabling efficient operations for farmers. For example, Umitron Corporation's real-time fish behaviour tracking and AI-based feeding decisions cut waste and transportation costs and improve feed conversion efficiency over traditional methods.

3.2 Strategies to Mitigate the Automotive Industry's Pollution

Automotive manufacturers, while having a smaller water footprint, still generate wastewater containing pollutants like grease and metals. To address this, Original Equipment Manufacturers (OEMs) globally are investing in innovative technologies to optimize water usage in their production processes. MNEs in the automotive sector are actively working to recycle and reuse water to minimize environmental impact and improve operational efficiency. This requires advanced treatment methods to remove solids and contaminants such as oil and grease from wastewater. The integration of a Reverse Osmosis (RO) system emerges as a highly recommended solution within this context, serving dual purposes of conserving water and minimizing paint wastage (Shivajee, 2022). By reducing the load on Effluent Treatment Plants (ETPs), RO systems contribute significantly to operational streamlining and resource conservation. Toyota, a global automotive leader, conducts thorough impact assessments in 22 facilities across North America, Asia, and Europe (Toyota, 2023), demonstrating its dedication to sustainable water management. Hyundai Motor similarly addresses water scarcity concerns in its factories, reflecting proactive environmental stewardship. Nissan adopts waterless car washing strategies, saving significant water per vehicle, while Maruti Suzuki implements water management techniques, highlighting the industry's collective commitment to resource conservation.

OEMs and MNEs in the automotive sector are actively adopting water conservation strategies to enhance operational efficiency and address environmental challenges. By investing in advanced treatment technologies, recycling measures, and innovative cleaning processes, they set new benchmarks for sustainable production. For instance, BMW Group's cascade systems enable intelligent reuse of process water, minimizing overall consumption and aligning with sustainable practices (BMW, 2021). Its innovative water recirculation system in car wash facilities reduces consumption during vehicle testing procedures, such as leak testing. This collective commitment underscores the industry's dedication to environmental sustainability and resource optimization.

Wastewater treatment holds promise for automotive MNEs as they can extract processed water from the treatment of wastewater. For example, Toyota has demonstrated a commitment to sustainable water management practices, with 95 percent of its water consumption for production being met through recycled and rainwater sources (Toyota, 2023). The company's investment in a modern wastewater treatment plant allows for the reuse of 70–85 percent of effluent for industrial and gardening needs, minimizing environmental impact by treating all effluent on-site. Similarly, Bosch has earmarked €10 million per year to reduce water withdrawal by 25 percent by 2025, reflecting its commitment to sustainable water management practices (Capgemini Research Institute, 2020).

Significant water usage occurs in the pre-treatment and cleaning stages before painting in automotive manufacturing. BMW Group's €161 million investment in 2018 emphasizes its commitment to environmentally friendly production methods, aiming to reduce electricity consumption, CO_2 emissions, and solvent use. Similarly, Honda's €67 million (8.1 billion yen) investment in 2018 covered environmental initiatives such as installing and certifying management systems, waste reduction, and ecosystem protection measures. The vehicle painting process is the most water-intensive process in an automotive company and has witnessed several innovations by the automobile MNEs to optimize water usage to minimize the impact of water sources and associated marine ecosystems. BMW's adoption of dry separation in the painting process represents an innovative approach to water conservation (BMW, 2021). By utilizing a special recirculation system and cardboard filters, BMW minimizes water usage, reducing environmental impact. Audi is also making strides in water conservation, with plans to reuse wastewater for obtaining new process water. Projects at Audi production sites aim to recycle hydro-rinse agents used in paint shops, further minimizing water consumption and environmental impact (Audi, 2022). Furthermore, BMW's utilization of six million litres of treated rainwater in water-deficit areas, such as its Chennai plant in India, provides up to 50 percent of the plant's water needs (BMW, 2021). Toyota also prioritizes water sustainability in India, meeting 95 percent of its production water demand through recycled and rainwater sources (Toyota, 2023).

3.3 Strategies for Pharmaceutical Companies to Mitigate Water Pollution

Pharmaceutical MNEs use various methods to treat water before releasing it into natural bodies. These include physical and chemical approaches. Coagulation, a common chemical method, involves adding iron or aluminium salts, such as aluminium sulphate or ferric chloride, to the water (Safe Drinking Water Foundation, 2017). Chemical compounds called coagulants carry a positive charge that neutralizes negatively charged debris in water, causing them to bind together or coagulate, a process known as flocculation. These larger particles settle at the bottom in sedimentation tanks. Coagulation, despite its complexity, is crucial for removing hydrophilic colloids from pharmaceutical wastewater. By using inorganic metallic salts and polymers, MNEs can effectively eliminate chromaticity and toxic organic matter, enhancing wastewater biodegradability.

Chlorination has emerged as an effective method for removing pharmaceuticals and toxic by-products from wastewater. Activated carbon, with its high adsorption capacity and stable properties, is also commonly used to eliminate natural and synthetic organic contaminants in pharmaceutical effluent treatment, helping to meet discharge standards (Sudhir et al., 2023). Ultrasonic degradation, based on ultrasonic cavitation, induces physical and chemical changes in wastewater, effectively breaking down organic compounds.

Electrochemical oxidation eliminates toxic pollutants, enhancing purification. Pharmaceutical MNEs employ microfiltration and ultrafiltration, membrane-based techniques, for wastewater treatment. Microfiltration separates impurities by passing water through a sieve-like membrane (Snyder et al., 2007). Ultrafiltration, on the other hand, utilizes membranes under pressure, allowing water and small molecules to pass through while trapping macromolecules and colloids. This technique is particularly useful for removing large molecular contaminants from wastewater. Pharmaceutical MNEs employ reverse osmosis following pre-treatment of wastewater, electrodialysis, and other methods for wastewater treatment, crucial for purifying water and meeting discharge standards. These techniques, alongside other physical and chemical treatments, enable effective wastewater management, minimizing environmental impact and promoting sustainable water practices.

For example, Novartis—a Swiss multinational pharmaceutical corporation—has installed a water recycling plant in one of its sites in Singapore to treat and recycle wastewater, using reverse osmosis technology (Novartis, 2020). The company has also implemented wastewater recovery projects in Germany and Slovenia to recover more than 31 thousand cubic meters of water. In 2020, the company had reused 8.6 million cubic meters of water, representing 16 percent of its total water withdrawal. The effluent load's chemical oxygen demand (COD) decreased from 2761 metric tons in 2016 to 1606.7 metric tons in 2020, a 42 percent reduction. This reduction enhances oxygen availability for marine life where effluent is released. Abbott—an American multinational medical devices and healthcare company—reported in 2022 that it recycled and reused 11 percent of the water used in its operations, while 26 percent of the facilities had water recycling activities in 2022 (Abbott, 2023). Advanced oxidative processes, physical-chemical methods, activated sludge, neutralization, and aerated water lagoons are employed for effluent water treatment. This treatment aids in preserving marine life and reducing chemical oxygen demand. Effluent discharge without treatment has declined since 2021, and emissions to water are monitored and reported per local regulations.

Pharmaceutical MNEs are also using AI in wastewater treatment. Researchers at the Korean Institute of Science and Technology have developed a technology to classify emerging trace substances based on their physicochemical properties and predict the concentrations of the substances using clustering and prediction-based AI tools (National Research Council of Science and Technology, 2024). The AI system boasts a prediction accuracy of about 0.75, enabling the tracing of 13 trace substances in wastewater. This surpasses the accuracy of existing AI techniques, which stands at 0.40 and relies on biological information. While similar technologies globally detect pollutants in water bodies from pharmaceutical companies, their adoption by pharmaceutical MNEs remains challenging.

3.4 *Strategies for the Oil and Gas Industry*

Oil and gas companies are responsible for mitigating water pollution by reducing toxic chemical concentrations in wastewater, decreasing water consumption, and preventing oil spills. The 'in-plant' system in refineries controls the concentration of pollutants in processing water through methods like stormwater separation and water reuse in cooling towers. Meanwhile, 'end-of-pipe' technology employs deep wastewater treatment to reduce pollutant concentrations in final effluents. Additionally, activated sludge removes hydrocarbons before discharge of wastewater into oceans (Radelyuk et al., 2021).

AI can help oil and gas projects minimize water usage in hydraulic fracturing, reducing stress on water resources and protecting marine ecosystems (Waqar et al., 2023). AI models also assist in minimizing hazardous waste at well-treatment jobs and reutilizing produced water for an appropriate recovery factor (Koroteev & Tekic, 2021). AI outperforms traditional aerial and field investigation methods in detecting oil spills using computational techniques on satellite images, enabling efficient hydrocarbon recovery (Singh et al., 2022). These optical remote sensing datasets are built by analysing the absorption of infrared light and sun-glint in the oil spill region and detecting thermal infrared wavelength characteristics between oil and marine water to assess the presence and thickness of oil spills over the water surface. When abnormal thermal emissivity is detected in ocean water, the system triggers an alarm. Advanced satellite sensors improve imaging techniques to capture detailed pictures of seas and oil spills, providing valuable data. AI systems, trained on historical data, offer rapid detection, enhancing efficiency. However, accurate detection requires robust, standardized datasets for effective AI model training (Huby et al., 2022). Figure 5 illustrates the application of AI in detecting oil spills and initiating appropriate response mechanisms.

Oil and gas MNEs are coming to the fore in experimenting with AI technologies to improve response times of oil spill clean-up. ExxonMobil—an American oil and gas multinational—has been using AI and cloud technologies since 2019 to optimize drilling operations, detect leakages autonomously, conduct predictive maintenance, and reduce response time to leakage events by automating decision-making (ExxonMobil, 2019). Similarly, Aker BP—a Norwegian oil exploration and development company—has invested in a radar-based oil spill detection technology developed by Vissim that utilizes image processing techniques for detecting oil spills with higher sensitivity (Durakovic, 2024). Vissim has also created a digital dashboard for Aker BP to integrate oil spill-related data from several detection sources, including radars, subsea sensors, and satellites. Additionally, British Petroleum has invested USD 5 million in geospatial analytics capabilities to gather spectral imagery of its operations from satellites, drones, and airplanes to detect

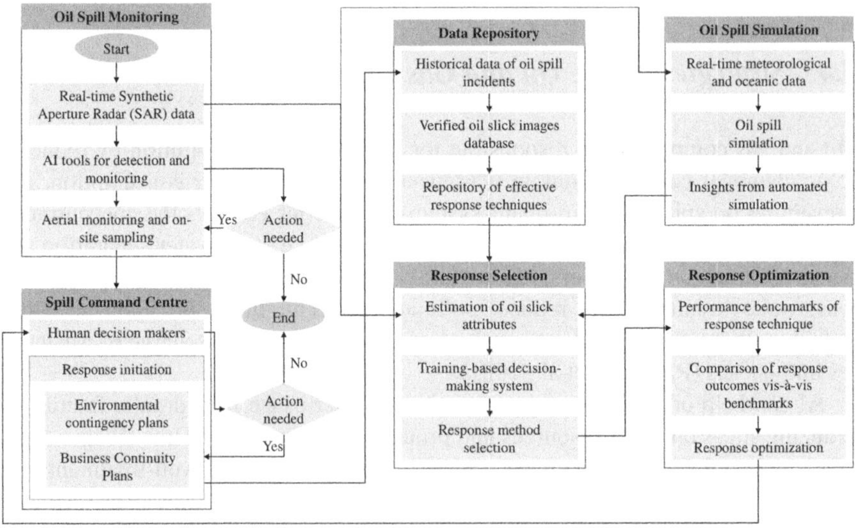

Fig. 5 Application of AI for oil spill detection and ocean pollution mitigation. (Source: Authors' elaboration adapted from Mohammadiun et al., 2021)

environmental changes, emissions, and leaks using machine learning algorithms (Ganta, 2020).

The deep-sea exploration process also affects marine ecosystems and threatens the life of species below water. In this regard, Shell is working with SparkCognition AI to generate images of the subsurface using fewer sweeps of seismic data for searching new oil and gas deposits, which can potentially help in deep-sea conservation (Inspenet, 2023). Using generative AI tools, Shell can generate images using seismic data scanners with increased efficiency and reduced disturbance to underwater ecosystems, as the time for exploration scans can be reduced from nine months to less than nine days. AI-driven paradigm shifts are transforming operations in oil-producing countries, impacting oil & gas MNEs. For example, Khurais, an oil field discovered in 1957, represents a shift in Saudi Arabia's oil production, utilizing drones, robotics, and AI for maintenance planning, satellite imagery analysis, and hazard detection. Drones scan production areas for inefficiencies and detect methane leaks (Gnana, 2024). The Saudi state-owned oil company, Saudi Aramco has developed an AI system to deploy drones in sub-surface terrain to capture images which will then be analysed by machine learning algorithms to detect oil spills and pollutants in water (Benito, 2024). Another system developed by Saudi Aramco features AI and machine learning to automate seismic data processing for real-time updates on subsurface reservoir models. By predicting oil spills and leakages and treating wastewater to reduce contaminant concentration, oil and gas MNEs can positively impact life below water.

3.5 Strategies for the Semiconductors Industry

In electronics manufacturing, back grinding and chemical mechanical polishing generate significant wastewater. Semiconductor MNEs use techniques like electrocoagulation, electro-adsorption, and coagulation-flocculation to treat wastewater and remove heavy metals from sludges and brine (Sadafale & Gaikwad, 2023). Coagulation-flocculation combines small particles into a heavier mass, removed as sludge, with over 90 percent efficiency in heavy metal removal. Electrocoagulation achieves nearly 99 percent efficiency in removing copper ions and fragments. Ion exchange effectively removes lead, nickel, cadmium, mercury, zinc, copper, and chromium by attaching metal ions to an immobile solid particle, replaced by a solid particle cation. Reverse osmosis, achieving over 90 percent efficiency in heavy metal removal, is instrumental in water treatment and reclamation. Advanced oxidation processes decompose natural micro-pollutants, while nanofiltration adsorbs heavy metals and organic matter, protecting marine ecosystems (Hsieh et al., 2023). Biomass sources like biochar remove heavy ions and microbial processes help immobilize metals in aqueous streams, reducing them to less soluble forms and precipitating chemical compounds, aiding in their removal from water.

Maximizing the reuse and recycling of water requires categorizing wastewater for applying the appropriate water treatment technologies. Different MNEs classify wastewater differently for treatment purposes. For example, Powerchip Semiconductor Manufacturing Corporation (PSMC) classifies wastewater into 20 different categories, including organic wastewater and waste acid solution (Wang et al., 2023). On the other hand, Samsung classified wastewater simply into three categories—acidic, alkaline, and fluoric wastewater. Taiwan Semiconductor Manufacturing Co., Ltd. (TSMC) Is known to classify wastewater into 38 different categories, including hydrofluoric acid, acidic, caustic, organic, and high-concentration liquid waste to name a few. United Microelectronics Corporation (UMC) also classifies wastewater into 27 categories, including solvent-based, inorganic, and others. Similarly, Winbond Electronics classifies wastewater for recycling purposes, and Yangtze Memory Technologies classifies wastewater into 29 categories. The classification system helps to determine the most appropriate wastewater treatment method. For example, PSMC uses RO systems for the wastewater generated in its plants, which is different from Samsung's method of using techniques, such as activated carbon filtration, acid-base neutralization, and coagulating sedimentation, besides using RO systems. Winbond uses membrane bioreactors for wastewater treatment. GlobalFoundries uses a combination of RO, activated carbon, catalytic treatment, and ceramic filtration before reusing wastewater in its operations. Deionization systems are used by Microchip Technology Inc., while RO systems are prevalent at semiconductor MNEs such as Micron, Infineon, UMC, and Western Digital. Sterilization of wastewater using ultraviolet (UV) radiations is also carried out at MNEs such as SK Hynix.

To minimize the flow of treated wastewater into marine resources, the semiconductor MNEs also strive to reuse or recycle the water according to its quality and requirements. For example, recycled water is reused in semiconductor manufacturing parks for the production of ultrapure water, irrigation of landscaped parks, cooling of the facility, and production of scrubber water that removes inorganic contaminants from wastewater (Wang et al., 2023). The reuse of water inside the semiconductor facility helps to reduce pressure on the available freshwater resources and conserves marine life from the onslaught of organic and inorganic contaminants that may otherwise enter into the marine ecosystems.

Effective water management by semiconductor MNEs supports net positive water goals, where they generate more water than they use. This involves enhancing manufacturing processes and engaging in external initiatives to enrich marine ecosystems. For example, in 2021, Intel had saved 9.3 billion gallons of water through its internal efficiency improvement methods leading to a 114 percent increase in water savings since 2019 (Intel, 2022). Intel's conservation efforts go beyond its factories. The company focuses on water restoration projects in the watersheds affected by its sites. In 2021, Intel-funded projects in Arizona restored 890 million gallons to the watershed, encompassing traditional conservation methods and initiatives to reshape local economies. For instance, one project incentivized farmers to transition to barley cultivation which requires less irrigation, conserving water, and alleviating stress on the underground water table (Intel, 2022). Additionally, Intel's investment in a local malt house reduced transaction costs associated with malting barley. This malted barley can then be sold to local breweries, reducing their reliance on out-of-state suppliers (Intel, 2022). In the Indian city of Bengaluru, Intel has taken steps to support the city's water resources. The city, once known as the 'City of Lakes', has faced extreme water stress and groundwater depletion over the past two decades due to rapid urban development. Intel has funded two water restoration projects in the region, which, upon full implementation, will restore over 100 million gallons of water annually between two lakes in the metropolitan city (Intel, 2022). Micron too has implemented several innovative solutions that help it to reclaim 75 percent of process water inside the facility (Chase, 2020). Moreover, its Singapore facility sources 98 percent of its water from reclaimed and desalination facilities, which helps to reduce its reliance and impact on freshwater reserves.

3.6 *Strategies for the Textile Manufacturing Sector*

Textile effluents vary based on the textile type and chemicals used in the process. 'Fast fashion' increases pressure on water resources. MNEs in textiles must treat wastewater to prevent dye and heavy metal contamination. The treatment choice depends on effluent composition, concentration, and characteristics of materials present in the effluents (Mostafa, 2015). The common process includes removing suspended solids, oil, and gritty materials. Bars and fine screens eliminate suspended materials like yarns and lint, protecting aquatic ecosystems. Settling removes suspended particles, and mechanical scrapping removes floating ones. Effluent acidity is neutralized to a pH of 5 to 9, and adsorption removes colours and pollutants like cyanides, phenols, dyes, and pesticides. Activated carbon is an effective adsorbent for dissolved organics. Treatment involves disinfection, filtration, sedimentation, and reverse osmosis before storing clean water.

Next, secondary treatment is applied to reduce dissolved and colloidal organic components, along with colour and stabilizers. This stage aims to remove a higher Biochemical Oxygen Demand (BOD) than can be achieved through sedimentation alone (Azanaw et al., 2022). Additionally, it significantly reduces phenol and oil content. Textile wastewater treatment commonly employs anaerobic methods, utilizing consortia of microorganisms (anaerobic sludge or biofilm) as active agents. Anaerobic processes offer advantages over aerobic methods, including the capacity to handle higher organic loads, aeration-free operation, and reduced sludge production. In anaerobic treatment, organic matter in the effluent undergoes oxidation by microorganism-coated surfaces of filter media, resulting in the production of carbon dioxide and water. To prepare treated effluent for reuse or to remove specific contaminants, textile effluents often undergo tertiary or advanced treatment. Common tertiary processes include the removal of waste organic colour components through adsorption and the elimination of dissolved particles through membrane filtration. Ultrafiltration, nanofiltration, ion exchange, and electrodialysis are additional wastewater treatment methods, but their detailed explanations are beyond this chapter's scope. Water reclamation and reuse alleviate strain on freshwater sources and limit pollutant entry into water bodies, safeguarding aquatic ecosystems from textile wastewater's adverse effects.

Louis Vuitton employs technologies to reuse treated wastewater and recover rainwater at its production sites, along with water recycling systems in distilleries and workshops (Moët Hennessy Louis Vuitton, 2023). H&M ensures proper wastewater treatment across its supply chain, integrates water costs into sourcing decisions, and sets targets for better water performance. In Bangladesh, H&M collaborates with the Alliance for Stewardship, supplier factories, other clothing brands, and local communities to enhance water management in Gazipur, Narayanganj, and Savar (H&M Group, 2024). The 'bottle2fashion' project with Danone AQUA in Indonesia recycled over 7.5 million PET bottles into polyester in 2021 of which 663,869 PET bottles were utilized, preventing plastic pollution and supporting marine life conservation (H&M, 2021). Levi Strauss reduced freshwater use in water-stressed areas by 14 percent from 2018 to 2021, using 15,030 million litres in 2021 (Levi Strauss & Co., 2023). This demonstrates how MNEs in the textile and fashion industries play a crucial role in conserving water resources and preserving aquatic life by mitigating the impact of dyes, chemicals, metals, and microplastics released in their processes.

3.7 Strategies for the Shipping Industry

The large volume of vessels in seawater stresses marine ecosystems due to shipborne waste and ballast water introducing foreign microorganisms. The shipping industry mitigates water pollution through waste reduction and ballast water treatment. Ships use compactors, incinerators, and food grinders to process waste during cruises longer than 3.7 days, allowing safe sea disposal and port handling of non-assimilable waste (Mobilik & Hassan, 2016). In Malaysia, maritime MNEs are prohibited from rinsing cargo residues with water and releasing them into the sea, protecting marine life.

Ballast water treatment minimizes the migration of foreign organisms to marine ecosystems, in line with the International Convention on the Control and Management of Ballast Water and Ship Sediments, 2004. Mechanical methods include filtration, sedimentation, floatation, and coagulation-flocculation (Bui et al., 2021). Chemical methods involve electrolysis, advanced oxidation, and chlorination. UV radiation inactivates harmful microorganisms and plankton. Water passes through porous screens and grooved discs, trapping substances. Coagulation, flocculation, and membrane filters cause plankton over 50 micrometres to settle and fuse. Magnetite powder induces flocculation, and a magnetic field creates plankton flocks for easy separation (Dachev et al., 2021). Ozone, an oxidizing agent, destroys harmful cells and reduces plankton by 71 to 99 percent with five to 10 hours of treatment (Bui et al., 2021). Deoxygenation lowers oxygen levels, killing microorganisms and preventing tank corrosion (Dachev et al., 2021). Ultrasonic noises break the cell walls of harmful organisms, killing them. Although expensive, these mechanical and chemical methods are essential for marine life sustainability.

Several shipping MNEs have strategies to mitigate water pollution and conserve marine life. Cosco Shipping Lines (CSL) —a Chinese international container transportation and shipping company—strictly adheres to wastewater discharge regulations, complying with domestic and international laws and conventions for ballast water and oil sewage management (Cosco Shipping Lines, 2023). CSL minimizes its environmental impact by primarily sourcing water from desalination to reduce pressure on freshwater ecosystems and using ballast water treatment devices to manage and treat ballast water, preventing contamination. To prevent environmental damage, CSL prohibits the intake of ballast water in restricted areas infested with pests and pathogens. Crew members are required to remove sediment from ballast water promptly, and specialized hull coatings and regular inspections prevent marine species from adhering to ships. Similarly, Maersk has installed ballast water treatment facilities on 93 percent of its fleet, aiming for 100 percent compliance by September 2024 (Maersk, 2024). Hapag-Lloyd conducts technical screenings to address the release of ballast water with non-indigenous species, hull biofouling, and noise and vibrations from shipping, ensuring active mitigation of environmental impacts (Hapag-Lloyd, 2024).

3.8 *Responsible Aquaculture Practices by MNEs*

Implementing responsible aquaculture practices by multinational companies is crucial for sustainable seafood production and marine ecosystem protection. MNEs in aquaculture must collaborate with certifiers, policymakers, regulators, and NGOs to bring about systemic changes. The World Economic Forum recommends large-scale aquaculture companies support small-scale actors in sustainably increasing productivity (Ajayi et al., 2023). MNEs can also promote low trophic and environmentally restorative aquaculture practices. It's essential for MNEs to engage with small-scale actors along the aquaculture value chains, as they are integral and can benefit socially, economically, and environmentally. Advocates can guide retailers and distributors on diversifying and responsibly procuring blue foods, including lower-trophic species. Additionally, AI can enhance global aquaculture planning and monitoring through platforms integrating satellite technology, ecological modelling, open-source data, and connectivity. Figure 6 illustrates the process used by AI for managing smart aquaculture.

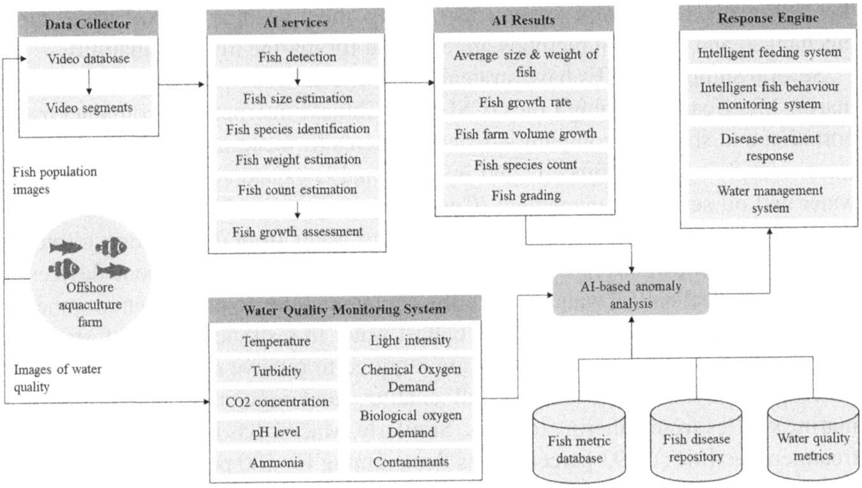

Fig. 6 Process flowchart of an AI-based aquaculture management system. (Source: Authors' elaboration)

3.9 Other Marine Conservation Projects by MNEs

MNEs are spearheading marine conservation efforts beyond their operational facilities. Coca-Cola Hellenic Bottling Company certifies its bottling plants with international water stewardship standards like Alliance for Water Stewardship or European Water Stewardship, assessing biodiversity impact, water ecosystems, and High Conservation Value Areas within the associated watersheds. The company also ensures compliance with the Environmental Management System ISO 14001 and treats 100% of its wastewater to support aquatic life (Martin & O'Neill, 2021). Coca-Cola's collaboration with WWF has established water reserves in Guatemala, protecting 2.2 billion cubic meters of freshwater, and conservation efforts in the Yangtze have boosted the finless porpoise population and reintroduced Père David's deer into the wild (Coca-Cola & WWF, 2020) (Table 2).

Apple's Austin campus uses recycled water for air-conditioning, wastewater reuse, condensate recovery, and stormwater capture, saving 60 million gallons of fresh water annually (Apple, 2023). Accenture and Emirates Nature-WWF are advancing Sustainable Blue Economy principles in the UAE (Economist Impact, 2023) and collaborating with the Australian Institute of Marine Science to use AI and analytics for reef conservation via an open-access tool, ReefCloud (Emirates Nature-WWF, 2024). In 2024, Bank of America completed the first debt-for-nature transaction in Continental Africa, refinancing $500 million of the Gabonese Republic's sovereign debt. Supported by The Nature Conservancy, this deal will generate $125 million for marine conservation over 15 years (Bank of America, 2023). The Gabon Blue Bond will fund $125 million in ocean conservation, aiding Gabon's goal to protect 30% of its lands, freshwater, and ocean by 2030. Reiterating

Table 2 Industry-specific marine sustainability strategies with rationale

Sector	Strategies	Rationale
Automotive	Dry separation technique in vehicle painting Integration of reverse osmosis system for wastewater Onsite effluent treatment Reducing the load on effluent treatment plants Waterless car washing strategies Water recirculation system in car wash facilities	Automotive companies can: Save water consumption costs Optimize vehicle painting processes Reuse treated water Comply with regulations Improve brand reputation
Fishing	AI-driven marine life monitoring system Collaboration with local fish farmers for training sessions Eco-labelling of sustainable fish products Sustainable credit for small-scale fish farmers	Improved water management boosts fishing companies' reputations. Ensures sustainable resources and strengthens community ties.
Oil & gas	AI for leakage detection and predictive maintenance Computational techniques for remote monitoring of oil fields Drones to detect oil spills and pollutants Monitoring using radars, subsea sensors, and satellites Stormwater separation and water reuse	Preventing spills ensures regulatory compliance, avoiding legal penalties and operational disruptions. It also saves costs related to spill response, remediation, and legal liabilities.

(continued)

Table 2 (continued)

Sector	Strategies	Rationale
Pharmaceutical	Coagulation and chlorination Electrochemical oxidation Microfiltration and Ultrafiltration Prediction-based AI tools to classify trace substances Ultrasonic degradation	Pharmaceutical companies can avoid hefty fines, legal action, and the potential shutdown of operations due to environmental incidents that can disrupt production. This can improve market position and potentially lead to better financial performance.
Semiconductor	Advanced oxidation Biomass to remove heavy metal ions Chemical coagulation-flocculation Ion exchange Membrane bioreactors Nanofiltration Reverse osmosis membrane technology Wastewater reuse	Preventing pollution ensures a reliable supply of high-quality water, crucial for maintaining product quality and production efficiency. It cuts costs related to water treatment and waste disposal. Better relations with local communities and governments enhance the company's social license to operate and expand prospects.
Shipping	Ballast water treatment Coagulation-flocculation Compactors, incinerators, and food grinders to process shipboard waste Magnetite powder Mechanical methods, such as filtration, sedimentation, floatation UV radiation to kill microorganisms Ultrasonic radiation to kill microorganisms	Shipping companies can avoid fines and enhance their reputation by reducing pollution. Clean marine environments mean fewer disruptions and damages to ships. Reduced damages, lower maintenance costs and extended vessel lifespan.
Textile	Adsorption to remove colours and organic pollutants Anaerobic methods to treat organic load Bars and fine screens to remove suspended pollutants Disinfection through filtration Electrodialysis Ion exchange Membrane filtration Reverse osmosis Sedimentation Ultrafiltration and nanofiltration Water reclamation and reuse	Failure to prevent water pollution can lead to hefty fines, legal penalties, and operational shutdowns. Efficient water management and pollution prevention reduce waste treatment costs and water usage, resulting in long-term savings. Cleaner production processes enhance efficiency and productivity, improving overall operational performance.

Source: Authors' elaboration

the connection between MNEs and Sustainable Development Goals (SDG 14), the next section analyses the financial viability of marine conservation projects for MNEs and the value of public-private partnerships in contributing to SDG 14.

4 Multinational Enterprises and Sustainable Development Goal 14

4.1 Economic Rationale for MNEs' Role

Investing in sustainability-led projects and implementing water management solutions can be costly and affect the profitability of MNEs. However, research by S&P Global found that companies involved in water-related activities have historically shown higher risk-adjusted returns than the global equity market (Cheng & Yi, 2020). McKinsey's study indicates that sustainable companies tend to be more profitable, with those having high environmental, social, and governance (ESG) ratings consistently outperforming the market (Bonini & Swartz, 2014). Additionally, Deutsche Bank's research found that companies with higher ESG ratings benefit from lower costs of debt and equity. Thus, investing in sustainability and spending on SDG 14 initiatives can enhance the financial strength of MNEs, justifying the initial costs of water management solutions.

Social and environmental externalities are intertwined with production decisions (Hart & Zingales, 2017), as reversing unsustainable practices, such as water pollution, is costlier than prevention. The study highlights that information asymmetry between company insiders and outsiders about environmental impacts means that externalities cannot be fully addressed by external rules alone. This necessitates proactive efforts by company leadership to reduce environmental footprints and conserve marine life. Effective sustainability management showcases an MNE's adaptability, strengthens its competitive market position, and enhances intangible assets (Schoenmaker & Schramade, 2019). Integrating sustainability into corporate strategy helps MNEs create long-term value and transition to a sustainable economy. Thus, sustainable business models address not only financial viability but also social and environmental concerns (Schoenmaker, 2020).

Indriastuti and Chariri (2021) discovered that sustainability investments positively impacted the financial performance of 132 Indonesian manufacturing companies. Implementing CSR projects reduced operating costs by replacing marketing expenses with CSR costs. Similarly, Pham et al. (2021) found that 116 listed Swedish companies investing in sustainability had higher earnings yield, ROA, and ROE, indicating better financial health compared to non-investing firms. These findings suggest that MNEs benefit from sustainability investments, supporting SDG 14 for conserving aquatic life. Enhanced environmental practices improve brand reputation and adaptability and integrating SDGs into corporate strategy provides both financial and non-financial benefits over time. MNEs can enhance marine ecosystem conservation through collaboration with non-corporate entities. Government agencies, NGOs, local communities, and regulators can bolster the effectiveness and scale of CSR initiatives. An ecosystem of stakeholders dedicated to ocean conservation can amplify MNEs' impact on SDG 14.

4.2 Partnerships with NGOs and Government Agencies

To conserve coastal marine biodiversity, MNEs can develop Public-Private Partnerships for the management of Marine Protected Areas (MPAs) (Blue Finance, 2018) with NGOs and government agencies. NGOs can play five types of roles, as Crosman (2013) defines:

- Advocate: Lobbying, litigating, and educating MNEs on marine conservation.
- Enabler: Empowering MNE stakeholders in local marine ecosystem management through capacity building and funding.
- Expert: Providing scientific input to MNEs on marine ecosystem conservation and enhancing organizational knowledge through original research and knowledge of conservation tools.
- Manager: Directly overseeing conservation program implementation and engaging in hands-on activities.
- Watchdog: Monitoring compliance with marine conservation agreements, pursuing legal action, and publicizing violations.

NGOs can collaborate with MNEs to manage Marine Protected Areas (MPAs) through Special Purpose Entities (SPEs), which function as separate legal bodies to manage risks (Blue Finance, 2018). These collaborations enhance sustainable fisheries globally. For instance, the Atlantic Seafood Company in Gambia partnered with the Fisheries Department and local fishers to implement sustainable practices, supporting 1500 fishers by providing crucial fish size distribution data (Marine Stewardship Council, 2013). In tourism, multinational hospitality enterprises can aid marine conservation. In Tanzania, tourism operators helped reduce dynamite fishing by raising safety concerns, prompting Navy intervention and increasing political support for regional conservation projects in the Mtwara and Tanga regions, working with fishing communities. Similarly, in El Nido, Philippines, Ten Knots Resort assisted in developing a marine park by enforcing measures against illegal fishing and pollution. These public-private partnerships demonstrate effective conservation efforts.

4.3 Community Engagement and Stakeholder Collaboration

Effective ocean conservation initiatives require local community involvement for sustainable solutions benefiting marine life. Integrated and collaborative multi-stakeholder approaches are essential for successful ocean management. This collaboration should transcend industry sectors, bridging public-private sector gaps. Companies must engage relevant stakeholders, explore partnerships, and define their roles for priority areas (Handoyo et al., 2022).

Conservation International—a multinational non-profit enterprise—has collaborated with Bechtel—a social enterprise—to install green-grey infrastructure for coastal conservation at two highly vulnerable locations in the Province of Iloilo, The Philippines. The local communities in these locations had been severely impacted by Typhoon Haiyan in 2013, among other storms. The green-grey projects at four pilot project sites combined mangrove restoration with breakwaters to improve the resilience of the local community, mangroves, and marine ecosystems (Handoyo et al., 2022). Another example comes from China, where Coca-Cola worked with WWF to construct wetlands for rural communities to cleanse household wastewater and mitigate the flow of contaminants into the Yangtze River and other waterways that could eventually end up in oceans otherwise (Coca Cola & World Wildlife Fund for Nature, 2020). In the Dongting subbasin, the Coca-Cola and WWF partnership prepared local communities for an impending fishing moratorium that aided the survival of finless porpoises in a natural reserve. Such partnerships among MNEs, international organizations, and local communities can usher in the rejuvenation of marine ecosystems. Figure 7 presents a framework for fostering collaboration among various stakeholders for SDG 14.

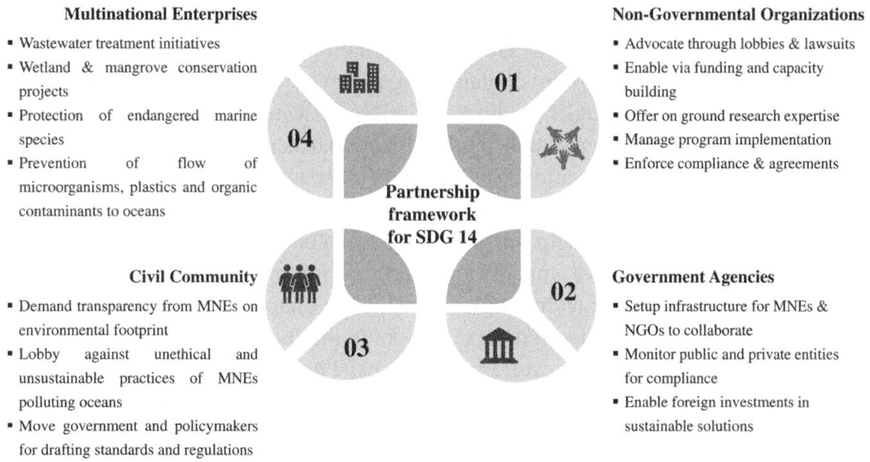

Fig. 7 Stakeholder Coordination Framework for SDG 14. (Source: Authors' elaboration)

4.4 *Accountability and Transparency in Reporting Environmental Impact*

Environmental accounting is crucial for stakeholder governance in MNEs as it demonstrates accountability for environmental impact (Wong et al., 2021). Environmental transparency involves making relevant information accessible to stakeholders, thereby supporting sustainable development (Aurelio et al., 2020). However, environmental accountability is not fully integrated into MNE practices due to the lack of mandatory standards and third-party auditing (Wong et al., 2021). It is recommended that MNEs ensure environmental data is interpretable, accessible, relevant, timely, and reliable. Transparency demands clear and coherent data, enabling governments to assess regulation needs and market efficacy. Reporting standards should cover data quality assurance, accuracy assessment, and verifiability. MNEs should share achieved targets with auditors, shareholders, and relevant stakeholders, taking responsibility for environmental impact. Effective information dissemination is vital for maintaining a positive social presence and communicating impact to the public. Adopting a comprehensive framework across sectors can facilitate environmental reporting and accountability, despite the absence of global regulations.

5 Conclusion

MNEs play a dual role in producing goods and services while also impacting both local and distant ecosystems. Across sectors like automotive, pharmaceuticals, healthcare, shipping, textiles, and tourism, MNEs contribute to marine pollution and threats to the existence of marine ecosystems. Marine pollution occurs in the form of industrial runoffs, wastewater discharge, microbial discharge, the introduction of metals and ions into the water, and chemical contamination of water bodies that eventually drain into oceans. However, technologies exist for wastewater management, pollution mitigation, contaminant removal, and predictive modelling, thereby contributing to SDG 14. Through the use of water treatment technologies, AI-led predictive intelligence, and public-private partnerships, MNEs can significantly impact ocean conservation. Stakeholders must actively engage in ocean conservation efforts through collaborative efforts and environmental stewardship.

The chapter intends to help sustainability officers and business executives in MNEs realize their negative impacts and start developing strategies for mitigating marine pollution using the technologies listed in the chapter. The results can guide existing and emerging MNEs in diverse sectors to implement existing technologies to contribute to SDG 14. However, limitations exist as the chapter does not explore the reasons deterring the adoption of the discussed sustainable solutions by MNEs. Further research is required to extend the findings of this chapter in understanding the drivers of technology adoption and enablers of cross-industry partnerships that can help MNEs collaborate on marine conservation efforts. The chapter concludes that MNEs bear accountability and responsibility to conserve marine ecosystems and must deploy the discussed technologies for marine sustainability.

References

Abbott. (2023). *Abbott CDP Water Security Questionnaire*. https://www.abbott.com/content/dam/corp/abbott/en-us/documents/pdfs/abbott-citizenship/2023-Abbott-CDP-Water-Security-Questionnaire.pdf

Ajayi, O., Beveridge, M., Frankfort, A., Hashemi, T., Kanyuuru, C. K., O'Neil, G., Nguyen, T., & Parkes, G. (2023). *The global sustainable aquaculture roadmap: Pathways for systemic change*. World Economic Forum. https://www3.weforum.org/docs/WEF_Global_Aquaculture_Roadmap_2023.pdf

Apple. (2023). *2023 Environmental Progress Report*. https://www.apple.com/environment/pdf/Apple_Environmental_Progress_Report_2023.pdf

Arsenault, C., & Le Billon, P. (2022). To fight illegal fishing in the Galapagos, Ecuador turns to Canadian satellite and sensing technology. *CBC*. https://www.cbc.ca/news/world/galapagos-illegal-fishing-satellite-technology-1.6469424

Asia Europe Environment Forum. (2020). *Garments sustainability in Asia and Europe*. https://asef.org/wp-content/uploads/2021/01/Policy-Brief-Stream3_Sustainable_Garments.pdf

Audi. (2022). *How Audi is reducing water consumption in production*. https://www.audi.com/en/sustainability/environment-resources/circular-economy/water-consumption.html

Aurelio, T., Luca, D. B., Gennaro, M., & Leoni, G. (2020). Environmental reporting, accountability and governance of local governments: An Italian multiple case study. *African Journal of Business Management, 14*(8), 229–242. https://doi.org/10.5897/AJBM2020.9027

Azanaw, A., Birlie, B., Teshome, B., & Jemberie, M. (2022). Textile effluent treatment methods and eco-friendly resolution of textile wastewater. *Case Studies in Chemical and Environmental Engineering, 6*, 100230. https://doi.org/10.1016/j.cscee.2022.100230

Bailey, K., Basu, A., & Sharma, S. (2022). The environmental impacts of fast fashion on water quality: A systematic review. *Water, 14*(7). https://doi.org/10.3390/w14071073

Bank of America. (2023). *Bank of America refinances Gabon sovereign debt for nature and ocean conservation*. https://newsroom.bankofamerica.com/content/newsroom/press-releases/2023/08/bank-of-america-refinances-gabon-sovereign-debt-for-nature-and-o.html

Beaumont, H. (2023). Oil wells guzzle precious California water. Next door, residents can't use the tap. *The Guardian*. https://www.theguardian.com/us-news/2023/jan/22/california-oil-wells-refineries-water-use-drought-fuller-acres

Benito, A. (2024). *How AI is reshaping Saudi Aramco's oil exploration and underwater operations strategy*. CIO. https://www.cio.com/article/2096554/how-ai-is-reshaping-saudi-aramcos-oil-exploration-and-underwater-operations-strategy.html.

Blue finance, International Coral Reef Initiative (ICRI). (2018). *Conservation investment blueprint: Public-private partnership for Marine Protected Areas*. Case Study of Blue finance in Dominican Republic. https://www.icriforum.org/wp-content/uploads/2019/12/ICRI%20Blueprint%20Public-Private%20Partnership%20for%20MPA%20by%20Blue%20Finance.pdf

BMW. (2021). *How the BMW Group saves water resources*. https://www.bmwgroup.com/en/news/general/2021/water.html

Bonini, S., & Swartz, S. (2014). Profits with purpose: How organizing for sustainability can benefit the bottom line. *McKinsey on Sustainability & Resource Productivity*. https://www.mckinsey.com/~/media/McKinsey/Business%20Functions/Sustainability/Our%20Insights/Profits%20with%20purpose/Profits%20with%20Purpose.ashx

Bui, V. D., Nguyen, Q. P., & Nguyen, D. T. (2021). A study of the ship ballast water treatment technologies & techniques. *Water Conservation & Management, 5*(2), 121–130. https://doi.org/10.26480/wcm.02.2021.121.130

Capgemini Research Institute. (2020). *The automotive industry in the era of sustainability*. https://www.capgemini.com/wp-content/uploads/2020/03/The-Automotive-Industry-in-the-Era-of-Sustainability.pdf.

Chase, M. (2020). *Innovation in water conservation*. Micron. https://in.micron.com/about/blog/company/innovations/innovation-in-water-conservation.

Cheng, T. & Yi, C. (2020). *Investing in water for a sustainable future*. S&P Dow Jones Indices. https://www.spglobal.com/spdji/en/documents/education/education-investing-in-water-for-a-sustainable-future.pdf

Coca-Cola Company and World Wildlife Fund for Nature. (2020). *The Coca-Cola Company & World Wildlife Fund (WWF) Global Partnership Report*. https://files.worldwildlife.org/wwfcmsprod/misc/coca_cola_impact_report/WWF_Coca-Cola_Partnership_Report_Final_v5.pdf

Cordes, E. E., Jones, D., Schlacher, T. A., Amon, D. J., Bernardino, A. F., Brooke, S., Carney, R., DeLeo, D. M., Dunlop, K. M., Escobar-Briones, E. G., Gates, A. R., Génio, L., Gobin, J., Henry, L., Herrera, S., Hoyt, S., Joye, M., Kark, S., Mestre, N. C., Metaxas, A., Pfeifer, S., Sink, K., Sweetman, A. K., & Witte, U. (2016). Environmental impacts of the deep-water oil and gas industry: A review to guide management strategies. *Frontiers in Environment Science, 4*. https://doi.org/10.3389/fenvs.2016.00058

Cosco Shipping Lines. (2023). *2022 Sustainability Report*. https://lines.coscoshipping.com/lines_resource/pdf/quality2022en.pdf

Council, N. R. (1996). *Stemming the tide: Controlling introductions of nonindigenous species by ships' ballast water*. National Academies Press.

Crosman, K. M. (2013). *The roles of non-governmental organizations in marine conservation*. University of Munich. https://deepblue.lib.umich.edu/bitstream/handle/2027.42/99557/Crosman_Roles_of_NGOs_in_Marine_Conservation_Final.pdf.

Dachev, Y., Tsvetkov, M., & Zlatev, V. (2021). Ship ballast water treatment. WSEAS transactions on environment and development. https://doi.org/10.37394/232015.2020.16.11

Durakovic, A. (2024). *Aker BP returns to Vissim for an upgraded oil spill detection solution*. Offshore Energy.. https://www.offshore-energy.biz/aker-bp-returns-to-vissim-for-upgraded-oil-spill-detection-solution/

Economist Impact. (2023). How the Australian Institute of Marine Science is protecting corals reefs using data and AI. *The Economist*. https://impact.economist.com/projects/enterprise-transformation/how-the-australian-institute-of-marine-science-is-protecting-corals-reefs-using-data-and-ai.html

Emirates Nature-WWF. (2024). *Emirates Nature-WWF teams up with Accenture to propel UAE's Sustainable Blue Economy*. https://www.emiratesnaturewwf.ae/en/press-release/940-emirates-nature-wwf-teams-up-with-accenture-to-propel-uaes-sustainable-blue-economy

ExxonMobil. (2019). *ExxonMobil to increase Permian profitability through digital partnership with Microsoft*. https://corporate.exxonmobil.com/news/news-releases/2019/0222_exxonmobil-to-increase-permian-profitability-through-digital-partnership-with-microsoft

Famiglietti, J., Sanyal, P., Xu, L., Bluhm, K., Garvey, K., & Ferris, D.M. (2022). *Global assessment of private sector impacts on water*. Global Institute for Water Security. https://valuingwaterinitiative.org/wp-content/uploads/2022/04/Ceres-Global-Water-Assessment-2022.pdf

Food and Agriculture Organization of the United Nations. (2024). *Proportion of fish stocks within biologically sustainable levels (Primary series)*. https://de-public-statsuite.fao.org/

Ganta, H. (2020). *BP Ventures invests in geospatial analytics firm for methane leaks*. Offshore Technology. https://www.offshore-technology.com/news/bp-ventures-invests-detect-methane-leaks.

Global Atlas of Environmental Justice. (2024a). *Taean national park oil spill, South Korea*. Environmental Justice Atlas. https://ejatlas.org/conflict/taean-national-park-oil-spill-south-korea.

Global Atlas of Environmental Justice. (2024b). *PFAS pollution by Arkema in 'chemical valley'*, Environmental Justice Atlas. https://ejatlas.org/conflict/pfas-pollution-by-arkema-in-chemical-valley-france

Global Atlas of Environmental Justice. (2024c). *TFA / PFAS pollution by Solvay, Salindres, France*. Environmental Justice Atlas. https://ejatlas.org/conflict/tfa-pfas-pollution-by-solvay-salindres-france

Global Atlas of Environmental Justice. (2024d). *Oil exploration and extraction in the Namibe Province, Angola*. Environmental Justice Atlas.https://ejatlas.org/conflict/oil-exploration-and-extraction-in-the-namibe-province-angola.

Global Atlas of Environmental Justice. (2024e). *Lithium mining in Salar del hombre Muerto, Argentina*. Environmental Justice Atlas. https://ejatlas.org/conflict/salar-del-hombre-muerto-litio-argentina

Gnana, J. (2024). *AI, drones and sustainability: Saudi Arabia's Khurais offers peek into future of oil production*. S&P Global Commodity Insights. https://www.spglobal.com/commodityinsights/en/market-insights/blogs/oil/031324-ai-drones-and-sustainability-saudi-arabias-khurais-offers-peek-into-future-of-oil-production

Government of Kerala. (2010). *Report and recommendations of the high power committee on the extent of damages caused by the Coca-Cola plant at Plachimada and claiming compensation*. https://groundwater.kerala.gov.in/wp-content/uploads/2018/04/coca-cola-kerala.pdf

Gruber, N., & Gregor, L. (2024). *Ocean acidification stripes*. https://oceanacidificationstripes.info/s/ph/basin/globalocean/entirebasin

H&M. (2021). *H&M is turning plastic waste from Indonesian islands into everyday fashion*. https://www2.hm.com/en_in/life/culture/inside-h-m/bottle-2-fashion-innovation.html

H&M Group. (2024). Water. https://hmgroup.com/sustainability/circularity-and-climate/water/

Handoyo, D., Meyer, N., Portfaix, A., Sivaprasad, D., Wijaya, V., Felber, G., Giron-Niva, A., Neo, G. H., Parkes, G., & Teleki, K. (2022). *What ocean sustainability means for business*. World Economic Forum. https://www3.weforum.org/docs/WEF_What_Ocean_Sustainability_Means_for_Business.pdf

Hapag-Lloyd. (2024). *2023 Sustainability Report*. https://www.hapag-lloyd.com/sustainability-report-2023/en/assets/downloads/HAPAG-LLOYD_Sustainability_Report_2023.pdf

Hart, O., & Zingales, L. (2017). Companies should maximize shareholder welfare not market value. *Journal of Law, Finance, and Accounting, 2*(2), 247–275.

Holcomb, S. (2023). *Oil Kills*. Oceana. https://oceana.org/blog/oil-kills.

Hsieh, S., Lin, P., Lin, I., Beck, D. E., & Lin, C. (2023). Assessing the contribution of semiconductors to the sustainable development goals (SDGs) from 2017 to 2022. *Heliyon, 9*(11), e21306. https://doi.org/10.1016/j.heliyon.2023.e21306

Huby, A., Sagban, R., & Alubady, R. (2022). Oil spill detection based on machine learning and deep learning: A review. In *Conference: 2022 5th International Conference on Engineering Technology and its Applications*. https://doi.org/10.1109/IICETA54559.2022.9888651

Iduk, U., & Samson, N. (2015). Effects and solutions of marine pollution from ships in Nigerian waterways. *International Journal of Scientific & Engineering Research, 6*(9).

Inspenet. (2023). *Shell dives into deep water with the help of AI for oil exploration*. Inspenet. https://inspenet.com/en/noticias/shell-dives-into-deep-water-with-the-help-of-ai-for-oil-exploration

Indriastuti, M., & Chariri, A. (2021). The role of green investment and corporate social responsibility investment on sustainable performance. *Cogent Business & Management, 8*(1). https://doi.org/10.1080/23311975.2021.1960120

Intel. (2022). *Intel achieves net positive water in 3 countries*. https://www.intel.com/content/www/us/en/newsroom/news/intel-achieves-net-positive-water-3-countries.html

Izmir. (2022, November 28). *Report accuses Swedish fashion chain of polluting water in Bangladesh*. Anadolu Agency. https://www.aa.com.tr/en/asia-pacific/report-accuses-swedish-fashion-chain-of-polluting-water-in-bangladesh/2750290

KEMI Swedish Chemicals Agency. (2014). *Chemicals in textiles—Risks to human health and the environment.* https://www.kemi.se/global/rapporter/2014/rapport-6-14-chemicals-in-textiles.pdf

Khan, M. (2023). New York state sues PepsiCo over plastic pollution in rivers. *Breaking News.* https://www.breakingnews.ie/world/new-york-state-sues-pepsico-over-plastic-pollution-in-rivers-1552393.html.

Koroteev, D., & Tekic, Z. (2021). Artificial intelligence in oil and gas upstream: Trends, challenges, and scenarios for the future. *Energy and AI, 3.* https://doi.org/10.1016/j.egyai.2020.100041

LaBrecque, S. (2023). *Rising microplastics in seas puts pressure on Tyre industry. Reuters..* https://www.reuters.com/sustainability/land-use-biodiversity/rising-microplastics-seas-puts-pressure-tyre-industry-2023-07-17/

Levi Strauss & Co. (2023). *2022 Sustainability Goals & Progress Update.* https://www.levistrauss.com/wp-content/uploads/2023/09/2022-LSCo.-Sustainability-Goals-Progress-Update.pdf

Maersk. (2024). *2023 Sustainability Report.* https://www.maersk.com/~/media_sc9/maersk/corporate/sustainability/files/resources/2024/2023-maersk-sustainability-report.pdf

Marine Stewardship Council. (2013). *Partnering for sustainable fisheries.* https://www.msc.org/docs/default-source/default-document-library/for-business/fishery-improvement-tools/msc-partnering-for-sustainable-fisheries.pdf

Martin, M., & O'Neill, S. (2021). *Coca-Cola HBC statement on fostering biodiversity.* Coca-Cola Hellenic Bottling Company. https://gr.coca-colahellenic.com/content/dam/cch/gr/documents/policies/2022/biodiversity-statement.pdf.

Melillo, G. (2022). Ford motor company sued over NJ tribal land pollution. *The Hill.* https://thehill.com/changing-america/sustainability/energy/3531324-ford-motor-company-sued-over-nj-tribal-land-pollution/.

Mills, E. L., Leach, J. H., Carlton, J. T., & Secor, C. L. (1993). Exotic species in the Great Lakes: A history of biotic crises and anthropogenic introductions. *Journal of Great Lakes Research, 19*(1), 1–54. https://doi.org/10.1016/S0380-1330(93)71197-1

Mobilik, J.M., & Hassan, R. (2016). Marine pollution threat from shipping activity towards ocean sustainability. In *Conference: Aquatic science colloquium 2016: Experiences sharing in aquatic science.*

Moët Hennessy Louis Vuitton. (2023). *LVMH announces an ambitious water efficiency plan to reduce the Group's overall water consumption footprint by 30%.* https://www.lvmh.com/news-documents/news/lvmh-announces-an-ambitious-water-efficiency-plan-to-reduce-the-groups-overall-water-consumption-footprint-by-30/

Mohale, H. P., Jawahar, P., Jayakumar, N., Oli, G. A., & Ravikumar, T. (2023). Application of deep learning (AI) in marine fisheries resource management. *Trends in Agricultural Science, 2*(9), 753–763. https://doi.org/10.5281/zenodo.8387275

Mohammadiun, S., Hu, G., Gharahbagh, A. A., Li, J., Hewage, K., & Sadiq, R. (2021). Intelligent computational techniques in marine oil spill management: A critical review. *Journal of Hazardous Materials, 419*(5), 126425. https://doi.org/10.1016/j.jhazmat.2021.126425

Mostafa, M. (2015). Waste water treatment in textile industries- The concept and current removal technologies. *Journal of Biodiversity and Environmental Sciences, 7*(1), 501–525.

Mühling, Ş. M. K. (2023). Utilizing artificial intelligence (AI) for the identification and management of Marine Protected Areas (MPAs): A review. *Journal of Geoscience and Environment Protection, 11*, 118–132. https://doi.org/10.4236/gep.2023.119008

Naik, G. (2023, April 5). *Inventor of blue jeans Levi Strauss & Co channels a new water-saving idea.* Business Standard. https://www.business-standard.com/world-news/inventor-of-blue-jeans-levi-strauss-co-channels-a-new-water-saving-idea-123042500974_1.html

National Research Council of Science and Technology (2024). *Quickly and easily predict emerging contaminant concentrations in wastewater with artificial intelligence.* https://www.eurekalert.org/news-releases/1031944

Niinimäki, K., Peters, G., Dahlbo, H., Perry, P., Rissanen, T., & Gwilt, A. (2020). The environmental price of fast fashion. *Nature Reviews Earth & Environment, 1*(4), 189–200. https://doi.org/10.1038/s43017-020-0039-9

Nixon, H., & Saphores, J. (2007). Impacts of motor vehicle operation on water quality in the US—Cleanup costs and policies. *Transportation Research Part D: Transport and Environment, 12*(8), 564–576. https://doi.org/10.1016/j.trd.2007.08.002

Novartis, A. G. (2020). Novartis environmental sustainability and occupational health and safety data supplement 2020. https://www.novartis.com/sites/novartis_com/files/novartis-environmental-sustainability-occupational-heatlth-safety-data-supplement-2020.pdf

Pham, D. C., Do, T. N. A., Doan, T. N., Nguyen, T. X. H., & Pham, T. K. Y. (2021). The impact of sustainability practices on financial performance: Empirical evidence from Sweden. *Cogent Business & Management, 8*(1). https://doi.org/10.1080/23311975.2021.1912526

Radelyuk, I., Tussupova, K., Klemeš, J. J., & Persson, K. (2021). Oil refinery and water pollution in the context of sustainable development: Developing and developed countries. *Journal of Cleaner Production, 302*, 126987. https://doi.org/10.1016/j.jclepro.2021.126987

Rakshit, P. (2022). *Auto industry help reduce water consumption with innovative solutions*. Car&Bike. https://www.carandbike.com/news/auto-industry-help-reduce-water-consumption-with-innovative-solutions-3200346.

Rather, M. A., Ahmad, I., Shah, A., Hajam, Y. A., Amin, A., Khursheed, S., Ahmad, I., & Rasool, S. (2024). Exploring opportunities of Artificial Intelligence in aquaculture to meet increasing food demand. *Food Chemistry: X, 22*. https://doi.org/10.1016/j.fochx.2024.101309

Sadafale, H., & Gaikwad, R.W. (2023). Emerging trends in wastewater treatment of semiconductor industry: A review. *Nature Environment and Pollution Technology, 22*, 2075–2081. https://doi.org/10.46488/NEPT.2023.v22i04.032

Safe Drinking Water Foundation. (2017). *Conventional water treatment: Coagulation and filtration*. https://www.safewater.org/fact-sheets-1/2017/1/23/conventional-water-treatment

Samal, K., Mahapatra, S., & Ali, M. H. (2022). Pharmaceutical wastewater as emerging contaminants (EC): Treatment technologies, impact on environment and human health. *Elsevier, Energy Nexus, 6*, 100076. https://doi.org/10.1016/j.nexus.2022.100076

Schoenmaker, D. (2020). *The impact economy: Balancing profit and impact*. Working Paper 2020/04, Bruegel. https://doi.org/10.2139/ssrn.3567026

Schoenmaker, D., & Schramade, D. (2019). Principles of sustainable finance. Oxford University Press. ISBN: 9780198826606.

Shivajee, V. (2022). Technological innovation for optimizing water consumption: A case study for sustainable painting process. *Journal of Research in Business and Management, 10*(8), 79–87.

Singh, U., Acharya, D., & Mishra, S. (2022). Oil spill detection & monitoring with artificial intelligence: A futuristic approach. In *Conference: Proceedings of the Workshop on Control and Embedded Systems*.

Snyder, S. A., Adham, S., Redding, A. M., Cannon, F. S., DeCarolis, J., Oppenheimer, J., Wert, E. C., & Yoon, Y. (2007). Role of membranes and activated carbon in the removal of endocrine disruptors and pharmaceuticals. *Desalination, 202*(1–3), 156–181. https://doi.org/10.1016/j.desal.2005.12.052

Sudhir, S. K., Bhatti, S., Godheja, J., Panda, S. & Haq, I. (2023). Treatment of pharmaceutical pollutants from industrial wastewater. In *Current Developments in Biotechnology and Bioengineering, Bioremediation of Endocrine Disrupting Pollutants in Industrial Wastewater*, pp. 1–16. https://doi.org/10.1016/B978-0-323-91902-9.00007-9

Tawari, C. C., & Davies, O. A. (2020). Impact of multinational corporations in fisheries development and management in Niger delta Nigeria. *Agriculture and Biology Journal of North America*.

Thornton, A. (2019). *How AI and satellites are used to combat illegal fishing*. Microsoft. https://news.microsoft.com/on-the-issues/2019/06/06/ocean-mind-illegal-fishing

Toyota. (2023). 2022–23 Sustainability Report. https://www.toyotabharat.com/documents/toyota-in-india/environment/tkm-sustainability-report-2023.pdf

Ullah, I., Nuta, F. M., Levente, D., Yiyu, B., Yihan, Z., Yi, C., Shah, M. H., & Kumar, R. (2023). Nexus between trade, industrialization, and marine pollution: A quantile regression approach. *Ecological Indicators, 155*, 110992. https://doi.org/10.1016/j.ecolind.2023.110992

Villard, A., Lelah, A., & Brissaud, D. (2015). Drawing a chip environmental profile: Environmental indicators for the semiconductor industry. *Journal of Cleaner Production, 86*, 98–109. https://doi.org/10.1016/j.jclepro.2014.08.061ff.hal-01063777

Wang, Q., Huang, N., Cai, H., Chen, X., & Wu, Y. (2023). Water strategies and practices for sustainable development in the semiconductor industry. *Water Cycle, 4*, 12–16. https://doi.org/10.1016/j.watcyc.2022.12.001

Waqar, A., Othman, I., Shafiq, N., & Mansoor, M. S. (2023). Applications of AI in oil and gas projects towards sustainable development: A systematic literature review. *Artificial Intelligence Review, 56*, 12771–12798. https://doi.org/10.1007/s10462-023-10467-7

Wild, S. (2008). Sustainability reporting in fishing industry management - Regulation versus voluntarism. *Australasian Accounting Business and Finance Journal, 2*(3), 10.14453/aabfj.v2i2.3.

Wilkinson, J. L., Boxall, A. B. A., Kolpin, D. W., & Teta, C. (2022). Pharmaceutical pollution of the world's rivers. *PNAS Environmental Sciences, 119*(8). https://doi.org/10.1073/pnas.2113947119

Wong, C. W. Y., Wong, C. Y., Boon-itt, S., & Tang, A. K. Y. (2021). Strategies for building environmental transparency and accountability. *Sustainability, 13*. https://doi.org/10.3390/su13169116

Woollacott, E. (2024). *How AI is being used to prevent illegal fishing*. British Broadcasting Corporation. https://www.bbc.com/news/business-68564249

Arijit Goswami is a seasoned professional in the technology services and consulting sector, bringing eight years of experience in advising multinational clients on emerging technologies for digital transformation. Currently, he leads client engagement and enterprise success programs at MarketsandMarkets. Arijit has spearheaded sustainability initiatives within his organization and volunteered on climate action projects in India. He holds a degree in electronics engineering from Panjab University and an MBA from the Indian Institute of Management Kozhikode. Additionally, he earned a certificate in sustainability leadership from the University of Exeter. Arijit has collaborated with the World Economic Forum's Global Shapers Community on climate change projects and has served as a fellow with UNESCO, UNRISD, UNAOC, and the World Bank's Global Youth for Climate Network on climate technology research projects. His diverse domain expertise spans financial services, automotive, and telecommunications sectors, with a strong interest in research and thought leadership on sustainable technologies.

MNEs, Climate Change, and Heart Diseases: AI for Global Health Equity

Hakan Lane, Charles J. Oakley, Tanzeem Hossain, Arijit Goswami, Rajan Kumar Sah, and Mohit Mishra

Abbreviations

ACS	Acute Coronary Syndrome
AI	Artificial Intelligence
AMI	Acute Myocardial Infarction
BRFSS	Behavioral Risk Factor Surveillance System
COP	Conference of Parties
COPD	Chronic Obstructive Pulmonary Disease
CVD	Cardiovascular Disease
CMNN	Communicable, Maternal, Neonatal and Nutritional
DMD	Density of Medical Doctors
DMN	Density of Nurses or Midwifery Personal
EHE	Extreme Heat Event

H. Lane
Johannes Gutenberg University Mainz, Mainz, Germany
e-mail: hlane@uni-mainz.de

C. J. Oakley (✉)
Liberty University, Lynchburg, VA, USA
e-mail: coakley10@liberty.edu

T. Hossain
University of Freiburg, Freiburg im Breisgau, Germany

A. Goswami
Markets and Markets, Bengaluru, Karnataka, India

R. K. Sah
Tribhuvan University, Kathmandu, Nepal

M. Mishra
Outpayce from Amadeus, Bengaluru, Karnataka, India

J. Bento, M. M. Torres (eds.), *Global Perspectives on Climate Change, Inequality, and Multinational Corporations*, CSR, Sustainability, Ethics & Governance, https://doi.org/10.1007/978-3-031-80797-8_4

GDP	Gross Domestic Product
GDPpc	Gross Domestic Product Per Capita
GHG	Greenhouse Gas
LEAB	Life Expectancy at Birth
NCD	Non-Communicable Disease
OR	Odds Ratio
PTSD	Post-Traumatic Stress Disorder
VBD	Vector-Borne Disease
WHO	World Health Organization

1 Introduction

Climate change presents a unique and multifaceted threat to human health, distinguished by its broad and diverse impacts (Costello et al., 2009). The rise in global temperatures and increased pollution are strongly correlated with the prevalence of both communicable and non-communicable diseases (NCDs). Specifically, climate-induced changes in weather patterns have altered the epidemiology of infectious diseases, resulting in the increased frequency and geographic spread of vector-borne diseases such as malaria and dengue (De Blois et al., 2015).

Additionally, climate-related events such as floods and droughts often lead to the contamination of drinking water supplies, subsequently triggering outbreaks of water-borne illnesses. Beyond these physical health challenges, climate change also exerts significant psychological stress, contributing to a rise in mental health issues such as anxiety, depression, and overall psychological distress, which in turn exacerbate physical health conditions (Bolan et al., 2024). The health impacts of climate change are complex and layered, encompassing both direct consequences—such as the immediate effects of extreme weather events and natural disasters—and indirect effects, including prolonged exposure to heat, increased incidence of vector-borne diseases, and other associated health risks. Crucially, these impacts are not uniformly distributed; rather, they disproportionately affect regions and populations based on socio-economic status, geographic location, and access to healthcare resources.

This chapter addresses the central research question: How can Artificial Intelligence (AI) contribute to mitigating health inequities on a global scale, particularly in the context of climate change? The exploration presented here posits that AI has significant potential to enhance health equity by providing remote diagnostics and treatment options in underserved regions. Furthermore, AI-driven predictive analytics can be instrumental in optimising the allocation of healthcare resources, thereby improving health outcomes in populations most vulnerable to the adverse effects of climate change. The chapter also examines how AI applications can mitigate the health risks associated with urban pollution, a key driver of climate-related health issues. Through an analysis of case studies and relevant literature, this study seeks to demonstrate the transformative role AI can play in addressing the intertwined challenges of climate change and global health disparities.

2 Literature Review

2.1 Climate Change and Human Health

Among all the factors that threaten human health in the current world, climate change is one of the most complex due to its range of influences and negative impact on health (Costello et al., 2009). Increased heat and pollution are found to be linked with various communicable and non-communicable diseases (NCD) (Friel et al., 2011; Wu et al., 2016). Besides these direct impacts on physical health, climate change also forces many people to migrate from vulnerable areas, impacting their mental health (Sri et al., 2023).

Humans are continuously contributing to climate change. Earth's average temperature increased by approximately 0.85 degrees Celsius in the twentieth century, and 1:% degree global mean surface temperature to the pre-industrial period. The rate of global warming has increased to over 0.18 degrees Celsius per decade over the last 30 years. The increasing threat of global warming severely threatens human health, notably the risk of heart disease through heat stress and air pollution. Other expected climate change features, such as more violent and extreme weather events and the rising sea level, are important for population health (De Blois et al., 2015). Figure 1 presents the impacts of climate change through extreme weather events.

Due to this variability in temperature, extreme heat events have become more frequent. For example, there were 70,000 deaths estimated to be attributable to these extreme conditions, with more than one-third of the deaths occurring in France, Italy, and Spain (Peters & Schneiderl., 2021), exacerbated by air pollution. The principal greenhouse gases are carbon dioxide (76%), methane (16%), and, to a

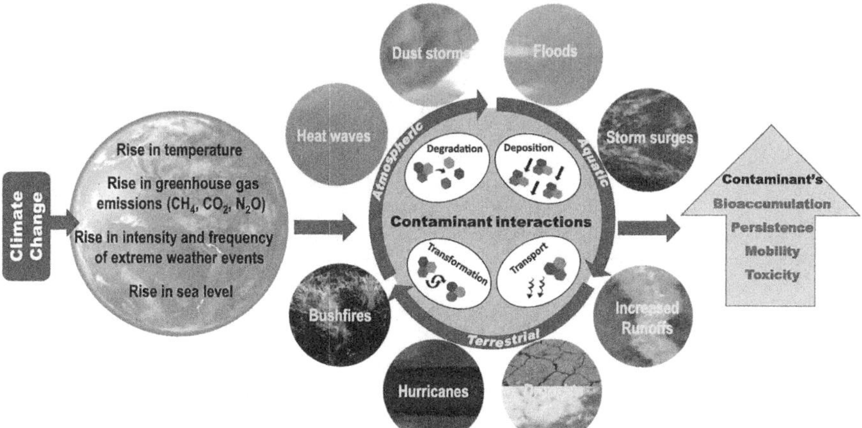

Fig. 1 Impacts of climate change on the fate of contaminants through extreme weather events. (Source: Bolan et al. (2024) under Open Access Creative Commons CC-BY licence. https://doi.org/10.1016/j.scitotenv.2023.168388)

limited extent, nitrous oxide (2%). Extreme heat waves can lead to dehydration, heat stroke, and an increased heart rate. As glaciers melt and ocean water expands, sea levels are rising, threatening coastal areas with flooding and erosion. Forced displacement due to sea level rise creates stress and mental health problems, both of which negatively affect our cardiovascular health (Shivanna, 2022), as portrayed in Fig. 2.

AI-driven health technologies could be used to address many of the challenges. For example, AI-driven solutions have helped reduce health workers' workload (Lozano et al., 2018). Although these interventions provide promise, many ethical, regulatory, and practical issues remain that require guidance before scale-up or widespread deployment (Wahl et al., 2018). A wealth of data is available to demonstrate how AI is being tested to address health challenges relevant to achieving SDGs. Many AI health solutions have shown promising preliminary results and may soon be used to supplement existing strategies for providing health services (Guo & Li, 2018).

The remaining issues include the development of AI-driven health interventions, how efficacy and effectiveness are assessed and reported, large-scale deployment planning and the ethical, regulatory, and economic standards and guidelines that will help protect the interests of communities in low- and middle-income countries (Schwalbe & Wahl, 2020).

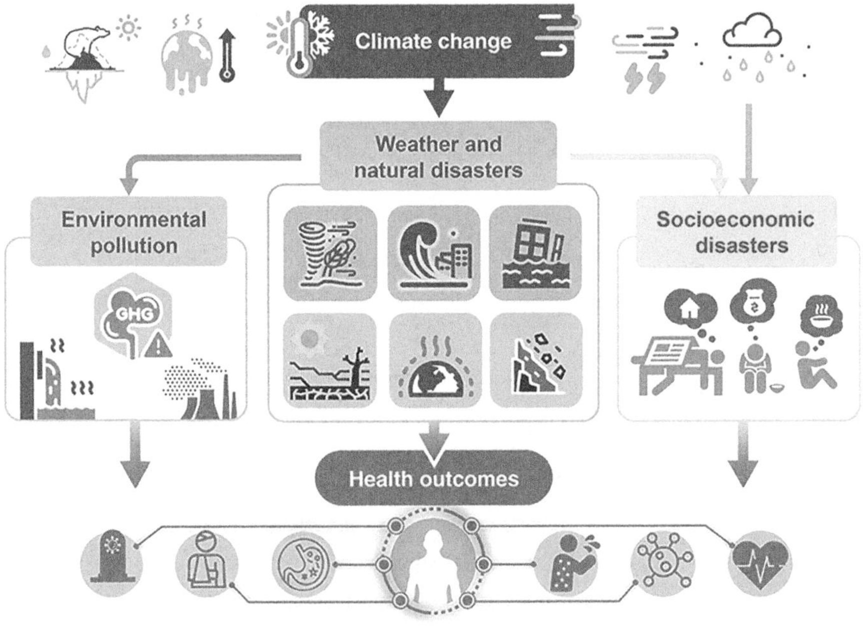

Fig. 2 Global climate change and human health. (Source: Zhao et al. (2022) under Creative Commons License. https://creativecommons.org/licenses/by/4.0/)

2.2 *Communicable Diseases*

Rising temperatures and extreme natural calamities have been reported to change the epidemiology of various infectious diseases (Wu et al., 2016). Vector-borne Diseases (VBD) like malaria and dengue are becoming more frequent in their endemic areas, while some of them are also spreading to new regions, as the vectors and the pathogens are finding suitable habitats there due to warmer temperatures (Caminade et al., 2019). Around 22% of dengue cases that occurred during large outbreaks in Hanoi, Vietnam, from 2008 to 2016 were found to be attributable to higher temperatures (Cheng et al., 2020). Food- and water-borne diseases are mostly associated with events like floods, cyclones, and droughts, where drinking water gets contaminated for prolonged periods and infects the affected communities (Cissé, 2019). A study in Bangladesh found that during rainfall, the risk for cholera disease increased 2 days after heat waves (Odds Ratio, OR = 1.53, 95% CI: 1.07–2.19), possibly due to drinking water contamination (Wu et al., 2018).

2.3 *Cardiovascular Disease (CVD) and Other Non-communicable Diseases (NCDs)*

Climate change has emerged as a significant driver, particularly regarding the rise of heart disease (Costello et al., 2009). Scientific evidence indicates a complex interplay between *The Struggle to Stay in Place* climate change and heart health, with factors such as air pollution, heat stress, and changes in food production contributing to the disproportionate burden of heart disease on vulnerable populations (Lowe, 2014). On top of climate change (Cicci et al., 2022), Brook et al. (2010) and Pope et al. (2009) both demonstrate the detrimental effects of air pollution on cardiovascular health, including increased rates of heart attacks, strokes, and other cardiovascular events.

Extreme Heat Events (EHE) and heat waves are significantly associated with CVD-related mortality, especially in people older than 65 (Cheng et al., 2019; Liu et al., 2022). According to de Bont et al. (2024), a 14.7% (95% CI, 10.3; 19.3) increase in all-cause daily mortality was caused by heat waves in multiple cities in India from 2008 to 2019. The increased frequency of cold spells due to climate change is also detrimental to heart health, increasing AMI by 9% for a 10 °C decrease in temperature (Claeys et al., 2017).

People with morbidity or pre-existing disease conditions are often more vulnerable, as dehydration during EHEs can lead to kidney diseases and is one of the most common causes of hospitalisation and mortality among patients with end-stage renal failure (Remigio et al., 2019; Luna-Cerón et al., 2024). Hypoglycemia in people with type-2 diabetes is also a common phenomenon during periods of EHEs (Bogar et al., 2022). Other extreme weather events such as tornadoes, thunderstorms, and hurricanes have become more frequent. Climatic events like wildfires and dust storms have become more frequent (Manisalidis et al., 2020) and also impose increased health risks (Makrufardi et al., 2023). Exposure to climate change-related natural hazards might lead to chronic NCDs like hypertension and thus deteriorate the overall quality of life (Ghosh et al., 2022; Zhao et al., 2022).

Short- and long-term exposure to polluted air is linked to increased risk of CVDs, Chronic Obstructive Pulmonary Diseases (COPD), and other lung diseases (Hansel et al., 2016; Kim et al., 2021; Chang et al., 2023). A study conducted on over half a million people in the U.S. demonstrated that the mortality from ischemic heart disease and stroke increased by 16% (95% CI 1.09–1.22) and 14% (95% CI 1.02–1.27), respectively, with each 10 $\mu g/m^3$ increase of $PM_{2.5}$ (Hayes et al., 2020). Various studies also suggest that air pollution might have influences on diseases such as asthma, rhinitis, and pollen allergies (D'Amato et al., 2015; Lee et al., 2023; Seastedt & Nadeau, 2023). Ruus and Mornoş (2022) summarised all the effects of meteorological factors and air pollution on CVDs in Fig. 3.

Extreme weather events such as storms, floods, wildfires, droughts, sea level rise, and heat are projected to make nearly a fifth of the planet too hot for humans by 2070. Due to these causes, millions of people are displaced as a result of political unrest, economic hardship, and conflict. Living in climate hotspots, where they lack the resources to adapt, refugees, internally displaced people, and stateless people are at the forefront of the climate emergency (Fig. 4). Regions of critical concern are the African Sahel, Middle East and North Africa, Central America's Dry Corridor, and South Asia (Agache et al., 2022).

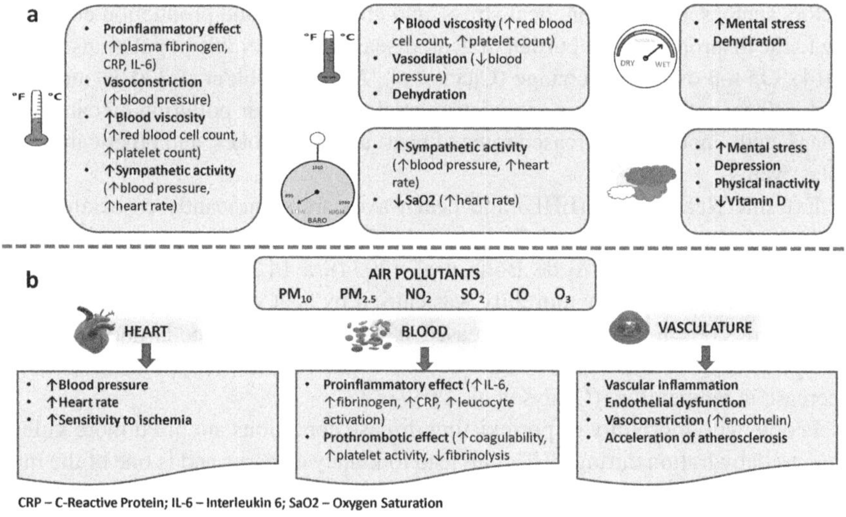

Fig. 3 Pathophysiological mechanisms of (**a**) meteorological factors and (**b**) air pollutants on acute coronary syndrome. (Source: Ruus and Mornoş (2022) under exclusive licence to Springer Science+Business Media, LLC, PMC Open Access Subset. https://www.ncbi.nlm.nih.gov/pmc/tools/openftlist/)

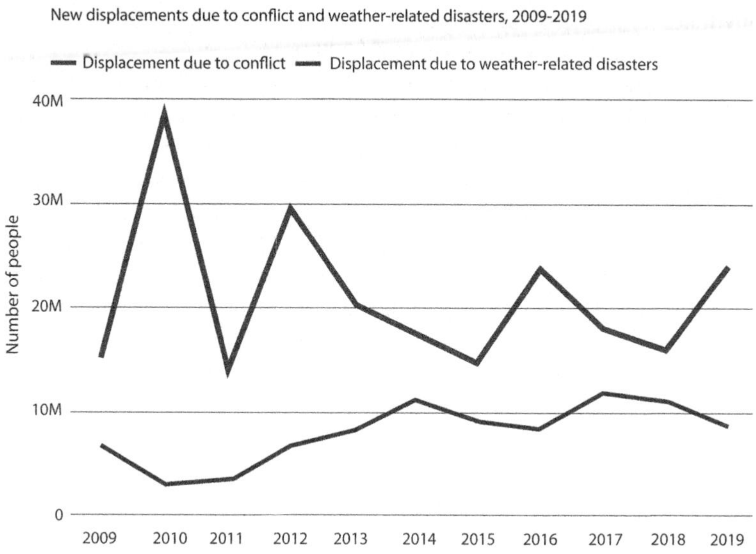

Fig. 4 Displaced on the frontlines of the climate emergency. (Source: UNHCR, The UN Refugee Agency. https://arcg.is/1fivTq0)

2.4 Mental Health and Climate Migration

Climate change and its consequences also led to adverse mental health outcomes such as depression and anxiety (Gawrych, 2022; Riaz et al., 2023). People affected by natural calamities like floods and cyclones suffer from post-traumatic stress disorders (PTSD), suicidal tendencies and substance abuse (Padhy ct al., 2015; Huang ct al., 2023; Walinski et al., 2023). Migration and displacement due to adverse climate is another important mental health stressor (Torres & Casey, 2017; Sri et al., 2023).

2.5 Multinational Corporations and Health Inequity

In recent decades, the role of multinational companies in exacerbating health inequity has come under increasing scrutiny. These companies wield significant influence over access to healthcare, pricing essential medicines, and formulating public health policies, often exerting substantial influence over global health outcomes through their operations in domains like food production, supply chains, and marketing strategies. As an example, diet is a strong risk factor for worldwide noncommunicable diseases (NCDs) (Global Burden of Disease Collaborators, 2020). The food industry's practices, behaviours, and commodities significantly impact global health equity (Buse et al., 2017). The proliferation of highly refined carbohydrates, cholesterol, and high-sugar diets leads to an increased prevalence of obesity and degenerative diseases (Popkin, 2002).

In food production and distribution, multinational corporations often prioritise profitability over nutritional quality and sustainability considerations (Monteiro et al., 2013; Swinburn et al., 2011). Multinational corporations, referred to as 'Big food', focus on specific nutritional issues and draw attention away from the broader health issues associated with processed food (Clapp & Scrinis, 2016). The global packaged food market in 2023 encompassed a whopping US$2.9 T (Financial Services Monitor Worldwide, 2023) and, therefore, wields considerable power and influence in policy governance. Convenience, sustainability, and shelf life are major factors in the popularity of packaged foods in emerging economic countries. Neoliberal economic policies have allowed multinationals to influence policies and regulatory governance (Falkner, 2003).

Marketing practices often glamorise unhealthy lifestyles while downplaying health risks, e.g., stating that a certain product has '100% of the recommended daily allowance of vitamin C' but failing to mention processed sugar or unhealthy fats. This kind of 'nutritional reductionism' creates a skewed consumer understanding of the product's impact on total body health (Jacobs & Tapsell, 2008; Clapp & Scrinis, 2016, p. 582).

Efforts to address these disparities and hold multinational corporations accountable for their impact on global health equity include universal basic income, early intervention education, more robust regulatory measures, promoting consumer education and empowerment, and fostering international cooperation to address cross-border health challenges ((Ruckert et al., 2017; Thornton et al., 2016; Kickbusch et al., 2016). Addressing the root causes of health inequities driven by multinational corporations can help policymakers and stakeholders work towards a more equitable and sustainable global health landscape. However, high reliance on public-private governance may create gaps in policy and equity, so international public health oversight is needed (Lie, 2019).

2.6 Spatial Impact on Health

Multinational corporations have a high global climate impact, using lax regulatory enforcement to maximise profit while neglecting climate sustainability. The environmental impacts go beyond their home borders (López et al., 2019), and 63% of worldwide carbon dioxide emissions and methane between 1751 and 2010 came from 90 entities (Heede, 2013).

Other consequences of climate change, such as elevated temperatures resulting in individual heat stress, pose additional challenges to heart health, particularly among vulnerable populations. Watts et al. (2015) document the increased incidence of heat-related illnesses and cardiovascular complications during extreme heat, underscoring the need for adaptive measures to protect public health. Vulnerable populations, including low-income communities and people of colour disproportionately bear the adverse effects of climate change-related heart disease. Hajat et al. (2015), and Costello et al. (2009) indicate that socio-economic factors and environmental stressors exacerbate disparities in environmental disease prevalence and outcomes, underscoring the need for targeted interventions to address health equity concerns.

This underscores the urgent need for comprehensive strategies to protect cardiovascular health in the face of environmental challenges. By addressing air pollution, heat stress, and changes in food production while prioritising health equity considerations, policymakers and public health professionals can work towards safeguarding heart health and promoting resilience in the face of climate change. In conclusion, evidence-based policy development represents a cornerstone in addressing health inequities exacerbated by climate change. By formulating informed policies and equitable interventions, governments can bolster resilience, protect vulnerable populations, and foster health equity amidst evolving environmental challenges, ensuring a sustainable future for future generations.

2.7 *Health Inequity*

When considering the inequality of climate change effects, it has been noted that the most impactful ramifications of climate change on health include extreme events, infectious diseases and impaired access to water and food (Ebi & Hess, 2020). Yarveysi et al. (2023) found that natural disaster risks are higher for poorer communities, and it was noted during floods that lower-income households suffered three times higher exposure than the upper echelons (Odersky & Löffler, 2024). The Germanwatch NGO has created the Global Climate Risk, which identifies which countries are at the most risk for climate change-related natural disasters (Fig. 5). The map clearly defines the highest risks for African countries and nations close to the Arctic.

When looking at infections, Ayorinde et al. (2023) revealed that people in lower income brackets had higher risks of tuberculosis, HIV, antimicrobial resistance and multiple other diseases through a thorough review of multiple studies in several countries. Finally, for unequal access to water, through various methods, a Gini coefficient between countries was calculated as 0.41 (Malakar & Mishra, 2017), and in some countries with a high degree of privatisation as high as 0.95 in Chile (Correa-Parra et al., 2020) or 0.7 in many parts of Brazil (Jesus et al., 2023).

3 Methodology

This grounded study reviews case studies from various continents and socio-cultural contexts to examine the implementation of AI tools in improving healthcare access and decision-making. The focus is on how AI can address health inequities

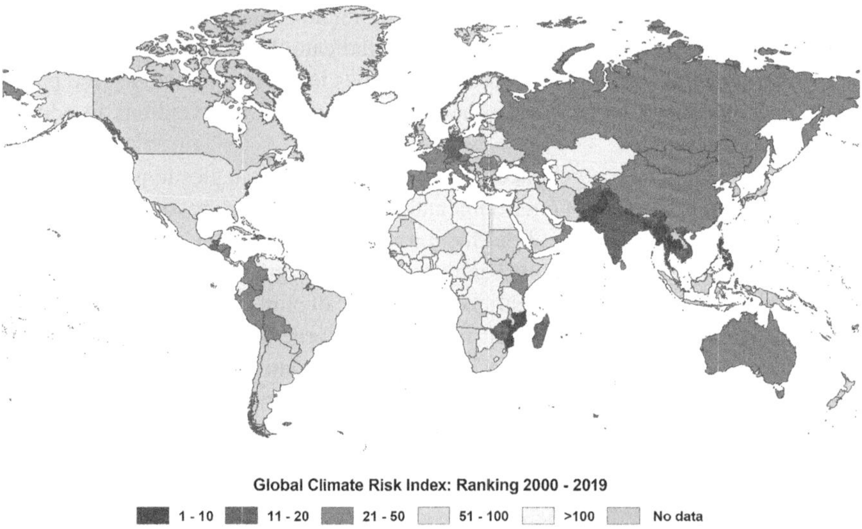

Fig. 5 Global Climate risk. (Source: Germanwatch. www.germanwatch.de)

exacerbated by climate change, particularly in communities vulnerable to cardiovascular diseases (CVD). The data sources include electronic health records, environmental monitoring, and predictive modelling to assess AI's impact on health outcomes.

Health inequities, characterised by systematic and avoidable disparities in health outcomes among different population groups, persist as a significant and multifaceted global challenge (Cunningham et al., 2022; O'Kane et al., 2024). Health inequity, defined as disparities in health outcomes, disproportionately affects marginalised low-income communities (O'Kane et al., 2024). The concept of equality in health is not fully defined and often depends on the context. Hoyer et al. (2022) explored the narratives and identified three main themes: (a) Maximising health, (b) equal healthcare, (c) absence of disparities. One of the most commonly used, the Gini coefficient for health starts by plotting the cumulative income on the x-axis and the cumulative incidence as y in a Lorenz curve. The Gini index is derived as the deviation from the equality diagonal, as shown in Fig. 6.

The Behavioural Risk Factor Surveillance System (BRFSS) from the Center of Disease Control, a nationwide health-related telephone survey system conducted by the Centers for Disease Control and Prevention (CDC), provided significant foundational data. The BRFSS collects data from U.S. residents regarding their health-related risk behaviours, chronic health conditions, and use of preventive services. Based on the BRFSS, we derive the Gini coefficients for several diseases in the United States. Table 1 reports values for several conditions, including access to healthcare and lifestyle factors.

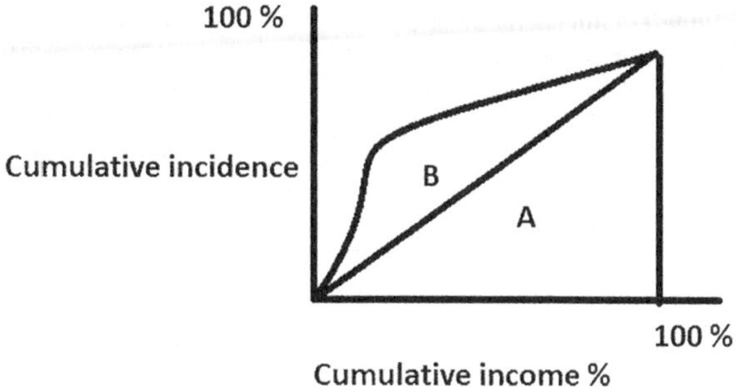

Fig. 6 Lorenz curve and Gini coefficient quantities. (Source: Authors' elaboration)

Similarly, the Gini coefficient can be used to measure the amount of health inequality between nations. Steinbeis et al. (2019) derived measures in each of the categories Communicable, Maternal, Neonatal and Nutritional (CMNN), Non-Communicable (NCD) and Injury-Related Diseases over the time 1990–2017 as portrayed in Fig. 7.

3.1 Understanding the Causes of Health Inequities

Health inequities emerge from a complex interplay of social, economic, and environmental determinants that systematically disadvantage specific segments of society (Friel et al., 2008; Hammonds & Reverby, 2019). Social factors profoundly influence individuals' health trajectories (Cunningham et al., 2022). These pathways include inadequate housing, food insecurity, and limited access to healthcare services (Marmot et al., 2008).

The GDP per capita is a key indicator of average economic well-being and prosperity. GDP per capita data was cross-referenced with health statistics from the World Health Organization (WHO) to examine disparities in global healthcare outcomes with a particular focus on life expectancy at birth (LEAB) and the availability of healthcare professionals.

The data reveal that LEAB is significantly higher in high-GDP per capita countries compared to their low-GDP per capita counterparts. Similarly, the number of medical doctors (DMD) and nursing and midwifery personnel (DMN) are both many times higher (Fig. 8). The higher availability of healthcare professionals in wealthier nations likely contributes to better health outcomes. Racism, discrimination, and colonial legacies exacerbate health inequities (Williams & Mohammed, 2008). Indigenous populations and racial minorities frequently confront systemic barriers hindering their access to quality healthcare services, culminating in elevated rates of morbidity and mortality (Farmer et al., 2019; Gonzales et al., 2016).

Table 1 Gini coefficients common non-communicable diseases United States 2023

Disease	Gini coefficient
CHD	0.36
High blood pressure	0.16
High cholesterol	0.25
Heart attack	0.45
Stroke	0.64
Diabetes	0.38
Not good physical health	0.66
Have any health insurance	0.09
Could not see a doctor because of the cost	0.73
Could not get medicines because of the cost	0.66
Do you have medicare?	0.38
Delayed getting medical care	0.41
Without health care coverage for the past 12 months	0.65
Smoking every day	0.29
Any exercise past 30 days	0.73

Note: As a reference, the income coefficient for the United States in 2013 was 0.41 (Source: Authors' elaboration)

4 Case Studies

4.1 Innovative Health

The overall health-related quality of life of people with heart failure in low- and middle-income countries is poor (Mulugeta et al., 2024), making the need for improved policies urgent in order to meet the United Nations 2025 health goals (Ruan et al., 2018). Governments must set a technology-based agenda to improve CVD prevention (Agbor et al., 2020). Although low- and middle-income countries have much lower GHG emissions per capita, health vulnerabilities range are in many cases much more pronounced (Campbell-Lendrum & Corvalán, 2007; Borg et al., 2021). A study in India on CVD found age-related sex differences in the vulnerability of the population to non-optimal temperatures (Shrikhande et al., 2023). Similar cases can be found that prove more impacts in low- and middle-income countries in aspects of both climate change and CVD.

Due to the higher level of wealth and technology, most examples of deployment have occurred in the West. In 2023, AI was employed by the National Health Service (NHS) in the U.K. to identify patients at risk of requiring hospitalisation, enabling community teams to intervene proactively and alleviate pressure on the elderly

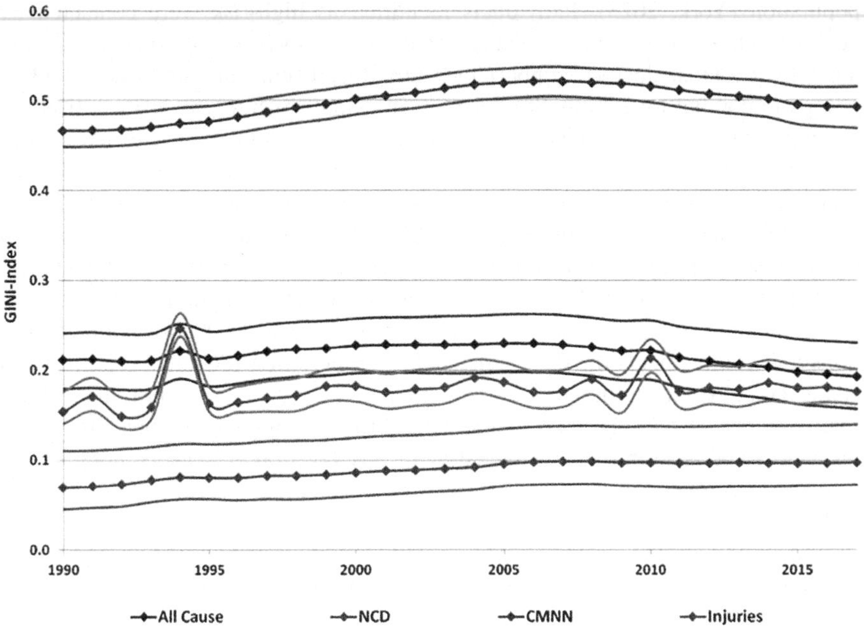

Fig. 7 Worldwide Gini coefficients between countries for different conditions 1990–2017. (Authors' elaboration)

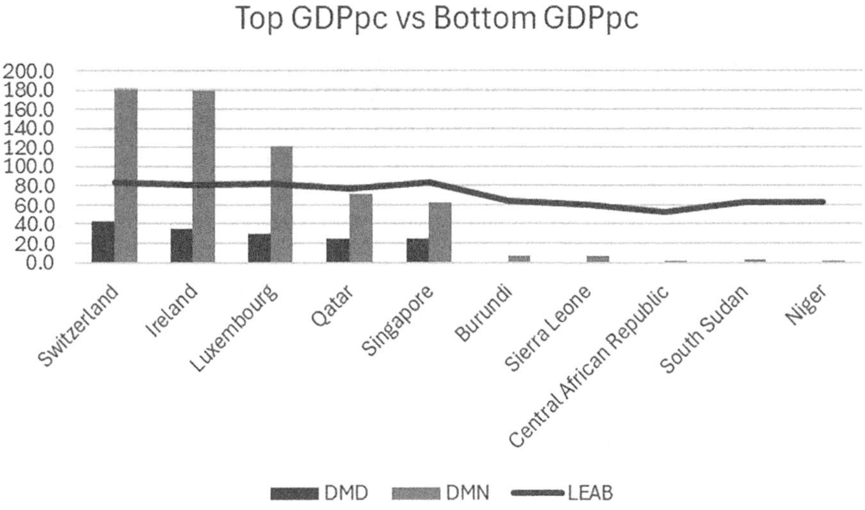

Fig. 8 Highest and lowest GDP per capita comparison chart. (Source: Authors' elaboration). *Note*: DMD Density of medical doctors, DMN density of nursing and midwifery personnel, LEAD Life expectancy at birth, GDPpc Gross domestic product per capita

population (York, 2023). Individuals identified as high-risk were contacted by health coaches, nurses, or G.P.s, who offered services such as food parcels, referring them to specialist doctors, support measures to prevent falls, or local volunteer groups to combat loneliness. Similarly, in Buckinghamshire, the NHS utilised AI technology integrated with electronic sensors on household appliances like kettles and fridges to detect changes in patients' dietary and drinking habits and react accordingly. A team of scientists from the Barcelona Supercomputing Center—Centro Nacional de Supercomputación (BSC-CNS) have developed an AI system that can be used to collect data on the concentration of different air pollutants and predict breaches of city pollution limits (Barcelona Supercomputing Center, 2023). Another centre in Germany collects data from more than 14,000 air pollution stations and complements the data with satellite information (Schultz, 2022). MNEs can leverage AI technology to forecast periods of heightened susceptibility in local river systems, such as instances of agricultural runoff (Singh & Pradhan, 2024). For instance, there is an initiative where strategically positioning sensors along a river enables comprehensive data collection on river conditions, precipitation patterns, and soil composition (Fisher, 2023). This proactive approach guides interventions, like advising farms to postpone fertiliser application, thereby mitigating potential environmental impacts, with a remarkable accuracy rate exceeding 90%.

4.2 Early Prediction of Heart Disease Through the Use of AI and Advanced Technologies

Cardiovascular diseases (CVDs) are the leading cause of death worldwide. In 2019, out of 17 million premature deaths (under the age of 70) from non-communicable diseases, 38% were due to CVDs. Acute coronary syndrome (ACS) where blood flow to the heart is restricted includes ST-elevation myocardial infarction (STEMI), non-ST elevation myocardial infarction (NSTEMI), and unstable angina (Byrne et al., 2024). When ACS was diagnosed with AI-based detection, sensitivity ranged from 64% to 100% and specificity ranged from 65% to 100%. (Chan et al., 2023). Although the overall quality of evidence was low due to the inclusion of case-control studies, the study informs the potential of using AI-assisted ACS detection for accurate diagnosis and prompt treatment for ACS. Similar cases have been found with cardiac arrhythmias, which can be found in anybody at any time without age biases. Cardiac arrhythmia is a condition in which the heart's normal rhythm is disrupted (Fenton et al., 2008). Paroxysmal Atrial Fibrillation (PAF) is the most common major cardiac arrhythmia. When PAF was predicted with the ML method with evaluation using the Atrial Fibrillation Prediction Database (AFPDB), an accuracy of 98% was reached (Ebrahimzadeh et al., 2018), proving why AI-based cardiac arrhythmias prediction has become more common (Attia et al., 2021).

4.3 *Wearables*

Smart wearables track workouts, sleep hygiene, cardiovascular health, arrhythmia (Riazul Islam et al., 2015). They increase awareness, facilitate home care and could potentially reduce clinical instability (Mortara et al., 2009). Better self-management and intervention approaches reduce healthcare costs by preventing unnecessary hospitalizations (Sana et al., 2020). This augments the physician-patient relationships, increases autonomy and reduces spending (Appelboom et al., 2014; Bonato, 2005). AI-assisted tolls can screen left ventricular dysfunction with 93% accuracy, lifesaving in many cases (Attia et al., 2022).

5 Results

The findings demonstrate that AI-driven health technologies can reduce the workload for healthcare workers, improve diagnostic accuracy, and facilitate remote monitoring of patients' health. In regions with limited healthcare infrastructure, AI-enabled tools provide essential support, offering health education and early disease detection. Successful AI applications in mitigating urban pollution and monitoring water quality are crucial for preventing water-borne diseases and improving overall health outcomes.

This study asked: *How can AI contribute to reducing inequality in global health, particularly in the context of climate change*? Using the propositions for this study (a) AI can enhance health equity by providing remote diagnostics and treatment in underserved areas. (b) AI-powered predictive analytics can optimise resource allocation and improve health outcomes. (c) AI applications in combating urban pollution can reduce health risks associated with climate change. This study can definitively show that AI offers transformative solutions to address health inequities by providing remote diagnostics, personalised treatment plans, and predictive analytics. AI-powered tools can enhance health equity by reaching underserved areas, optimising resource allocation, and improving health outcomes. However, challenges such as data privacy, limited technology access, and potential biases in AI algorithms need to be addressed.

6 Discussion, Limitations and Future Work

6.1 *Impact of Health Inequities on Population Groups*

The relative inequality is worst for CMNNs, most likely due to a number of factors. CMNN diseases are often more prevalent in low-income and marginalised populations, where access to healthcare services, clean water, sanitation, and nutrition may be limited. They can spread more easily in crowded or unsanitary conditions, further exacerbating health disparities. On the other hand, NCDs are often associated with lifestyle factors such as diet, physical activity, and tobacco use, which can be influenced by socio-economic status but are not solely determined by it. Individuals with higher incomes may have better access to healthcare, education, and resources to prevent and manage NCDs, leading to less pronounced health inequalities compared to CMNN diseases.

The ramifications of health inequities are particularly acute among marginalised and vulnerable population groups, including Indigenous peoples, racial minorities, women, and individuals experiencing poverty (World Health Organization, 2006). For example, Indigenous populations endure disproportionately higher burdens of chronic diseases such as diabetes and cardiovascular ailments due to historical and contemporary social and economic marginalisation (Farmer et al., 2019; Gonzales et al., 2016). Similarly, women confront unique health inequities related to reproductive health, access to maternal care, and experiences of gender-based violence, all of which contribute to diminished health outcomes relative to men (World Health Organization, 2006). Moreover, health inequities intersect with various axes of oppression, including gender, sexuality, disability, and immigration status, further compounding disparities in health outcomes (Hankivsky, 2012).

6.2 Healthcare

Equitable healthcare access is paramount in mitigating health disparities. Vulnerable communities often encounter barriers to healthcare access exacerbating their health burdens (Gulliford et al., 2002), and policy should address systemic disparities. This involves targeted interventions for healthcare infrastructure in underserved areas, expanded access to primary care and community-based health initiatives (Shi et al., 2021), and concerted efforts to address social determinants of health (Friel et al., 2008).

Policies should prioritise the participation of marginalised communities in decision-making processes related to healthcare delivery. Community engagement and empowerment are essential for fostering trust, building resilience, and ensuring that healthcare services meet the diverse needs of local populations. Community-based participatory research (CBPR) emphasises collaborative partnerships between community health stakeholders, researchers, and community organisations to solve local issues (Berge et al., 2009). Partnerships between policymakers, healthcare providers, community organisations, and advocacy groups are critical for advancing health equity initiatives (Chang et al., 2017; Cunningham et al., 2022), together with equitable access to healthcare services (Hammonds & Reverby, 2019; Lavizzo-Mourey et al., 2021).

6.3 Multinational Corporations and Health Inequity

In recent decades, the role of multinational companies in exacerbating health inequity has come under increasing scrutiny. These companies wield significant influence over access to healthcare, pricing essential medicines, and formulating public health policies, often exerting substantial influence over global health outcomes through their operations in domains like food production, supply chains, and marketing strategies. As an example, diet is a strong risk factor for worldwide non-communicable diseases (NCDs) (Global Burden of Disease Collaborators, 2020). The food industry's practices, behaviours, and commodities significantly impact global health equity (Buse et al., 2017). The proliferation of highly refined carbohydrates, cholesterol, and high-sugar diets leads to an increased prevalence of obesity and degenerative diseases (Popkin, 2002).

In food production and distribution, multinational corporations often prioritise profitability over nutritional quality and sustainability considerations (Monteiro et al., 2013; Swinburn et al., 2011). Multinational corporations, referred to as 'Big food', focus on specific nutritional issues and draw attention away from the broader health issues associated with processed food (Clapp & Scrinis, 2016). The global packaged food market in 2023 encompassed a whopping US$2.9 T (Financial Services Monitor Worldwide, 2023) and, therefore, wields considerable power and influence in policy governance. Convenience, sustainability, and shelf life are major factors in the popularity of packaged foods in emerging economic countries. Neoliberal economic policies have allowed multinationals to influence policies and regulatory governance (Falkner, 2003).

Marketing practices often glamorise unhealthy lifestyles while downplaying health risks, e.g., by stating that a certain product has '100% of the recommended daily allowance of vitamin C' but failing to mention processed sugar or unhealthy fats. This kind of 'nutritional reductionism' creates a skewed consumer understanding of the product's impact on total body health (Jacobs & Tapsell, 2008; Clapp & Scrinis, 2016, p. 582).

Efforts to address these disparities and hold multinational corporations accountable for their impact on global health equity include universal basic income, early intervention education, more robust regulatory measures, promoting consumer education and empowerment, and fostering international cooperation to address cross-border health challenges (Ruckert et al., 2017; Thornton et al., 2016; Kickbusch et al., 2016). Addressing the root causes of health inequities driven by multinational corporations can help policymakers and stakeholders work towards a more equitable and sustainable global health landscape. However, high reliance on public-private governance may create gaps in policy and equity, so international public health oversight is needed (Lie, 2019).

6.4 Spatial Impact on Health

Multinational corporations have a high global climate impact, using lax regulatory enforcement to maximise profit while neglecting climate sustainability. The environmental impacts go beyond their home borders (López et al., 2019), and 63% of worldwide carbon dioxide emissions and methane between 1751 and 2010 came from 90 entities (Heede, 2013). Other consequences of climate change, such as elevated temperatures resulting in individual heat stress, pose additional challenges to heart health, particularly among vulnerable populations. Watts et al. (2015) document the increased incidence of heat-related illnesses and cardiovascular complications during extreme heat, underscoring the need for adaptive measures to protect public health.

The adverse effects of climate change-related heart disease are disproportionately borne by vulnerable populations, including low-income communities and people of colour. Hajat et al. (2015) and Costello et al. (2009) indicate that socioeconomic factors and environmental stressors exacerbate disparities in environmental disease prevalence and outcomes, underscoring the need for targeted interventions to address health equity concerns. This underscores the urgent need for comprehensive strategies to protect cardiovascular health in the face of environmental challenges. By addressing air pollution, heat stress, and changes in food production while prioritising health equity considerations, policymakers and public health professionals can work towards safeguarding heart health and promoting resilience in the face of climate change. In conclusion, evidence-based policy development represents a cornerstone in addressing health inequities exacerbated by climate change. By formulating informed policies and equitable interventions, governments can bolster resilience, protect vulnerable populations, and foster health equity amidst evolving environmental challenges, ensuring a sustainable future for future generations.

7 Moving Forward—Practical Applications

7.1 Health Service Adaptation

AI holds the potential to ensure health equity and bring quality healthcare to vulnerable populations and rural areas where healthcare penetration is low. Multinational healthcare enterprises can leverage AI to expand access to their services in marginalised regions. AI algorithms can analyse medical images from C.T. scans, X-rays, and MRI scans to provide diagnostic reports of high quality for vulnerable populations and develop a personalised treatment plan. The use of AI systems for rural healthcare increases health equity as rural patients gets quality treatment in real time, instead of travelling long distances to reach geographically dispersed hospitals. Figure 9 illustrates the application of AI for automated medical diagnostics.

The use of AI in combating pollution using predictive monitoring and big data analytics positions is a potential tool for MNEs to leverage for sustainability efforts. Figure 10 illustrates the use of AI in monitoring Air Quality Indices (AQI), which estimate concentrations of nitrous oxide, sulphur dioxide, carbon dioxide, and particulate matter. AI mitigates urban soil pollution by predicting contamination risks using historical data. Sensors and drones monitor soil quality in real-time, detecting pollutants promptly. AI optimises remediation strategies by analysing soil composition and pollutant data, ensuring efficient clean-up efforts. It optimises soil remediation strategies by analysing vast datasets on soil composition, pollutant types, and remediation techniques, ensuring efficient and cost-effective remediation efforts. AI tools for urban planning incorporate soil pollution data to prevent further contamination. This allows cities to designate green spaces and landfills strategically and AI enables cities to promote sustainable urban development through data analysis and predictive modelling.

It can be deduced that AI holds immense promise in bolstering health equity, yet its application in healthcare also raises significant concerns. Central to this discussion is the imperative that AI initiatives actively address rather than perpetuate existing disparities faced by vulnerable populations. Failure to tailor AI solutions to local contexts and needs risks exacerbating health inequities (PATH, 2023). This challenge is compounded by the presence of unintended biases in AI algorithms, limitations in the availability and accuracy of local training data, and deficiencies in privacy protocols, all of which can propagate inaccuracies and inequities in diagnosis and treatment. Moreover, a critical issue lies in the skewed representation within health data used to train Algorithms, often derived from predominantly white populations and lacking ethno-racial information. Health data biases against BIPOC communities can skew AI algorithms. The inclusion of ethno-racial data is mishandled, leading to suboptimal outcomes (Gurevich et al., 2023). Over-reliance on AI without adequate capacity-building and oversight may erode the human touch and scientific rigour in healthcare delivery (PATH, 2023), necessitating a delicate balance to ensure AI technologies truly advance health equity. AI reliance without oversight of biased algorithms, privacy breaches etc. (OECD, 2024) risks diminishing healthcare quality. The potential should be maximised through equity, transparency, and accountability. Comprehensively addressing health inequities demands a

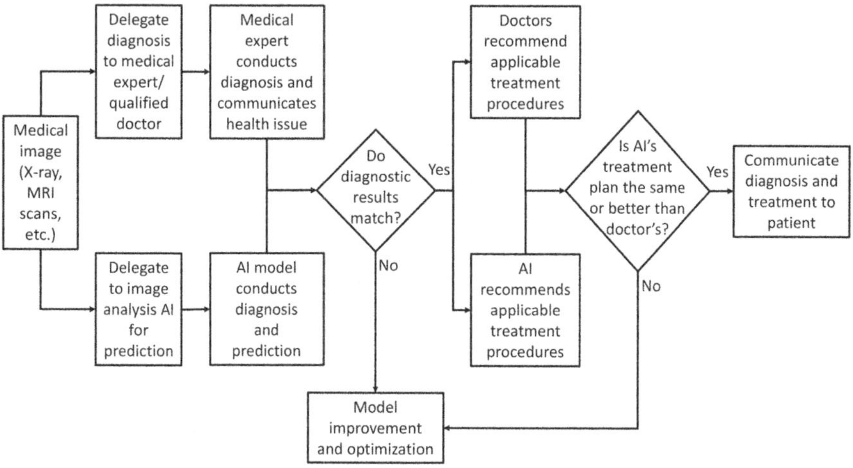

Fig. 9 Use of AI for medical diagnostics and model optimization. (Source: Authors' elaboration)

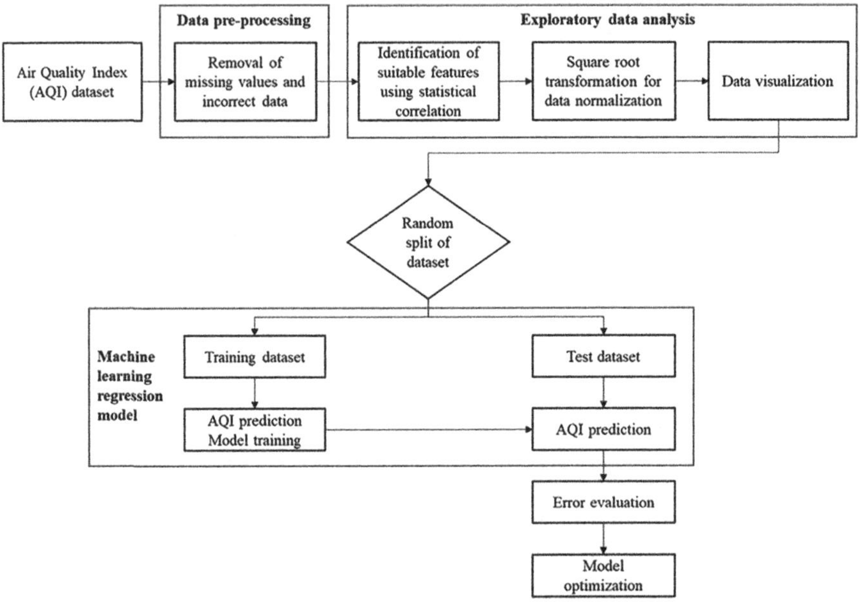

Fig. 10 Application of AI for monitoring air health. (Source: Authors' elaboration)

concerted effort to dismantle the structural barriers and systemic injustices perpetuating health outcomes disparities. By acknowledging the multifaceted nature of health inequities and embracing inclusive and equitable approaches to intervention, societies can work towards achieving health equity for all individuals, irrespective of their background.

7.2 Healthcare Equity Through AI

Ensuring equitable access to healthcare resources and services is paramount in addressing the health challenges exacerbated by spatial isolation. Low-income communities and marginalised groups often face barriers to healthcare access and experience disproportionate health burdens related to environmental changes. Policy efforts to promote health equity in the context of climate change must address systemic disparities and prioritise the needs of the underserved (Friel et al., 2008). This includes improving healthcare infrastructure in underserved areas, expanding access to primary care services, and implementing community-based health initiatives. Addressing social and economic disparities in healthcare access to promote health equity is paramount (Costello et al., 2009; Popkin, 2002; Ruckert et al., 2017).

Policies must prioritise the participation of marginalised communities in decision-making processes related to healthcare delivery and climate adaptation. Community engagement and empowerment are essential for fostering trust, building resilience, and ensuring that healthcare services meet the diverse needs of local populations. Partnerships between policymakers, healthcare providers, community organisations, and advocacy groups are critical for advancing health equity initiatives in the context of spatial isolation.

Addressing health inequities exacerbated by multinational corporations requires a multifaceted approach integrating health equity considerations into reducing power asymmetry, climate adaptation, and mitigation strategies. By prioritising equitable access to healthcare resources and services, policymakers can enhance resilience, promote community well-being, and mitigate the disproportionate health impacts of climate change on vulnerable populations.

7.3 Policy Development

Policy development is a critical avenue for addressing the exacerbation of health inequities induced by the expanding impacts of climate change. These policies should be underpinned by empirical evidence and informed by research on the interplay of environmental, social, and economic aspects (Costello et al., 2009; Kickbusch et al., 2016) and centre on safeguarding the most at-risk populations. Low-income communities, indigenous groups, the elderly, and marginalised demographics often bear the brunt of climate and nutritionally induced health threats due to limited access to healthcare, socio-economic disparities, heightened exposure to environmental hazards, and scarcity of nutritional options (Popkin, 2002; Watts et al., 2015).

Addressing health inequities necessitates policies that bolster public health infrastructure and foster resilience. Early warning systems for extreme weather events, access to clean water and sanitation, and bolstering healthcare delivery systems in marginalised regions are critical components. Policies must acknowledge and confront the systemic barriers that hinder equitable access to healthcare resources and services. This entails investing in underserved areas, bolstering healthcare infrastructure, and expanding access to primary care services. Frameworks should seamlessly integrate health equity considerations across various sectors, including urban planning, agriculture, and disaster management (Kickbusch et al., 2016; Lowe, 2014). Adopting a holistic approach enables governments to implement measures that mitigate immediate health impacts and tackle underlying social determinants of health, fostering equitable outcomes for all.

7.4 Artificial Intelligence and Global Health

Adequate literature exists to substantiate that AI helps to reach out to a larger patient base, helping medical practitioners to increase coverage and grow their geographical presence while collecting more voluminous health data to improve their services. For rural areas, remote patient monitoring and mobile health can be beneficial. Smartphone applications can be turned into mobile medical kits for remote monitoring of patient's health conditions, early predictability of diseases, timely screening, and improved adherence to the medication routine (Pahune, 2024). For disease surveillance and outbreak prediction, AI-powered diagnostic tools can assist healthcare providers in correctly diagnosing diverse diseases, chronic conditions, and infectious ailments. In vulnerable regions where sufficient qualified doctors are unavailable, AI-enabled clinical decision support machines can aid doctors in deciding the contextually appropriate medical treatment, thereby augmenting human intelligence and reducing diagnostic errors (Ibid).

Vulnerable sections of society also suffer from language barriers which can prevent them from accessing remote healthcare services. AI-based applications by healthcare MNEs can help these patients choose their preferred language of medical instruction and access healthcare advice online, provided their region has robust Internet connectivity (Pahune, 2024). MNEs can distribute AI-powered health education tools to rural populations via smartphones to bridge the knowledge gap in healthcare. With increased coverage of healthcare services using AI-based tools and with an increasing number of people sharing their health data with consent, healthcare MNEs can grow their knowledge via data sets.

Urban regions also house vulnerable and marginalised communities that are exposed to polluted ecosystems in their living areas, while well-earning people are also exposed to pollutants in the air, water, and land. MNEs in the healthcare and pharmaceutical industry deal with issues of diverse pollutants. Ninety-nine percent of the world's population breathes polluted air that exceeds the pollution limits recommended by the World Health Organization (WHO) (2024a, 2024b). However, with the use of AI, local governments can keep a check on the magnitude of pollution at different times of the day and recommend remedial action.

MNEs use AI technology to combat sanitary water outflows from leaking pipes, preserve clean water, and run public health surveillance through wastewater-based epidemiology to detect high-risk pathogens (Fluid Analytics, 2024). MNEs in the data analytics industry can leverage AI's advanced analytical capabilities to analyse complex and extensive datasets and provide precise early warnings for water-related health hazards (Schenker, 2024). Machine learning algorithms enhance prediction models, improving the accuracy of alerts.

Similarly, MNEs in retail, pharmaceutical, food, and other consumer product industries pollute water by directly discharging produced wastewater into flowing or stagnant water bodies. AI-driven technologies play a crucial role in safeguarding aquatic ecosystems by continuously monitoring and managing the effects of water pollutants on diverse species. They facilitate various functions, including damage assessment and resource optimization, leading to the implementation of efficient recovery strategies (Frontiers, 2024).

8 Limitations and Future Work

This chapter confines its examination of AI in health equity to business use cases, specifically focusing on how multinational enterprises (MNEs) can leverage the technology to achieve Sustainable Development Goal 3 (SDG3). It does not delve into the detailed mechanisms of machine learning systems, algorithms, and the necessary guardrails that MNEs must consider prior to AI system implementation. Furthermore, the discussion on AI applications for mitigating urban pollution is exploratory in nature and does not assess the feasibility or viability of such systems for MNEs. Further research into the feasibility of AI systems in the various contexts discussed in this chapter is recommended.

Limitations of the study include the need for a more detailed exploration of AI systems' technical aspects and feasibility in various contexts. Future research should focus on these areas to enhance AI's effectiveness in global health. Future work should explore the development of AI-driven health interventions, assess their effectiveness, and establish ethical and regulatory guidelines to protect communities in low- and middle-income countries. The chapter also suggests investigating how AI can support public health adaptation strategies, such as modifying healthcare infrastructure to cope with increased demand due to climate change.

9 Conclusion

The Anthropocene epoch, marked by emissions predominantly driven by multinational enterprises, has inflicted substantial harm on ecosystems and human health, with disproportionate impacts on lower- and middle-income populations. In this context, AI developed through strategic public-private partnerships, presents a promising avenue for constructing resilient and equitable health systems. The success of AI applications, ranging from diagnostic tools to predictive analytics, underscores its transformative potential in addressing the complex challenges at the intersection of climate change and health.

This study contributes to the ongoing discourse on the climate-health nexus by emphasising AI's critical role in mitigating health inequalities and offering practical solutions for adapting healthcare services to the evolving impacts of climate change. Effectively containing and managing the escalating public health threats posed by climate change necessitates coordinated policy actions at local, regional, and international levels. Innovative data-gathering efforts, as suggested by Barteit et al. (2023), are essential for quantifying the vulnerabilities and risks associated with various diseases, enabling targeted interventions. At the community level, effective risk communication and education are crucial for raising awareness among vulnerable populations, as highlighted by Kreslake et al. (2016). Additionally, ensuring adequate resources and support for healthcare professionals, as emphasised by Friel et al. (2011), is vital for enhancing the capacity of health systems to respond to climate-related challenges.

Health inequities arise from a complex interplay of social, economic, and environmental determinants that systematically disadvantage specific segments of society (Friel et al., 2008; Hammonds & Reverby, 2019). Social determinants, including inadequate housing, food insecurity, and limited access to healthcare services, profoundly shape individuals' health trajectories (Cunningham et al., 2022; Marmot et al., 2008). Addressing these inequities requires a comprehensive approach that integrates AI-driven innovations with broader efforts to address the underlying social determinants of health.

References

Agache, I., Sampath, V., Aguilera, J., Akdis, C. A., Akdis, M., Barry, M., Bouagnon, A., Chinthrajah, S., Collins, W., Dulitzki, C., Erny, B., Gomez, J., Goshua, A., Jutel, M., Kizer, K. W., Kline, O., LaBeaud, A. D., Pali-Schöll, I., Perrett, K. P., & Nadeau, K. C. (2022). Climate change and global health: A call to more research and more action. *Allergy, 77*(5), 1389–1407. https://doi.org/10.1111/all.15229

Agbor, V., Ntusi, N., & Noubiap, J. (2020). An overview of heart failure in low- and middle-income countries. *Cardiovascular Diagnosis and Therapy, 10*(2), 244–251. https://doi.org/10.21037/cdt.2019.08.03

Appelboom, G., Camacho, E., Abraham, M. E., Bruce, S. S., Dumont, E. L., Zacharia, B. E., D'Amico, R., Slomian, J., Reginster, J. Y., Bruyère, O., & Connolly, E. S. (2014). Smart wearable body sensors for patient self-assessment and monitoring. *Archives of Public Health, 72*(1), 28. https://doi.org/10.1186/2049-3258-72-28

Attia, Z. I., Harmon, D. M., Behr, E. R., & Friedman, P. A. (2021). Application of artificial intelligence to the electrocardiogram. *European Heart Journal, 42*, 4717. https://doi.org/10.1093/eurheartj/ehab649

Attia, Z. I., Harmon, D. M., Dugan, J., Manka, L., Lopez-Jimenez, F., Lerman, A., Siontis, K. C., Noseworthy, P. A., Yao, X., Klavetter, E. W., Halamka, J. D., Asirvatham, S. J., Khan, R., Carter, R. E., Leibovich, B. C., & Friedman, P. A. (2022). Prospective evaluation of smartwatch-enabled detection of left ventricular dysfunction. *Nature Medicine, 28*(12), 2497–2503. https://doi.org/10.1038/s41591-022-02053-1

Ayorinde, A., Esan, O. B., Buabeng, R., Taylor, B., & Salway, S. (2023). Ethnic inequities in maternal health. *BMJ (Clinical Research ed.), 381*, 1040. https://doi.org/10.1136/bmj.p1040

Barcelona Supercomputing Center. (2023). *A pioneering artificial intelligence method to fight urban air pollution*. Phys. Retrieved April 18, 2024, from https://phys.org/news/2023-04-artificial-intelligence-method-urban-air.html

Barteit, S., Colmar, D., Nellis, S., Thu, M., Watterson, J., Gouwanda, D., Bärnighausen, T., & Su, T. T. (2023). Developing climate change and health impact monitoring with eHealth at the South East Asia Community Observatory and Health and Demographic Surveillance Site, Malaysia (CHIMES). *Frontiers in Public Health, 11*. https://doi.org/10.3389/fpubh.2023.1153149

Berge, J. M., Mendenhall, T. J., & Doherty, W. J. (2009). Using community-based participatory research (CBPR) to target health disparities in families. *Family Relations, 58*(4), 475–488. https://doi.org/10.1111/j.1741-3729.2009.00567.x

Bogar, K., Brensinger, C. M., Hennessy, S., Flory, J. H., Bell, M. L., Shi, C., Bilker, W. B., & Leonard, C. E. (2022). Climate change and ambient temperature extremes: Association with serious hypoglycemia, diabetic ketoacidosis, and sudden cardiac arrest/ventricular arrhythmia in people with type 2 diabetes. *Diabetes Care, 45*(11), e171–e173. https://doi.org/10.2337/dc22-1161

Bolan, S., Padhye, L. P., Jasemizad, T., Govarthanan, M., Karmegam, N., Wijesekara, H., Amarasiri, D., Hou, D., Zhou, P., Biswal, B. K., Balasubramanian, R., Wang, H., Siddique, K. H., Rinklebe, J., Kirkham, M., & Bolan, N. (2024). Impacts of climate change on the fate of contaminants through extreme weather events. *Science of the Total Environment, 909*, 168388. https://doi.org/10.1016/j.scitotenv.2023.168388

Bonato, P. (2005). Advances in wearable technology and applications in physical medicine and rehabilitation. *Journal of Neuroengineering and Rehabilitation, 2*, 2. https://doi.org/10.1186/1743-0003-2-2

Borg, F. H., Greibe Andersen, J., Karekezi, C., Yonga, G., Furu, P., Kallestrup, P., & Kraef, C. (2021). Climate change and health in urban informal settlements in low- and middle-income countries—A scoping review of health impacts and adaptation strategies. *Global Health Action, 14*(1), 1908064. https://doi.org/10.1080/16549716.2021.1908064

Brook, R. D., Rajagopalan, S., Pope, C., Brook, J. R., Bhatnagar, A., Diez-Roux, A. V., Holguin, F., Hong, Y., Luepker, R. V., Mittleman, M. A., Peters, A., Siscovick, D., Smith, S. C., Whitsel, L., & Kaufman, J. D. (2010). Particulate matter air pollution and cardiovascular disease. *Circulation, 121*(21), 2331–2378. https://doi.org/10.1161/cir.0b013e3181dbece1

Buse, K., Tanaka, S., & Hawkes, S. (2017). Healthy people and healthy profits? Elaborating a conceptual framework for governing the commercial determinants of non-communicable diseases and identifying options for reducing risk exposure. *Globalization and Health, 13*(1), 34. https://doi.org/10.1186/s12992-017-0255-3

Byrne, R. A., Rossello, X., Coughlan, J. J., Barbato, E., Berry, C., Chieffo, A., Claeys, M. J., Dan, G.-A., Dweck, M. R., Galbraith, M., Gilard, M., Hinterbuchner, L., Jankowska, E. A., Jüni, P., Kimura, T., Kunadian, V., Leosdottir, M., Lorusso, R., Pedretti, R. F. E., & Zeppenfeld, K. (2024). 2023 ESC Guidelines for the management of acute coronary syndromes. *European Heart Journal: Acute Cardiovascular Care, 13*(1), 55–161. https://doi.org/10.1093/ehjacc/zuad107

Caminade, C., McIntyre, K. M., & Jones, A. E. (2019). Impact of recent and future climate change on vector-borne diseases. *Annals of the New York Academy of Sciences, 1436*(1), 157–173. https://doi.org/10.1111/nyas.13950

Campbell-Lendrum, D., & Corvalán, C. (2007). Climate change and developing-country cities: Implications for environmental health and equity. *Journal of Urban Health: Bulletin of the New York Academy of Medicine, 84*(Suppl 1), 109–117. https://doi.org/10.1007/s11524-007-9170-x

Chan, P. Z., Ramli, M. A. I. B., & Chew, H. S. J. (2023). Diagnostic Test Accuracy of artificial intelligence-assisted detection of acute coronary syndrome: A systematic review and meta-analysis. *Computers in Biology and Medicine, 167*, 107636. https://doi.org/10.1016/j.compbiomed.2023.107636

Chang, H. Y., Jung, C. K., Woo, J. I., Lee, S., Cho, J., Kim, S. W., & Kwak, T. Y. (2019). Artificial intelligence in pathology. *Journal of Pathology and Translational Medicine, 53*(1), 1–12. https://doi.org/10.4132/jptm.2018.12.16. Epub 2018 Dec 28. PMID: 30599506; PMCID: PMC6344799.

Chang, A. Y., Ogbuoji, O., Atun, R., & Verguet, S. (2017). Dynamic modeling approaches to characterize the functioning of health systems: A systematic review of the literature. *Social Science & Medicine, 194*, 160–167. https://doi.org/10.1016/j.socscimed.2017.09.005

Chang, J. H., Lee, Y. L., Te Chang, L., Chang, T. Y., Hsiao, T. C., Chung, K. F., Ho, K. F., Kuo, H. P., Lee, K. Y., Chuang, K. J., & Chuang, H. C. (2023). Climate change, air quality, and respiratory health: A focus on particle deposition in the lungs. *Annals of Medicine, 55*(2). https://doi.org/10.1080/07853890.2023.2264881

Cheng, J., Bambrick, H., Yakob, L., Devine, G., Frentiu, F. D., Toan, D., Thai, P. Q., Xu, Z., & Hu, W. (2020). Heatwaves and dengue outbreaks in Hanoi, Vietnam: New evidence on early warning. *PLoS Neglected Tropical Diseases, 14*(1), 1–15. https://doi.org/10.1371/journal.pntd.0007997

Cheng, J., Xu, Z., Bambrick, H., Prescott, V., Wang, N., Zhang, Y., Su, H., Tong, S., & Hu, W. (2019). Cardiorespiratory effects of heatwaves: A systematic review and meta-analysis of global epidemiological evidence. *Environmental Research, 177*, 108610. https://doi.org/10.1016/j.envres.2019.108610

Cicci, K. R., Maltby, A., Clemens, K. K., Vicedo-Cabrera, A. M., Gunz, A. C., Lavigne, É., & Wilk, P. (2022). High temperatures and cardiovascular-related morbidity: A scoping review. *International Journal of Environmental Research and Public Health, 19*(18). https://doi.org/10.3390/ijerph191811243

Cissé, G. (2019). Food-borne and water-borne diseases under climate change in low- and middle-income countries: Further efforts needed for reducing environmental health exposure risks. *Acta Tropica, 194*, 181–188. https://doi.org/10.1016/j.actatropica.2019.03.012

Claeys, M. J., Rajagopalan, S., Nawrot, T. S., & Brook, R. D. (2017). Climate and environmental triggers of acute myocardial infarction. *European Heart Journal, 38*(13), 955–960. https://doi.org/10.1093/eurheartj/ehw151

Clapp, J., & Scrinis, G. (2016). Big food, nutritionism, and corporate power. *Globalizations, 14*(4), 578–595. https://doi.org/10.1080/14747731.2016.1239806

Correa-Parra, J., Vergara-Perucich, J. F., & Aguirre-Nuñez, C. (2020). Water privatization and inequality: Gini coefficient for water resources in Chile. *Water, 12*(12), 3369. https://doi.org/10.3390/w12123369

Costello, A., Abbas, M., Allen, A., Ball, S., Bell, S., Bellamy, R., Friel, S., Groce, N., Johnson, A., Kett, M., Lee, M., Levy, C., Maslin, M., McCoy, D., McGuire, B., Montgomery, H., Napier, D., Pagel, C., Patel, J., & Patterson, C. (2009). Managing the health effects of climate change: Lancet and University College London Institute for Global Health Commission. *The Lancet, 373*(9676), 1693–1733. https://doi.org/10.1016/S0140-6736(09)60935-1

Cunningham, R., Polomano, R. C., Wood, R. M., & Aysola, J. (2022). Health systems and health equity: Advancing the agenda. *Nursing Outlook, 70*(6), S66–S76. https://doi.org/10.1016/j.outlook.2022.05.010

D'Amato, G., Holgate, S. T., Pawankar, R., Ledford, D. K., Cecchi, L., Al-Ahmad, M., Al-Enezi, F., Al-Muhsen, S., Ansotegui, I., Baena-Cagnani, C., Baker, D. J., Bayram, H., Bergmann, K. C., Boulet, L. P., Buters, J., D'Amato, M., Dorsano, S., Douwes, J., Finlay, S. E., & Annesi-Maesano, I. (2015). Meteorological conditions, climate change, new emerging factors, and asthma and related allergic disorders. A statement of the World Allergy Organization. *World Allergy Organization Journal, 8*(1), 25. https://doi.org/10.1186/s40413-015-0073-0

De Blois, J., Kjellstrom, T., Agewall, S., Ezekowitz, J. A., Armstrong, P. W., & Atar, D. (2015). *The Effects of Climate Change on Cardiac Health Cardiology 131*(4) 209–217. https://doi.org/10.1159/000398787

de Bont, J., Nori-Sarma, A., Stafoggia, M., Banerjee, T., Ingole, V., Jaganathan, S., Mandal, S., Rajiva, A., Krishna, B., Kloog, I., Lane, K., Mall, R. K., Tiwari, A., Wei, Y., Wellenius, G. A., Prabhakaran, D., Schwartz, J., Prabhakaran, P., & Ljungman, P. (2024). Impact of heatwaves on all-cause mortality in India: A comprehensive multi-city study. *Environment International, 184*, 108461. https://doi.org/10.1016/j.envint.2024.108461

Ebi, K. L., & Hess, J. J. (2020). Health risks due to climate change: Inequity in causes and consequences. *Health Affairs (Project Hope), 39*(12), 2056–2062. https://doi.org/10.1377/hlthaff.2020.01125

Ebrahimzadeh, E., Kalantari, M., Joulani, M., Shahraki, R. S., Fayaz, F., & Ahmadi, F. (2018). Prediction of paroxysmal Atrial Fibrillation: A machine learning based approach using combined feature vector and mixture of expert classification on HRV signal. *Computer Methods and Programs in Biomedicine, 165*, 53–67. https://doi.org/10.1016/j.cmpb.2018.07.014

Falkner, R. (2003). Private environmental governance and international relations: Exploring the links. *Global Environmental Politics, 3*(2), 72–87. https://doi.org/10.1162/152638003322068227

Farmer, H. R., Wray, L. A., & Thomas, J. R. (2019). Do race and everyday discrimination predict mortality risk? Evidence from the health and retirement study. *Gerontology and Geriatric Medicine, 5*, 233372141985566. https://doi.org/10.1177/2333721419855665

Fenton, F., Cherry, E., & Glass, L. (2008). Cardiac arrhythmia. *Scholarpedia, 3*(7), 1665. https://doi.org/10.4249/scholarpedia.1665

Financial Services Monitor Worldwide. (2023). *Packaged food global market report 2023*. SyndiGate Media. https://link.gale.com/apps/doc/A754821784/GBIB?u=vic_liberty&sid=bookmark-GBIB&xid=4f456597

Fisher, J. (2023). *A.I. to stop water pollution before it happens*. BBC. Retrieved April 18, 2024, from: https://www.bbc.com/news/science-environment-65913940

Fluid Analytics. (2024). *Public-health surveillance through wastewater-based epidemiology (WBE)*. https://fluidanalytics.ai

Friel, S., Bowen, K., Campbell-Lendrum, D., Frumkin, H., McMichael, A. J., & Rasanathan, K. (2011). Climate change, non-communicable diseases, and development: The relationships and common policy opportunities. *Annual Review of Public Health, 32*, 133–147. https://doi.org/10.1146/annurev-publhealth-071910-140612

Friel, S., Marmot, M., McMichael, A. J., Kjellstrom, T., & Vågerö, D. (2008). Global health equity and climate stabilization: A common agenda. *The Lancet, 372*(9650), 1677–1683. https://doi.org/10.1016/s0140-6736(08)61692-x

Gawrych, M. (2022). Climate change and mental health: A review of current literature. *Psychiatria Polska, 56*(4), 903–915. https://doi.org/10.12740/PP/OnlineFirst/131991

Ghosh, A. K., Demetres, M. R., Geisler, B. P., Ssebyala, S. N., Yang, T., Shapiro, M. F., Setoguchi, S., & Abramson, D. (2022). Impact of hurricanes and associated extreme weather events on cardiovascular health: A scoping review. *Environmental Health Perspectives, 130*(11), 116003. https://doi.org/10.1289/EHP11252

Global Burden of Disease Collaborators. (2020). Global burden of 369 diseases and injuries in 204 countries and territories, 1990–2019: A systematic analysis for the global burden of disease study 2019. *The Lancet, 396*, 1204–1222. https://doi.org/10.1016/S0140-6736(20)30925-9

Gonzales, K. L., Noonan, C., Goins, R., Henderson, W. G., Beals, J., Manson, S. M., Acton, K. J., & Roubideaux, Y. (2016). Assessing the everyday discrimination scale among American Indians and Alaska natives. *Psychological Assessment, 28*(1), 51–58. https://doi.org/10.1037/a0039337

Gulliford, M., Figueroa-Munoz, J., Morgan, M., Hughes, D., Gibson, B., Beech, R., & Hudson, M. (2002). What does 'access to health care' mean? *Journal of Health Services Research & Policy, 7*(3), 186–188. https://doi.org/10.1258/135581902760082517

Guo, J., & Li, B. (2018). The application of medical artificial intelligence technology in rural areas of developing countries. *Health Equity, 2*(1), 174–181. https://doi.org/10.1089/heq.2018.0037. PMID: 30283865; PMCID: PMC6110188.

Gurevich, E., Hassan, B. E., & Morr, C. E. (2023). Equity within A.I. systems: What can health leaders expect? *Healthc Manage Forum, 36*(2), 119–124. https://doi.org/10.1177/08404704221125368

Hajat, A., Hsia, C., & O'Neill, M. S. (2015). Socio-economic disparities and air pollution exposure: A global review. *Current Environmental Health Reports, 2*(4), 440–450. https://doi.org/10.1007/s40572-015-0069-5

Hammonds, E. M., & Reverby, S. M. (2019). Toward a historically informed analysis of racial health disparities since 1619. *American Journal of Public Health, 109*(10), 1348–1349. https://doi.org/10.2105/ajph.2019.305262

Hankivsky, O. (Ed.). (2012). *An intersectionality-based policy analysis framework*. Institute for Intersectionality Research and Policy, Simon Fraser University.

Hansel, N. N., McCormack, M. C., & Kim, V. (2016). The effects of air pollution and temperature on COPD. *COPD: Journal of Chronic Obstructive Pulmonary Disease, 13*(3), 372–379. https://doi.org/10.3109/15412555.2015.1089846

Hayes, R. B., Lim, C., Zhang, Y., Cromar, K., Shao, Y., Reynolds, H. R., Silverman, D. T., Jones, R. R., Park, Y., Jerrett, M., Ahn, J., & Thurston, G. D. (2020). PM2.5 air pollution and cause-specific cardiovascular disease mortality. *International Journal of Epidemiology, 49*(1), 25–35. https://doi.org/10.1093/ije/dyz114

Heede, R. (2013). Tracing anthropogenic carbon dioxide and methane emissions to fossil fuel and cement producers, 1854–2010. *Climatic Change, 122*(1–2), 229–241. https://doi.org/10.1007/s10584-013-0986-y

Hoyer, D., Dee, E., O'Leary, M. S., Heffernan, M., Gelfand, K., Kappel, R., & Fromknecht, C. Q. (2022). How do we define and measure health equity? The state of current practice and tools to advance health equity. *Journal of Public Health Management and Practice: JPHMP, 28*(5), 570–577. https://doi.org/10.1097/PHH.0000000000001603

Huang, W., Gao, Y., Xu, R., Yang, Z., Yu, P., Ye, T., Ritchie, E. A., Li, S., & Guo, Y. (2023). Health effects of cyclones: A systematic review and meta-analysis of epidemiological studies. *Environmental Health Perspectives, 131*(8), 86001. https://doi.org/10.1289/EHP12158

Jacobs, D. R., & Tapsell, L. C. (2008). Food, not nutrients, is the fundamental unit in nutrition. *Nutrition Reviews, 65*(10), 439–450. https://doi.org/10.1111/j.1753-4887.2007.tb00269.x

Jesus, F. S. M. d., Monteiro, A. M. V., & Tomasella, J. (2023). Spatial inequalities in access to safe drinking water in an upper-middle-income country: A multi-scale analysis of Brazil. *Water, 15*(8), 1620. https://doi.org/10.3390/w15081620

Kickbusch, I., Allen, L., & Franz, C. (2016). The commercial determinants of health. *The Lancet Global Health, 4*(12), e895–e896. https://doi.org/10.1016/s2214-109x(16)30217-0

Kim, S. Y., Kim, S. H., Wee, J. H., Min, C., Han, S. M., Kim, S., & Choi, H. G. (2021). Short and long term exposure to air pollution increases the risk of ischemic heart disease. *Scientific Reports, 11*(1), 5108. https://doi.org/10.1038/s41598-021-84587-x

Kreslake, J. M., Price, K. M., & Sarfaty, M. (2016). Developing effective communication materials on the health effects of climate change for vulnerable groups: A mixed methods study. *BMC Public Health, 16*(1). https://doi.org/10.1186/s12889-016-3546-3

Lavizzo-Mourey, R. J., Besser, R. E., & Williams, D. R. (2021). Understanding and mitigating health inequities—Past, current, and future directions. *New England Journal of Medicine, 384*(18), 1681–1684. https://doi.org/10.1056/nejmp2008628

Lee, A. S., Aguilera, J., Efobi, J. A., Jung, Y. S., Seastedt, H., Shah, M. M., Yang, E., Konvinse, K., Utz, P. J., Sampath, V., & Nadeau, K. C. (2023). Climate change and public health: The effects of global warming on the risk of allergies and autoimmune diseases: The effects of global warming on the risk of allergies and autoimmune diseases. *EMBO Reports, 24*(4), e56821. https://doi.org/10.15252/embr.202356821

Lie, A. (2019). Power in global nutrition governance. *Global Governance, 25*(2), 277–303. https://doi.org/10.1163/19426720-02502006

Liu, J., Hansen, A., Zhang, Y., Driscoll, T., Morgan, G., Bi, P., Liu, J., Varghese, B. M., Hansen, A., Zhang, Y., Driscoll, T., Morgan, G., Dear, K., Gourley, M., & Capon, A. (2022). Heat exposure and cardiovascular health outcomes: A systematic review and meta-analysis. *Articles Lancet Planet Health, 6*, 484–495. https://doi.org/10.1016/S2542-5196(22)00117-6

López, L.-A., Cadarso, M.-Á., Zafrilla, J., & Arce, G. (2019). The carbon footprint of the U.S. multinationals' foreign affiliates. *Nature Communications, 10*(1). https://doi.org/10.1038/s41467-019-09473-7

Lowe, M. (2014). Obesity and climate change mitigation in Australia: Overview and analysis of policies with co-benefits. *Australian and New Zealand Journal of Public Health, 38*(1), 19–24. https://doi.org/10.1111/1753-6405.12150

Lozano, R., Fullman, N., Abate, D., et al. (2018). Measuring progress from 1990 to 2017 and projecting attainment to 2030 of the health-related Sustainable Development Goals for 195 countries and territories: a systematic analysis for the Global Burden of Disease Study 2017. *Lancet, 392*, 2091–2138.

Luna-Cerón, E., Pherez-Farah, A., Krishnan-Sivadoss, I., & Guerrero-Beltrán, C. E. (2024). Molecular challenges and opportunities in climate change-induced kidney diseases. *Biomolecules, 14*(3). https://doi.org/10.3390/biom14030251

Makrufardi, F., Manullang, A., Rusmawati Ningtyas, D., Lin, S. C., Chuang, H. C., & Chung, K. F. (2023). Extreme weather and asthma: A systematic review and meta-analysis. *European Respiratory Review, 32*(168), 230019. https://doi.org/10.1183/16000617.0019-2023

Malakar, K., & Mishra, T. (2017). Application of Gini, Theil and concentration indices for assessing water use inequality. *International Journal of Social Economics, 44*(10), 1335–1347. https://doi.org/10.1108/IJSE-01-2016-0017

Manisalidis, I., Stavropoulou, E., Stavropoulos, A., & Bezirtzoglou, E. (2020). Environmental and health impacts of air pollution: A review. *Frontiers in Public Health, 8*. https://doi.org/10.3389/fpubh.2020.00014

Marmot, M., Friel, S., Bell, R., Houweling, T., & Taylor, S. (2008). Closing the gap in a generation: Health equity through action on the social determinants of health. *The Lancet, 372*(9650), 1661–1669. https://doi.org/10.1016/s0140-6736(08)61690-6

Monteiro, C. A., Moubarac, J., Cannon, G., Ng, S. W., & Popkin, B. (2013). Ultra-processed products are becoming dominant in the global food system. *Obesity Reviews, 14*(S2), 21–28. https://doi.org/10.1111/obr.12107

Mortara, A., Pinna, G. D., Johnson, P., Maestri, R., Capomolla, S., La Rovere, M. T., Ponikowski, P., Tavazzi, L., & Sleight, P. (2009). Home telemonitoring in heart failure patients: The HHH study (Home or Hospital in Heart Failure). *European Journal of Heart Failure, 11*(3), 312–318. https://doi.org/10.1093/eurjhf/hfp022

Mulugeta, H., Sinclair, P. M., & Wilson, A. (2024). Health-related quality of life of people with heart failure in low- and middle-income countries: A systematic review and meta-analysis. *Quality of Life Research, 33*(5), 1175–1189. https://doi.org/10.1007/s11136-023-03563-2

Odersky, M., & Löffler, M. (2024). Differential exposure to climate change? Evidence from the 2021 floods in Germany. *The Journal of Economic Inequality, 22*, 551. https://doi.org/10.1007/s10888-023-09605-6

O'Kane, M., Agrawal, S., Binder, L., Dzau, V., Gandhi, T. K., Harrington, R., Mate, K., McGann, P., Meyers, D., Rosen, P., Schreiber, M., & Schummers, D. (2024). *An equity agenda for the field of health care quality improvement (NAM Perspectives)* [Discussion Paper]. National Academy of Medicine. https://doi.org/10.31478/202109b.

OECD. (2024). *A.I. in health: huge potential risks*. Organization for Economic Cooperation and Development. https://www.oecd.org/health/AI-in-health-huge-potential-huge-risks.pdf

Padhy, S. K., Sarkar, S., Panigrahi, M., & Paul, S. (2015). Mental health effects of climate change. *Indian Journal Of Occupational And Environmental Medicine, 19*(1), 3–7. https://doi.org/10.4103/0019-5278.156997

Pahune, S. (2024). *A brief overview of how A.I. enables healthcare sector rural development*. Preprint retrieved from https://doi.org/10.13140/RG.2.2.16675.63525

PATH. (2023). *The power of A.I. for health equity*. https://www.path.org/our-impact/articles/the-power-of-ai-for-health-equity/

Peters, A., Schneider, A. (2021). Cardiovascular risks of climate change. *Nature Reviews Cardiology 18*(1) 1–2. https://doi.org/10.1038/s41569-020-00473-5

Pope, C., Ezzati, M., & Dockery, D. W. (2009). Fine-particulate air pollution and life expectancy in the United States. *New England Journal of Medicine, 360*(4), 376–386. https://doi.org/10.1056/nejmsa0805646

Popkin, B. M. (2002). An overview on the nutrition transition and its health implications: The Bellagio meeting. *Public Health Nutrition, 5*(1a), 93–103. https://doi.org/10.1079/phn2001280

Remigio, R. V., Jiang, C., Raimann, J., Kotanko, P., Usvyat, L., Maddux, F. W., Kinney, P., & Sapkota, A. (2019). Association of extreme heat events with hospital admission or mortality among patients with end-stage renal disease. *JAMA Network Open, 2*(8), E198904. https://doi.org/10.1001/jamanetworkopen.2019

Riaz, M. M. A., Nayyer, B., Lal, A., Nawaz, F. A., & Zil-e-Ali, A. (2023). Climate change and mental health: A call to action to include mental health and psychosocial support services (MHPSS) in the Pakistan flood crisis. *BJPsych International, 20*(3), 56–58. https://doi.org/10.1192/bji.2023.13

Riazul Islam, S. M., Kwak, D., Humaun Kabir, M., Hossain, M., & Kwak, K.-S. (2015). The internet of things for health care: A comprehensive survey. *IEEE Access, 3*, 678–708. https://doi.org/10.1109/ACCESS.2015.2437951

Ruan, Y., Guo, Y., Zheng, Y., Huang, Z., Sun, S., Kowal, P., Shi, Y., & Wu, F. (2018). Cardiovascular disease (CVD) and associated risk factors among older adults in six low-and middle-income countries: Results from SAGE Wave 1. *BMC Public Health, 18*(1), 778. https://doi.org/10.1186/s12889-018-5653-9

Ruckert, A., Huynh, C., & Labonté, R. (2017). Reducing health inequities: Is universal basic income the way forward? *Journal of Public Health, 40*(1), 3–7. https://doi.org/10.1093/pubmed/fdx006

Ruus, A. A., & Mornoş, C. (2022). The impact of meteorological factors and air pollutants on acute coronary syndrome. *Current Cardiology Reports, 24*(10), 1337–1349. https://doi.org/10.1007/s11886-022-01759-5

Sana, F., Isselbacher, E. M., Singh, J. P., Heist, E. K., Pathik, B., & Armoundas, A. A. (2020). Wearable devices for ambulatory cardiac monitoring. *Journal of the American College of Cardiology, 75*(13), 1582–1592. https://doi.org/10.1016/j.jacc.2020.01.046

Schenker, J. L. (2024). *How A.I. and robotics are helping cities tackle urban water pollution.* The Innovator. https://theinnovator.news/how-ai-and-robotics-are-helping-cities-tackle-urban-water-pollution/

Schultz, M. (2022). *How can we use A.I. to fight air pollution?* Open Access Government. https://www.openaccessgovernment.org/how-can-we-use-ai-to-fight-air-pollution/137726/

Schwalbe, N., & Wahl, B. (2020). Artificial intelligence and the future of global health. *Lancet, 395*(10236), 1579–1586. https://doi.org/10.1016/S0140-6736(20)30226-9. PMID: 32416782; PMCID: PMC7255280.

Seastedt, H., & Nadeau, K. (2023). Factors by which global warming worsens allergic disease. *Annals of Allergy, Asthma and Immunology, 131*(6), 694–702. https://doi.org/10.1016/j.anai.2023.08.610

Shi, F., Xia, L., Shan, F., Song, B., Wu, D., Wei, Y., Yuan, H., Jiang, H., He, Y., Gao, Y., Sui, H., & Shen, D. (2021). Large-scale screening to distinguish between COVID-19 and community-acquired pneumonia using infection size-aware classification. *Physics in Medicine & Biology, 66*(6), 065031. https://doi.org/10.1088/1361-6560/abe838

Shivanna, K. R. (2022). Climate change and its impact on biodiversity and human welfare. *Proceedings of the Indian National Science Academy. Part A, Physical Sciences, 88*(2), 160–171. https://doi.org/10.1007/s43538-022-00073-6

Shrikhande, S. S., Pedder, H., Röösli, M., Dalvie, M. A., Lakshmanasamy, R., Gasparrini, A., Utzinger, J., & Cissé, G. (2023). Non-optimal apparent temperature and cardiovascular mortality: The association in Puducherry, India between 2011 and 2020. *BMC Public Health, 23*, 291. https://doi.org/10.1186/s12889-023-15128-6

Singh, S. K., & Pradhan, B. (2024). Artificial Intelligence Applications in Addressing Water Contamination for Environmental Sustainability. Frontiers in Environmental Science. Retrieved May 28, 2024, from https://www.frontiersin.org/research-topics/61043/artificial-intelligence-applications-in-addressing-water-contamination-for-environmental-sustainability

Sri, A., Bhugra, D., Persaud, A., Tribe, R., Gnanapragasam, S., Castaldelli-Maia, J. M., Torales, J., & Ventriglio, A. (2023). Global mental health and climate change: A geo-psychiatry perspective. *Asian Journal of Psychiatry, 84*, 103562. https://doi.org/10.1016/j.ajp.2023.103562

Steinbeis, F., Gotham, D., von Philipsborn, P., & Stratil, J. M. (2019). Quantifying changes in global health inequality: The Gini and Slope Inequality Indices applied to the Global Burden of Disease data, 1990–2017. *BMJ Global Health, 4*(5), e001500. https://doi.org/10.1136/bmjgh-2019-001500

Swinburn, B. A., Sacks, G., Hall, K. D., McPherson, K., Finegood, D. T., Moodie, M. L., & Gortmaker, S. L. (2011). The global obesity pandemic: Shaped by global drivers and local environments. *The Lancet, 378*(9793), 804–814. https://doi.org/10.1016/s0140-6736(11)60813-1

Thornton, R. J., Glover, C. M., Cené, C. W., Glik, D. C., Henderson, J. A., & Williams, D. R. (2016). Evaluating strategies for reducing health disparities by addressing the social determinants of health. *Health Affairs, 35*(8), 1416–1423. https://doi.org/10.1377/hlthaff.2015.1357

Torres, J. M., & Casey, J. A. (2017). The centrality of social ties to climate migration and mental health. *BMC Public Health, 17*(1), 600. https://doi.org/10.1186/s12889-017-4508-0

Walinski, A., Sander, J., Gerlinger, G., Clemens, V., Meyer-Lindenberg, A., & Heinz, A. (2023). The effects of climate change on mental health. *Deutsches Arzteblatt International, 120*(8), 117–124. https://doi.org/10.3238/arztebl.m2022.0403

Watts, N., Adger, W., Agnolucci, P., Blackstock, J., Byass, P., Cai, W., Chaytor, S., Colbourn, T., Collins, M., Cooper, A., Cox, P. M., Depledge, J., Drummond, P., Ekins, P., Galaz, V., Grace, D., Graham, H., Grubb, M., Haines, A., & Costello, A. (2015). Health and climate change: Policy responses to protect public health. *The Lancet, 386*(10006), 1861–1914. https://doi.org/10.1016/s0140-6736(15)60854-6

Wahl, B., Cossy-Gantner, A., Germann, S, & Schwalbe, N. R. (2018). Artificial intelligence (AI) and global health: How can AI contribute to health in resource-poor settings? BMJ Global Health 3(4) e000798. https://doi.org/10.1136/bmjgh-2018-000798

Williams, D. R., & Mohammed, S. A. (2008). Discrimination and racial disparities in health: Evidence and needed research. *Journal of Behavioral Medicine, 32*(1), 20–47. https://doi.org/10.1007/s10865-008-9185-0

World Health Organization. (2006). *European strategies for tackling social inequities in health: Levelling up Part 2*. World Health Organization. https://iris.who.int/bitstream/handle/10665/107791/89384?sequence=1

World Health Organization. (2024a). *World health statistics 2022*. https://data.who.int/indicators/i/AC597B1

World Health Organization. (2024b). *Air pollution*. https://www.who.int/health-topics/air-pollution

Wu, J., Yunus, M., Ali, M., Escamilla, V., & Emch, M. (2018). Influences of heatwave, rainfall, and tree cover on cholera in Bangladesh. *Environment International, 120*, 304–311. https://doi.org/10.1016/j.envint.2018.08.012

Wu, X., Lu, Y., Zhou, S., Chen, L., & Xu, B. (2016). Impact of climate change on human infectious diseases: Empirical evidence and human adaptation. *Environment International, 86*, 14–23. https://doi.org/10.1016/j.envint.2015.09.007

Yarveysi, F., Alipour, A., & Moftakhari, H. (2023). Block-level vulnerability assessment reveals disproportionate impacts of natural hazards across the conterminous United States. *Nature Communications, 14*, 4222. https://doi.org/10.1038/s41467-023-39853-z

York, M. (2023). *Artificial intelligence to identify vulnerable people that need preventative care this winter*. AT Today. https://attoday.co.uk/artificial-intelligence-to-identify-vulnerable-people-that-need-preventative-care-this-winter/

Zhao, Q., Yu, P., Mahendran, R., Huang, W., Gao, Y., Yang, Z., Ye, T., Wen, B., Wu, Y., Li, S., & Guo, Y. (2022). Global climate change and human health: Pathways and possible solutions. *Eco-Environment and Health, 1*(2), 53–62. https://doi.org/10.1016/j.eehl.2022.04.004

Hakan Lane holds a position as a Postdoc researcher at the University of Mainz in Germany within the curAIHeart project where machine learning is used as a tool for heart disease prevention. He has worked on previous projects about sustainable behaviour and data processing and has been a member of several environmental activist groups.

Charles J. Oakley holds a PhD in Public Administration from Liberty University and an MPA from George Mason University. With extensive experience collaborating with major public agencies, he specializes in change management and crisis response. His research explores public policy, moral panic, and punctuated equilibrium theory, with a growing focus on the implications of artificial intelligence in public administration and global health governance.

Tanzeem Hossain is a research professional with a focus on public health and climate change. She has a Master's degree in Global Urban Health from the University of Freiburg, Germany. With a background in Disaster Management, her professional experience includes working as a Data Manager for the Government of Bangladesh and the World Health Organization on an emergency COVID-19 program.

Arijit Goswami is a Strategy Consultant at MarketsandMarkets, with over 8 years of experience in the intersection of AI, automotive and financial services industry. He holds an MBA from Indian Institute of Management Kozhikode and a bachelors in Engineering from Panjab University. He has previously worked with UNESCO, UNAOC and World Bank Group's GYCN to study the intersection of AI, digital technologies and climate change, and has published papers on AI and ethics.

Rajan Kumar Sah is a passionate writer, researcher, and engineer from Nepal. He holds a Bachelor's degree in Civil Engineering from Thapathali Campus, Tribhuvan University. His professional experience encompasses building designs and site supervision works in the construction of world-class high-rise hotels in Nepal. He has been part of numerous international organisations and youth groups. His interests span across artificial intelligence, structural engineering, environmental engineering, architecture, occupational safety, climate change, and SDGs.

Mohit Mishra, a Computer Science and Engineering graduate from Lovely Professional University, thrives in software development and machine learning. His experience at Amadeus Software Labs and Omdena showcases his expertise in optimising performance and tackling social challenges. He is passionate about leveraging technology for positive impact and holds diverse interests spanning AI, sustainability, and social good. His comprehensive technical skill set includes proficiency in various programming languages, data science frameworks, and cloud technologies.

Inequality in South African Banking: Executive Directors' Remuneration and Company Performance

Suzanne Kieviet and Henriette Elsabe Scholtz

1 Introduction

Ghauri and Cooke (2022) posit that sustainable development is a central issue for the world economy today. Additionally, the contexts of emerging markets and developing countries have particular significance, because they impose complex constraints on the achievement of the Sustainable Development Goals (SDGs). South Africa is an emerging market corporate governance pioneer with a well-developed governance framework (the King Report on Corporate Governance) which provides world-class guidance on corporate disclosure. Simultaneously, the country also has the highest wage inequality in the world, with a Gini coefficient of 0.63 (ILO, 2018). Despite South Africa being an emerging market corporate governance pioneer, the significant economic instability and growing wage inequality make it foreseeable that the remuneration policies and practices of companies listed on the Johannesburg Stock Exchange (JSE) receive considerable attention from stakeholders.

Recent scandals involving multinationals such as Steinhoff and Tongaat Hulett again focused stakeholder attention on the size and structure of executive remuneration packages against a backdrop of grotesque wage inequality. In the field of corporate governance, interest in reporting excessive executive directors' remuneration (EDR) is longstanding and aligning shareholders' and management's interests

All authors certify that they have no affiliations with or involvement in any organisation or entity with any financial interest or non-financial interest in the subject matter or materials discussed in this manuscript. No funding was received for conducting this study.

S. Kieviet (✉) · H. E. Scholtz
Stellenbosch University, Stellenbosch, South Africa
e-mail: kieviets@sun.ac.za; hscholtz@sun.ac.za

J. Bento, M. M. Torres (eds.), *Global Perspectives on Climate Change, Inequality, and Multinational Corporations*, CSR, Sustainability, Ethics & Governance,
https://doi.org/10.1007/978-3-031-80797-8_5

remains a contentious area (Meng, 2020; Vieito et al., 2008). Verster (2022) debates that the dilemma of EDR involves balancing the need to fairly compensate and incentivise executives while ensuring that their pay aligns with the long-term interests of stakeholders. Thus, the question is asked: is there a relationship between EDR and company performance in the banking sector of South Africa, and if so, is this fair given South Africa's inequality and socio-economic status? Although the relationship between pay and performance has been researched previously, few studies focused specifically on South Africa's banking sector (Deysel & Kruger, 2015; Van Blerck, 2012). Those that examined the pay-performance relationship in South Africa only considered total EDR (Deysel & Kruger, 2015; Van Blerck, 2012). In contrast, this study acknowledges the individual incentive components.

This chapter empirically examines whether the relatively high increases in bank executive remuneration align with bank performance and shareholder value. It also considers if this remuneration is fair to all stakeholders given South Africa's inequality and socio-economic status. This study further endeavours to deepen the understanding of how the SDGs can be achieved within an emerging economy which has, on the one hand, a well-developed governance framework, whilst, on the other, the highest wage inequality in the world. The study examined whether there is a link between EDR and the performance of South African banks. To achieve the objective, EDR was regressed on numerous accounting and market-based performance measures to determine if EDR is linked to company performance. EDR is considered in the context of growth in company performance over a 12-year period.

The motivation of this study is twofold: to explore whether high remunerations offered to the executive directors of South African banks have added value to the shareholders and to understand the socio-economic factors in an emerging economy such as South Africa that significantly impact how companies and multinational entities can reduce inequalities through sustainable and ethical business practices. It provides important insights that will enlighten future discourses on the effectiveness of disproportionate EDR packages and their long-term relation with increased market performance and sustained accounting performance.

Furthermore, a better understanding of the socio-economic factors at play will enhance the impact that multinational companies can have on reducing inequalities through sustainable and ethical business practices. The company assessment was limited to the annual reports of the banks listed on the JSE for a 12-year period. Based on prior research, specific market-based and accounting-based performance measures were used in the study. Nevertheless, this study is useful to gain insight into how sustainable development goals can be achieved to reduce inequalities in South Africa. The question of how multinational companies can drive economic inclusivity and therefore reduce inequality through sustainable and ethical business practices is best answered by understanding the socio-economic circumstances at play in a country such as South Africa. South Africa is specifically selected, as it is an emerging market corporate governance pioneer, with significant economic instability and growing wage inequality.

2 South Africa's Socio-Eeconomic Challenges

Le Pere and Ikome (2009) studied prospects and challenges for economic development in Africa and concluded that despite the great promise and positive effects of democracy, growth and good governance, economic conditions in Africa will remain turbulent, subject to difficult circumstances and different inconsistencies. This depicts especially to, inter alia, terms of trade, growth and employment, capital flows and the political environment. These are too variable and unstable to achieve poverty reduction targets set in the millennium development goals and, if not judiciously managed at institutional and political levels, imply further marginalisation of Africa in the world economy. They opine that Africa's challenges and prospects can be disaggregated into two continuums: domestic and international. At a domestic level, Le Pere and Ikome (2009) opine that a bold vision in policy design and implementation is critically needed, specifically regarding monetary and fiscal discipline, good governance, macroeconomic stability, promoting entrepreneurship and effective institutions. Van Niekerk (2020) concludes that for economic inclusivity to realistically reduce global inequality, its full integration into three areas is necessary: business models, public policy and community development. De Schutter et al. (2012) furthermore emphasise the significant role of foreign direct investment in promoting economic development by shaping business policies in developing economies. Another example of a much-needed policy design and implementation is the education sector.

Cummings and Altbach (1997) argue that education plays a crucial role in a country's development. Mlambo and Adetiba (2019) discovered that South Africa's educational system has failed to produce sufficient graduates with the necessary skills to drive the country's development and economic growth. Moreover, many students studying in fields requiring critical skills often consider emigrating immediately after graduation due to politically driven employment policies that do not prioritise merit. Considering remuneration, Meng (2020) argues that overpaying executive directors exacerbates income inequality, while underpaying them could lead to issues with brain drain. Mlambo and Adetiba (2019) infer that urgent interventions are needed in South Africa to address brain drain, exacerbated by poor economic growth, political uncertainty and crime. Failure to prioritise and contain this issue could jeopardise the country's developmental objectives. Mpendulo and Mang'unyi (2018) concur that the government should enact economic reforms that would promote entrepreneurship and the private sector, since educational achievements have an association with entrepreneurship, which could assist in reducing unemployment and developing South Africa as a high-income country. According to the Quarterly Labour Force Survey (Q1 of 2024), nearly 60% of 15- to 24-year-olds in South Africa are unemployed, and for the entire working-age population, the unemployed constitutes 32.9% of the population (Stats SA, 2024). As of 2018, approximately 34% of the population in South Africa received social grants as an income support system (SASSA, 2018). Statistics South Africa (Stats SA, 2014) ascertains that more than 20% of the population lives below the food poverty line, with 45.5% living in moderate poverty. Zikhali (2021), however, argues that social grants do not provide a sustainable way by which poverty can be tackled. These scholars posit that strategies such as empowering people through education, capacity building and popular participation in social and economic activities are ways to achieve sustainable livelihoods and poverty reduction in South Africa.

As part of the contextual analysis, the prevailing socio-economic dynamics in South Africa are examined, with a specific focus on how it can substantially impact the ability of businesses to mitigate inequalities through the adoption of sustainable and ethical practices that prioritise social responsibility. Also, empirical analyses are used to verify if the perceived relatively high and apparent increases in bank executive directors' remuneration over the years relate to bank performance. Mans-Kemp (2021) found that there is a need for further guidance to ensure that quantum pay disclosures become more comparable across sectors. Van Zyl and Mans-Kemp (2022a, 2022b) add that the government and the Institute of Directors Southern Africa should be cognisant of South Africa's socio-economic challenges when reflecting on the potential to implement international remuneration best practices in the local context.

3 Literature Review

This section discusses the theoretical framework used in this study, followed by a literature review on the pay-performance relationship and accounting and market-based performance. We also present the research hypothesis in this section. The theoretical framework creates opposing expectations regarding the pay-performance ratio. A positive pay-performance ratio is expected by the optimal contract theory. According to agency theory, CEOs are often selfish and prioritise their personal wealth creation instead of acting in the best interests of shareholders, known as the agency problem (Jensen & Meckling, 1976). To effectively address the agency problem, the optimal contract theory proposes that compensation contracts should be properly designed, consisting of different compensation components linked to company performance (Weisbach, 2007:421).

In contrast, managerial power theory suggests a negative or no pay-performance ratio (Bebchuk et al., 2002). This theory proposes that powerful CEOs can influence the composition of their compensation packages to enhance their wealth without being intrinsically incentivised to create sustainable value for the company (Bebchuk & Fried, 2003). Therefore, managerial power theory claims that the greater the power of a CEO, the higher the compensation, and the less likely that the compensation will be linked to company performance (Bebchuk et al., 2002).

3.1 Pay-Performance Relationship

Interest in reporting excessive EDR is longstanding and aligning shareholders' and management's interests remains a contentious area in corporate governance (Meng, 2020; Vieito et al., 2008). Jones (2023) concurs that investor faith in the banking industry has declined following the collapse of supposedly 'too big to fail' companies like Steinhoff, SVB Bank, Signature Bank, Credit Suisse, Enron, Tongaat Hulett and Parmalat. Although all companies encounter potential conflicts of interest among stakeholders, this issue is exacerbated in the banking industry due to its inherent nature. The existence of asset-liability issues, high debt-to-equity ratios, and, among others, deposit insurance may lead to greater conflicts. Macey and O'Hara (2003) argue that bank directors play a more crucial role in company success compared to directors of industrial companies. King and Levine (1993) highlight the significant role banks play in economic growth, particularly in emerging economies. South Africa's banking system is known for its stringent regulation, among the most rigorous in the Western world (Singleton & Verhoef, 2010).

In South Africa, the various King Reports have offered governance guidance since 1994 to companies listed on the JSE. The emolument of directors proves to be a particularly intricate corporate governance consideration, and investors and practitioners increasingly query the size and compilation of director emolument packages and disclosure thereof in JSE-listed companies' remuneration reports (Van Zyl & Mans-Kemp, 2022a, 2022b). In answer to such critique, the King IV Report provides more definitive director remuneration recommendations than King III. According to the King IV recommendation, listed companies should disclose the remuneration report in three parts, consisting of a background statement, remuneration policy and implementation report from 2017, to enhance their accountability. Companies should further reveal how they would respond to a substantial (25% or more) non-binding shareholder vote cast against their remuneration policy and/or implementation report. Moreover, King IV emphasises that EDR should be fair and responsible, in the context of overall employee pay, and linked to performance metrics (Institute of Directors Southern Africa, 2016). Although a plethora of disclosure requirements are welcomed, Van Zyl and Mans-Kemp (2022a, 2022b) found that remuneration reports are often onerous to read and time-consuming and recommend that information should be reported in an uncomplicated manner. Despite the onerous disclosure requirements, Bussin et al. (2023) add that remuneration governance should be refined through the incorporation of retrospective CEO remuneration disclosure to augment pay-performance sensitivity. Specifically, Meng (2020) recommends that South African companies enforce an effective remuneration structure and that an executive supervision division will be meaningful and significant. Furthermore, it is proposed that vesting and holding periods for long-term incentives should be at least 5 years. These emolument policies are composed with the long-term survival of the company in mind, and companies should have arrangements in place that will enable them to recover or withhold variable pay when it is appropriate to do so. These recommendations are supported by Magnan et al. (2010)

who provided the following four recommendations. Firstly, EDR must be sufficient to attract high-calibre individuals and to compensate them for their responsibilities. Secondly, EDR must be set in a transparent and objective way, bearing clear benchmarks reflecting the most conceivable talent markets. Thirdly, a significant percentage of EDR must be 'locked in' for the long term (approximately 5–10 years). Lastly, the EDR must not be based upon the realisation of short-term goals, but rather be based on the long-term success of the company while not encouraging excessive risk-taking. Van Zyl and Mans-Kemp (2022a, 2022b) conclude that it is arduous to practically address stakeholders' conflicting information needs. For instance, while certain individuals would request more detail in remuneration reports, other parties might demand a more abbreviated version. These scholars suggested that the Institute of Directors Southern Africa could offer more guidance to further standardise remuneration reporting in the future.

Vieito et al. (2008) state that executive directors need to be incentivised to enhance shareholder value rather than pursuing their own interests. Agency theory decrees that an agency problem exists when an agent, such as the executive director, has an agenda which rivals with the interest of the shareholders (Jensen & Meckling, 1976; Shortland & Perkins, 2023). Jensen and Meckling (1976), however, argue that incentives, such as cash bonuses and share options, can be used as performance incentives to align managers' interests with those of shareholders. Verster (2022) contends that the debate typically steers towards capping executive directors' remuneration. From an incentivisation perspective, equity holders have unlimited upside potential, raising the question of whether the stewards of their capital should enjoy the same upside. Shortland and Perkins (2023) further argue that the dilemma of EDR involves balancing the need to fairly compensate and incentivise executives while ensuring that their pay aligns with the long-term interests of shareholders. Thus, the question arises: is there a relationship between EDR and company performance in South Africa's banking sector?

Although the link between pay and performance is a topic that's been researched in the past, very few studies focused on specifically the banking sector in South Africa (Deysel & Kruger, 2015; Van Blerck, 2012). Those who did report on the pay-performance relationship in South Africa only considered total EDR (Deysel & Kruger, 2015; Van Blerck, 2012), whereas this study acknowledges the individual incentive components. Some prior studies have reported an absence of a relationship or a weak relationship between EDR and company performance in the banking sector (Ahmed & Ndayisaba, 2017; Lawrence, 2020; Meng, 2020; Omoregie & Kelikume, 2017; Yusuf & Abubakar, 2017), while others have found a positive, although sometimes small, statistically significant relation (Appah et al., 2020; Meng, 2020).

Given that South Africa has the highest wage inequality in the world, it is foreseeable that the remuneration policies and practices of companies listed on the JSE receive considerable attention from stakeholders (Padia & Callaghan, 2020). Mans-Kemp (2021) concurs that there is a need for further guidance to ensure that quantum pay disclosures become more comparable across sectors. Van Zyl and Mans-Kemp (2022a, 2022b) add that the government and the Institute of Directors

Southern Africa should be cognisant of South Africa's socio-economic challenges when reflecting on the potential to implement international remuneration best practices in the local context.

3.2 Accounting- and Market-Based Performance

Several methods have been used in previous studies to measure company performance, which can be divided into two broad categories, namely accounting- and market-based performance measures. Market-based performance measures are defined as those based on share prices (Al-Matari et al., 2014) and provide a direct measure of shareholder wealth by incorporating a forward-looking aspect (Steyn, 2015). Tobin's Q is calculated as the market value of equity plus the book value of debt divided by the book value of total assets (Iress, 2023). The Tobin's Q ratio indicates the business's value and helps estimate whether it is over- or undervalued (Lin et al., 2017). Total shareholder return measures the return that a shareholder earns from an investment, considering both capital appreciation and dividends (Čupić & Todorović, 2011).

While the financial statements from publicly traded companies are subject to stringent regulatory scrutiny, which lends credibility and reliability to the analysis, it is crucial to recognise that accounting-based performance measures are inherently susceptible to the following limitations and constraints: firstly it is historic and therefore backward-looking, which offers no guarantee of future performance; management may become focused purely on short-term earnings management (Dechow et al., 2011). Secondly, accounting profits can be manipulated by adjusting discretionary amounts (Lambert & Sponem, 2005) and lastly temporary disequilibrium effects may be present, including variations in tax legislation (Blaylock et al., 2012).

Despite the limitations, accounting-based performance measures are still widely accepted and used, but the types of metrics applied have changed over the years. More traditional measures are included in this study, for example return on equity and headline earnings per share. Newer benchmarks, such as the economic value-added benchmark, which considers the excess value created by management after subtracting the expected return, are not included. The reason for excluding economic value-added as a proxy for company accounting performance is that it is not always easy to compare a company with its peers, given the differences in the cost of capital expectations of the various companies. Several studies have combined performance metrics in regression analyses (e.g., El-Sayed, 2013; Kirsten & Du Toit, 2018; Krasnikova, 2014). In accordance with King III and wider literature, this approach is justified for two main reasons. First, no single metric fully captures company performance (Al-Matari et al., 2014). Second, to prevent manipulation and ensure robust decision-making, King III recommends combining accounting and market-based metrics for the remuneration of executive directors (Institute of Directors Southern Africa, 2012). South African shareholders also advocate for including return, earnings growth and market-based benchmarks in executive directors' remuneration decisions (Bussin et al., 2023).

The majority of studies found evidence supporting optimal contract theory and reported a positive pay-performance relationship between total EDR and accounting-based performance measures (e.g., Talmor & Wallace, 2001; Deysel & Kruger, 2015; Ahmed & Ndayisaba, 2017; Chou & Buchdadi, 2018). Moreover, some studies also provide empirical support for the managerial power management theory, suggesting that a significant proportion of executive directors, particularly CEOs, tend to prioritise their personal interests over the broader interests of the company and its shareholders. More specifically, a negative relationship was reported between total EDR and accounting-based performance (e.g., Omoregie & Kelikume, 2017; Mohamed, 2020). Omoregie and Kelikume (2017), for instance, included 12 banks in Nigeria over a nine-year period and reported a negative relationship between return on equity and total EDR. Omoregie and Kelikume (2017) posit that their findings could be explained in that most of the activities of boards of Nigerian banks are largely hidden from the shareholders.

The literature predominantly reported a significant positive relationship between cash-based EDR (including short-term EDR) and accounting-based performance measures (Sigler & Porterfield, 2001; Talmor & Wallace, 2001; Van Blerck, 2012; Słomka-Gołębiowska, 2016). For example, Słomka-Gołębiowska (2016) examined all listed banks on the Polish Stock Exchange between 2005 and 2013 and found that the cash compensation of executive directors is dependent on bank performance, implying that directors are not paid for poor performance. Optimal contract theory is further supported by Leonard (1990) who studied 439 large U.S. corporations between 1981 and 1985 and concluded that companies with long-term incentive plans enjoy significantly larger increases in return on equity than companies without such plans. Studies examining the relationship between long-term EDR and accounting-based performance have also reported predominantly positive results. Becher et al. (2005) and Słomka-Gołębiowska (2016) report a positive relationship between long-term EDR and return on equity. Becher et al. (2005), for example, compared the EDR of banks and non-banks from 1992 to 1999 in the United States. Their findings indicate that banks that adopt a high degree of equity-based compensation exhibit a significant relationship with accounting-based performance, augmented corporate governance engagement and accelerated growth without a corresponding escalation in risk.

Most studies have found a statistically significant positive relationship between total and long-term EDR and shareholder return as market-related performance measures, supporting optimal contrast theory (Burhop, 2004; Doucouliagos et al., 2007; Ahmed & Ndayisaba, 2017). Doucouliagos et al. (2007) investigated Australian banks over the period spanning 1992–2005 and concluded that the positive relationship observed implies that Australian banks are managed in a manner consistent with the optimisation of shareholder value. The results for the relationship between EDR and Tobin's Q were mixed. Becher et al. (2005) and Chou and Buchdadi (2018) reported no relationship, while Van Blerck (2012) found a negative relationship. Based on the findings of most of the literature, we posit a positive relationship between EDR and company performance for our main research hypothesis.

4 Model and Methods

The objective of this study is to determine if the perceived relatively substantial increases in EDR over the years are related to bank performance. The study examines the relationship between different forms of EDR as a function of company performance. For this purpose, the following empirical model specification is adopted:

$$\mathrm{EDR}_{tj} = \beta_0 + \beta_1 \mathrm{HEPS}_{tj} + \beta_2 \mathrm{ROE}_{tj} + \beta_3 \mathrm{TSR}_{tj} + \beta_4 \mathrm{TQ}_{tj} + \beta_5 \mathrm{FS}_{tj} + \gamma_t + \varepsilon_{tj} \quad (1)$$

where subscript t is the year, j is the company, γ_t are the year's fixed effects, and ε is the regression residual (error term).

This study contributes to the body of research on the topic of executive compensation by examining the relationship between short-term, long-term and total EDR and both accounting-based and market-based company performance for all listed banks in South Africa. EDR is considered in the context of growth in company performance over a 12-year period. EDR is measured by short-term executive directors' remuneration (EDRS), long-term executive directors' remuneration (EDRL) and total executive directors' remuneration (EDRT). Company performance is divided into market-based performance and accounting-based performance. The former is measured as total shareholder return (TSR) and Tobin's Q (TQ). The latter is measured as headline earnings per share (HEPS) and return on equity (ROE). A control variable that captures firm size (FS) is added to the equation. Table 1 provides a summary of all variables used in this study.

This study employed a quantitative research approach, using panel data regressions to test the research hypothesis outlined before. To report the descriptive statistics, the mean, minimum, maximum, median and standard deviation values were obtained, using the raw data. The Pearson's Correlation coefficients were calculated to determine the correlation coefficients. This study used longitudinal data and panel data regressions are therefore necessary. Following the panel data tests suggested by Menard (2008:230–240), the F-test and Hausman test were conducted to determine the best panel data regression model. The fixed effects multivariate regression model was identified as the optimal model to investigate the relationship between executive directors' remuneration and firm performance. The statistical analysis was performed using a macro written in the programming language R and executed in Statistica, a statistical analysis program. This macro includes a factor that automatically accounts for the different years included in the study.

Endogeneity occurs when an independent variable is correlated with the error term in a regression model (Wooldridge, 2009:838). Roberts & Whited (2013:496) caution that many researchers do not adequately address this issue. The

Table 1 Definition of variables (Source: Authors' elaboration)

Acronym	Variable	Definition	Studies	Expected sign
A. Dependent variable				
EDRS	Short-term executive directors' remuneration	Natural log of annual cash bonus awards paid to executive directors (thousand South African Rand)	Lawrence (2020)	–
EDRL	Long-term executive directors' remuneration	Natural log of the value of shares and share options awarded to executive directors (thousand South African Rand)	Becher et al. (2005), Meng (2020), Omoregie and Kelikume (2017)	–
EDRT	Total executive directors' remuneration	Natural log of the basic salary, fringe benefits, annual cash bonus awards and value of shares and share options awarded to executive directors (thousand South African Rand)	Ahmed and Ndayisaba (2017), Chou and Buchdadi (2018), Deysel and Kruger (2015), Lee and Isa (2015), Meng (2020)	–
B. Dependent variable				
HEPS	Headline earnings per share	Ratio between profit or loss generated by operations and investments divided by the weighted number of average outstanding ordinary shares (%)	Doucouliagos et al. (2007), Deysel and Kruger (2015)	Positive
ROE	Return on equity	Ratio between profit after tax divided by owners' interest (%)	Meng (2020), Omoregie and Kelikume (2017)	Positive
TSR	Total shareholder return	Ratio between share price at year-end plus (dividend yield times share price at the beginning of the year) to share price at the beginning of the year (%)	Ahmed and Ndayisaba (2017), Doucouliagos et al. (2007), Meng (2020)	Positive
TQ	Tobin's Q	Ratio between equity (market value) and total assets (at replacement cost) (%)	Chou and Buchdadi (2018), Van Blerck (2012)	Positive
FS	Firm size	Natural log of total assets as shown in the statement of financial position of each company (million South African Rand)	Ahmed and Ndayisaba (2017), Chou and Buchdadi (2018), Deysel and Kruger (2015)	Positive

Durbin-Wu-Hausman test is commonly used in the literature to detect endogeneity, as seen in studies by Ataay (2018) and El-Sayed (2013). To conduct the Durbin-Wu-Hausman test, an instrumental variable is required (Roberts & Whited, 2013:513). This test is performed to confirm the suitability of the panel data regressions. This study focuses on the six South African banks that were listed on the JSE for the 2009–2022 reporting periods, namely Absa Bank Limited, Capitec Bank Limited, Finbond Group Limited, FirstRand Bank Limited, Nedbank Group Limited and Standard Bank of SA Limited.

The EDR data and total asset data were manually collected from the annual reports of the banks for 2009–2022. The annual reports of companies were obtained from the website of each of the banks. Data regarding headline earnings per share and return on equity were obtained from the financial ratio function within the Iress database. It is a provider of stock market and research data in South Africa available at: https://researchdomain.iress.co.za). Tobin's Q and total shareholder return data were sourced from the financial models function on the Iress database. The next section provides the descriptive statistics, followed by the correlation analysis, diagnostic tests, and the presentation and discussion of the research findings.

5 Results

5.1 *Descriptive Statistics*

As can be seen in Table 2, 84 company-year observations were studied. The descriptive statistics indicate that the average total salary earned by an executive director during the research period was 22.2 million South African Rand, with executive directors on average earning 32% from short-term incentives and 42% from long-term incentives. Shareholders, on average, earned 15.06 South African cents per share they held in South African listed banks. The mean for total shareholder returns was 22.41%, indicating an average return of 22% for the banks between 2009 and 2022. The average Tobin's Q was 1.14, indicating that, on average, the market value of banks was still larger than the book value of assets. A large variation was observed

Table 2 Summary statistics of all model variables (Source: Authors' elaboration)

Variable	Mean	Minimum	Maximum	Median	Standard deviation
EDRS	7160	0	21,447	6143	5693
EDRL	9408	0	48,266	8158	9006
EDRT	22,213	1082	60,010	20,015	15,219
HEPS	1506	−24	17,037	1113	2141
ROE	14.29	−31.47	35.43	14.73	10.58
TSR	22.41	−56.45	536.36	10.85	74.67
TQ	1.14	0.57	2.87	0.99	0.43
FS	204,897	4150	1,793,201	1,517,470	4,494,061

regarding the size of the banks. The smallest bank reported assets of 4.1 million South African Rand, while the largest bank's assets amounted to 4.5 trillion South African Rand. Descriptive statistics regarding the companies' dependent, independent and control variables were considered before estimating the regressions and are the actual figures (not adjusted for inflation). Table 2 presents these descriptive statistics for the raw data over the period 2009–2021.

5.2 *Correlation Analysis*

Multiple regression models require the absence of multicollinearity between the performance and the board structure of the companies tested. A serious multicollinearity problem is predicted by Arceneaux and Huber (2007) if the coefficient is greater than or equal to 0.7. The Pearman's correlation was used to examine relationships between the data elements and all coefficients were lower than 0.7 (Table 3). This correlation is particularly valuable as it enables a more granular examination of the relationship between EDR and company performance across a range of values, thereby facilitating a more nuanced understanding than would be feasible using regression equations alone. The highest correlation is the correlation between the size of the company (FS) and total EDR at 0.596, which is lower than 0.7. All coefficients between dependent and independent variables in Table 3 are as expected, except in the following instances: significant and negative correlations were found between EDRS and ROE, between EDRS and TQ, between EDRL and HEPS and between EDRT and TSR. Furthermore, an absence of a significant relationship was found between EDRS and HEPS; EDRS and TSR, EDRL and ROE, EDRL and TSR, EDRL and TQ, EDRT and TQ.

Table 3 Pearson's correlation matrix (Source: Authors' elaboration)

	EDRS	EDRL	EDRT	HEPS	ROE	TSR	TQ	FS
EDRS	1	0.512***	−0.606***	0.213**	0.144	−0.382***	0.059	0.207**
EDRL		1	−0.638***	0.380***	−0.031	−0.236	0.039	0.012
EDRT			1	0.379***	0.209**	−0.382***	0.059	0.207**
HEPS				1	0.157	−0.094	0.120	0.407***
ROE					1	0.026	0.179*	0.407***
TSR						1	−0.074	−0.119
TQ							1	−0.174*
FS								1

Note: *** Significant at the 0.01 level; ** Significant at the 0.05 level; * Significant at the 0.10 level

5.3 Diagnostic Tests

Diagnostic tests were performed on the data to test that the data meets the satisfactory levels of normality, homoscedasticity and multicollinearity. The normality plot suggested slight non-normality in the dataset. Moreover, the Breusch-Pagan test confirmed the presence of heteroskedasticity in all the regression models (all significant at the 1% level). The Huber-White method of using a robust covariance matrix was used to resolve the heteroskedasticity (MacKinnon & White, 1985:320). The results presented in Table 4 were reported after the Huber-White adjustments were made for heteroskedasticity. The underlying assumptions were also tested; the findings were investigated for multicollinearity by calculating a variance inflation factor (VIF) for each variable. All the VIFs tested are below 2. The tolerance factors are all well above 0.1 and below 10 for VIFs, suggesting that multicollinearity poses no problem when interpreting the regression results. The model specification test indicated that the explanatory variables were correctly specified. Thus, the Wald test, statistically significant (0.00), indicates that the model is fit.

5.4 Findings

Table 4 reports the results of the regression analysis which were conducted to determine whether the EDR is related to the bank performance of the 6 listed South African banks. The main interest in Table 4 is the sign and the significance of the independent and control variable tested. The regressions performed are significant

Table 4 Regression results (Source: Authors' elaboration)

	Model 1		Model 2		Model 3	
	Coefficients	*p*-value	Coefficients	*p*-value	Coefficients	*p*-value
HEPS	0.10	0.65	0.46	0.00***	0.01	0.58
ROE	0.13	0.08*	0.17	0.08*	0.51	0.60
TSR	−0.14	0.06*	−0.00	0.43	−0.06	0.55
TQ	0.37	0.17	0.13	0.02**	0.55	0.03**
FS	0.58	0.00***	−0.68	0.26	0.56	0.00***
R-Squared	0.29		0.38		0.47	
Adjusted R-Squared	0.14		0.20		0.40	
Sample size	84		84		84	
F-statistic	5.58***		5.47***		5.73	
F-test for FE	3.36***		1.59		2.80*	
Hausman test	2.93***		1.29		7.62*	

Note: Dependent variable is EDRS (Model 1), EDRL (Model 2) and EDRT (Model 3); *** Significant at the 0.01 level; ** Significant at the 0.05 level; * Significant at the 0.10 level.; FE denotes fixed effects

at the 1% significance levels. Some 14% (EDRS), 20% (EDRL) and 40% (EDRT) of the variation in EDR was explained by the independent and control variables.

A significant and positive relationship is found between the short-term remuneration awarded to the executive directors (EDRS) of the banks and ROE. The results thereby offer support for optimal contract theory, providing evidence to accept the main research hypothesis. This result indicates that executive directors of banks are remunerated by way of short-term bonuses for generating income for the equity invested by shareholders. By tying the short-term bonuses awarded to executive directors to ROE, the company aligns executives' personal financial goals with the firm's financial health and shareholder interests, ensuring that both parties benefit from improved performance. Empirically, the findings agree with the findings of Talmor and Wallace (2001). Contrary to what was expected from the main research hypothesis statement, a significant and negative relationship is reported between the EDRS and TSR. The negative relationship between EDRS and TSR indicates that these variables move in opposite directions and could be interpreted in that executive directors receive more short-term remuneration if TSR decreases. The result could be explained in that TSR could be influenced by market volatility, which is beyond the control of executive directors.

Moreover, when executive directors are remunerated with short-term incentives, they may be inclined to prioritise the attainment of short-term financial objectives, which may not necessarily be congruent with actions that generate sustainable, long-term value for shareholders. Van Zyl and Mans-Kemp (2022a, 2022b) suggest that in an economy facing rising unemployment, declining economic activity and the real threat of brain drain, adopting business-friendly policies that include reasonable EDR (prioritising the long-term survival of the company) should be regarded as a fair and prudent approach. Therefore, the government and the Institute of Directors Southern Africa should be mindful of South Africa's socio-economic challenges when considering the adoption of international remuneration best practices in the local context. No evidence could be found to reject the null hypothesis for the relationship between EDR short-term remuneration and HEPS and TQ. The proposals of optimal contract theory, which suggest linking executive remuneration contracts to bank performance, could be undermined by an absence of a relationship. An examination of the remuneration reports of the individual banks unveiled that most remuneration committees used HEPS as a performance measure when granting short-term incentives to executive directors.

In support of the main research hypothesis, the results reported in Table 4 reveal a significant and positive relationship between the long-term incentives awarded to executive directors and accounting-based performance (HEPS and ROE) and a statistically significant relationship with TQ as market-based performance measure. The results indicate that the long-term remuneration awarded to executive directors motivates them to focus on long-term strategies to increase the company's profitability and enhance the effectiveness of the company's equity and market value. The results provide evidence to support the optimal contract theory and could be interpreted to indicate that executive directors are rewarded for enhancing shareholder value over the long term. No statistically significant relationship was reported between EDRL and TSR. The banks' remuneration reports, however, showed that certain banks provide long-term remuneration to executive directors based on TSR. As explained above, TSR may be outside the control of executive directors as it is influenced by many external factors (such as market conditions and industry trends).

The results reported in Table 4 offer evidence to accept the main research hypothesis and report a significant and positive relationship between total EDR and Tobin's Q (on the 5% level) as a market-based performance measure. Empirically, the results agree with the findings of Rahman (2016). The results suggest that the total remuneration offered to executive directors of banks is effectively aligned with enhancing the company's market value. This alignment suggests that the remuneration structure motivates executives to focus on strategies that improve market performance, investor perceptions, and overall company valuation. The results support the principles of optimal contract theory indicating that the remuneration contract offered to the executive directors is linked to company performance. No evidence could be found to reject the null hypothesis between EDRT and HEPS, ROE and TSR. These findings agree with the findings of Meng (2020) as well as Yusuf and Abubakar (2017). The absence of the relationship can be explained by the fact that the EDRT includes the basic remuneration awarded to executive directors, other benefits offered, short-term bonuses (EDRS) and share-based remuneration (EDRL). EDRT may be influenced by a wide range of factors, including market conditions, industry trends, and company-specific factors. For example, Jacobs (2024) found that the number of companies listed on the JSE has decreased by more than half over the past 30 years. Additionally, Jeffery (2019) opines that South Africa's business environment is not attractive and not encouraging, with the country experiencing a difficult economic growth period and an unpredictable political system.

Understanding the relationships between executive directors' remuneration and company performance helps in designing executive remuneration packages consisting of short-term and long-term components that effectively motivate executives to achieve desirable company performance outcomes. By tying compensation to performance measures, companies can ensure that executives are incentivised to work towards maximising shareholder value. Some additional tests were performed to ensure the robustness of the results. Following El-Sayed (2013) endogeneity is assessed by re-estimating the regression models, using lagged variables in the two-stage least-squares regression models. The Durbin-Wu-Hausmann test was performed. Following Ataay (2018) and El-Sayed (2013) lagged performance variables were used as instruments. The Durbin-Wu-Hausmann test indicated the absence of endogeneity, confirming the suitability of the panel data regressions.

6 Conclusion

Despite the King Reports providing a robust governance framework, South Africa has the highest wage inequality globally and faces increasing wage disparity and economic volatility. Consequently, stakeholders are increasingly critical of the remuneration offered to executive directors in South Africa. While the relationship between company performance and corporate governance has been explored in academic literature, there has been limited research on the relationship between EDR and company performance in South Africa's banking sector. Given the crucial role banks play in any economy, this study aimed to determine if there is a relationship between EDR and company performance in the South African banking sector, and if so, is it fair given the country's inequality and socio-economic status? This study fills a gap in existing literature, by analysing this relationship, using a sample of all listed South African banks from 2009 to 2021.

Using panel data analysis, evidence shows that there is a positive and statistically significant relationship between the short-term bonuses awarded to executive directors and return on equity, long-term remuneration and headline earnings, return on equity and Tobin's Q, and total remuneration and Tobin's Q. This is consistent with the findings by Jensen and Meckling (1976) who posit that incentives, such as cash bonuses and share options can be used as performance incentives to align managers' interests with those of shareholders. The widening of inequality in the wage gap, however, is explained by understanding the interplay between the availability of skilled individuals (exacerbated by the brain drain) and the demand for high-calibre executive directors in companies. It is put forward that national development priorities should connect with international education agendas, as a more educated populace is a more employed populace. Furthermore, given an economy facing rising unemployment, declining economic activity, and the real threat of brain drain, adopting business-friendly policies that include reasonable EDR (prioritising the long-term survival of the company) should be considered fair practice. Therefore, the government and the Institute of Directors Southern Africa should be cognisant of South Africa's socio-economic challenges when considering the adoption of international remuneration best practices in the local context.

On a global front, the Sustainable Development Goals simultaneously call for reducing inequalities within and among countries. By gaining insight into South Africa's unique socio-economic challenges, the findings of this study should assist multinational corporations in decision-making, with the aim of achieving sustainable development goals. It is suggested that multinational entities should be aware of the socio-economic and political boundaries in which they operate, as their influence on developing economies should not be underestimated. Future researchers can focus on designing economic policies that will support and enhance an encouraging business environment. Researchers can also consider designing a sound education system to produce enough graduates with the required skills to spur South Africa's economic growth and development. Future research may also investigate the aptness and application of politically charged employment policies and include an investigation of other sectors of the JSE, as well as how multinational entities contribute towards creating sustainable business strategies. This study is limited to South African banks and cannot be generalised to other companies listed on the JSE. The unique socio-economic context of South Africa makes comparisons to other developing countries complex. Future research could focus on the banking sector in other developing countries or select another sector of listed JSE companies for analysis. This study only included popular accounting-based and market-based performance variables to measure firm performance. Future research could incorporate additional financial measures (such as return on equity and abnormal share returns) and non-financial measures (such as environmental, social, and governance activity measures). This study did not investigate the disclosure of executive directors' remuneration. Corporate governance requirements and international financial reporting standards (IFRS 2 share-based payment) have been effective so far, but detailed disclosure of directors' remuneration is still gradually improving. Future studies could investigate the corporate governance factors influencing executive directors' remuneration. Continued future developments in international financial reporting standards will aid future studies. Lastly, it is suggested that the accountability of executive directors' remuneration to shareholders could be promoted by the compensation policy.

References

Ahmed, A. D., & Ndayisaba, G. A. (2017). Do regulatory standards help align ED compensation and banks performance association? *The Journal of Developing Areas, 51*(4), 127–142.

Al-Matari, E. M., Al-Swidi, A. K., & Fadzil, F. H. B. (2014). The measurements of firm performance's dimensions. *Asian Journal of Finance & Accounting, 6*(1), 24–49.

Appah, E., Tebepah, S. F., & Awuji, C. E. (2020). Directors' compensation and financial performance of deposit money banks in Nigeria. *World Journal of Finance and Investment Research, 5*(1), 61–76.

Arceneaux, K., & Huber, G. A. (2007). What to do (and not do) with multicollinearity in state politics research. *State Politics & Policy Quarterly, 7*(1), 81–101.

Ataay, A. (2018). Performance sensitivity of executive pay: The role of ownership structure, board leadership structure and board characteristics. *Economic Research-Ekonomska Istraživanja, 31*(1), 1152–1168.

Bebchuk, L. A., Fried, J., & Walker, D. (2002). Managerial power and rent extraction in the design of executive compensation. *The University of Chicago Law Review, 69*(3), 751.

Bebchuk, L. A., & Fried, J. M. (2003). Executive compensation as an agency problem. *Journal of Economic Perspectives, 17*(3), 71–92.

Becher, D. A., Campbell, T. L., II, & Frye, M. B. (2005). Incentive compensation for bank directors: The impact of deregulation. *The Journal of Business, 78*(5), 1753–1778.

Blaylock, B., Shevlin, T., & Wilson, R. J. (2012). Tax avoidance, large positive temporary book-tax differences, and earnings persistence. *The Accounting Review, 87*(1), 91–120.

Burhop, C. (2004). Executive remuneration and firm performance: The case of large German Banks, 1854–1910. *Business History, 46*(4), 525–543.

Bussin, M., Wöcke, A., & Deysel, B. (2023). The impact of remuneration governance on chief executive officer overpayment. *South African Journal of Economic and Management Sciences, 26*(1), 1–10.

Chou, T. K., & Buchdadi, A. D. (2018). Executive's compensation, good corporate governance, ownership structure, and firm performance: A study of listed banks in Indonesia. *Journal of Business and Retail Management Research, 12*(3), 79–91.

Cummings, W. K., & Altbach, P. G. (Eds.). (1997). *The challenge of Eastern Asian education: Implications for America*. State University of New York Press.

Čupić, M., & Todorović, M. (2011). Total shareholder return-Decomposition, internal equivalent and alternatives. In *Problems of competitiveness of contemporary economies* (pp. 379–392).

Dechow, P. M., Ge, W., Larson, C. R., & Sloan, R. G. (2011). Predicting material accounting misstatements. *Contemporary Accounting Research, 28*(1), 17–82.

De Schutter, O., Swinnen, J., & Wouters, J. (Eds.). (2012). *Foreign direct investment and human development* (pp. 1–24). Routledge.

Deysel, B., & Kruger, J. (2015). The relationship between South African ED compensation and company performance in the banking industry. *Southern African Business Review, 19*(1), 137–169.

Doucouliagos, H., Haman, J., & Askary, S. (2007). Directors' remuneration and performance in Australian banking. *Corporate Governance: An International Review, 15*(6), 1363–1383.

El-Sayed, N. M. (2013). *An examination of executive directors' remuneration in FTSE 350 companies*. Unpublished doctoral dissertation. University of Exeter Business School.

Ghauri, P. N., & Cooke, F. L. (2022). MNEs and United Nations' sustainable development goals. In *The new frontiers of international business: Development, evolving topics, and implications for practice* (pp. 329–344). Springer.

ILO. (2018). *Global wage report 2018/19: What lies behind gender pay gaps*. International Labour Organization, International Labour Office. http://www.ilo.org

Institute of Directors Southern Africa (IoDSA). (2012). *Practice notes: A guide to the application of King III: Remuneration*. Johannesburg.

Institute of Directors Southern Africa (IoDSA). (2016). *King report on corporate governance for South Africa*. Johannesburg.

Iress. (2023). *Research domain user guide*. Johannesburg.

Jacobs, S. (2024). *JSE delistings start to bite*. Daily Investor. http://www.dailyinvestor.com/

Jeffery, A. (2019). A new empowerment strategy to liberate the poor. *The Policy Bulletin of the IRR (South African Institute for Race Relations), 3*(42), 4–16.

Jensen, M., & Meckling, W. (1976). Theory of the firm: Managerial Behavior, agency costs and ownership structure. *Journal of Financial Economics, 3*(4), 305–360.

Jones, M. (2023). *Banking turmoil pounding investor confidence, surveys show*. Reuters. http://www.reuters.com

King, R. G., & Levine, R. (1993). Finance and growth: Schumpeter might be right. *The Quarterly Journal of Economics, 108*(3), 717–737.

Kirsten, E., & Du Toit, E. (2018). The relationship between remuneration and financial performance for companies listed on the Johannesburg Stock Exchange. *South African Journal of Economic and Management Sciences, 21*(1), 1–10.

Krasnikova, N. (2014). *Executive remuneration, financial performance and corporate governance in UK and Spanish listed firms*. Unpublished doctoral dissertation. Ashton University.

Lambert, C., & Sponem, S. (2005). Corporate governance and profit manipulation: A French field study. *Critical Perspectives on Accounting, 16*(6), 717–748.

Lawrence, E. (2020). The impact of director's remuneration on firms' performance evidence: A study of Nigerian banking sector. *Tansian University Journal of Arts, Management & Social Sciences, 7*(1), 129–150.

Lee, S. P., & Isa, M. (2015). Directors' remuneration, governance and performance: The case of Malaysian banks. *Managerial Finance, 41*(1), 26–44.

Le Pere, G., & Ikome, F. (2009). Challenges and prospects for economic development in Africa. *Asia-Pacific Review, 16*(2), 89–114.

Lin, X., Wang, C., Wang, N., & Yang, J. (2017). Investment, Tobin's q, and interest rates. *Journal of Financial Economics, 130*(3), 620–640.

Leonard, J. S. (1990). Executive pay and firm performance. *ILR Review, 43*(3), 13-S.
Macey, J. R., & O'Hara, M. (2003). The corporate governance of banks. *Economic Policy Review, 9*(1), 91–107.
MacKinnon, J. G., & White, H. (1985). Some heteroskedasticity-consistent covariance matrix estimators with improved finite sample properties. *Journal of Econometrics, 29*(3), 305–325.
Magnan, M., St-Onge, S., & Gélinas, P. (2010). Director compensation and firm value: A research synthesis. *International Journal of Disclosure and Governance, 7*(1), 28–41.
Mans-Kemp, N. (2021). *The growing wage gap, and what it means for executive pay*. The Conversation Africa. https://www.businesslive.co.za/
Menard, S. (2008). *Handbook of longitudinal research: Design, measurement, and analysis*. Amsterdam.
Meng, D. (2020). *A comparison analysis of ED compensation related to shareholders value: South Africa versus China holding banks*. Unpublished master's thesis. University of the Western Cape.
Mlambo, V. H., & Adetiba, T. C. (2019). Brain drain and South Africa's socio economic development: The waves and its effects. *Journal of Public Affairs, 19*(4), e1942.
Mohamed, Y. (2020). *Association between executive directors' remuneration and earnings management among banks in Kenya*. Unpublished master's thesis. Strathmore University.
Mpendulo, G., & Mang'unyi, E. E. (2018). Exploring relationships between education level and unemployment. *Journal of Social Sciences (COES&RJ-JSS), 7*(2), 86–102.
Omoregie, O. K., & Kelikume, I. (2017). Executive compensation and banking sector performance: Evidence from Nigeria. *The Journal of Developing Areas, 51*(2), 1–15.
Padia, N., & Callaghan, C. W. (2020). Executive director remuneration and company performance: Panel evidence from South Africa for the years following King III. *Personnel Review, 50*(3), 829–844.
Rahman, M. M. (2016). Determinants of CEO compensation: Empirical evidence from listed banks of Bangladesh. *Journal of Business Studies, 1*(1), 127–161.
Roberts, M. R., & Whited, T. M. (2013). Chapter 7: Endogeneity in empirical corporate finance. In G. M. Constantinides, M. Harris, & R. M. Stulz (Eds.), *Handbook of the economics of finance* (Vol. 2, Part A, pp. 493–572). Elsevier.
Shortland, S., & Perkins, S. J. (2023). Quantum accountability: When does enough become too much in top pay decision-making? *Employee Relations, 45*(6), 1455–1475.
Sigler, K. J., & Porterfield, R. (2001). CEO compensation: Its link to bank performance. *American Business Review, 19*(2), 110–114.
Singleton, J., & Verhoef, G. (2010). Regulation, deregulation, and internationalisation in South Africa and New Zealand banking. *Business History, 52*(4), 536–563.
Słomka-Gołębiowska, A. (2016). The effect of remuneration committee independence on the pay–performance relationship: Evidence from the banking industry in Poland. *Eastern European Economics, 54*(1), 71–89.
SASSA. (2018). *Social grants*. http://www.sassa.gov.za/index.php/social-grants
Stats SA. (2014). *Poverty trends in South Africa: An examination of absolute poverty between 2006 and 2011*. Statistics South Africa.
Stats SA. (2024). *Quarterly labour force survey (QLFS) Q4:2023*. https://www.statssa.gov.za/publications/P0211/Presentation%20QLFS%20Q4%202023.pdf
Steyn, G. F. (2015). *The relationship between ED compensation and future share returns in South Africa*. Unpublished master's thesis. University of the Western Cape.
Talmor, E., & Wallace, J. S. (2001). *A unified analysis of executive pay: The case of the financial sector*. https://ssrn.com/abstract=215630
Van Blerck, G. T. (2012). *The relationship between executive remuneration at financial institutions and Economic Value Added*. Unpublished master's thesis. University of Pretoria.
Van Niekerk, A. J. (2020). Inclusive economic sustainability: SDGs and global inequality. *Sustainability, 12*(13), 5427.

Van Zyl, M., & Mans-Kemp, N. (2022a). Insider perspectives on director remuneration governance deliberations. *Personnel Review, 52*(3), 799–816.

Van Zyl, M., & Mans-Kemp, N. (2022b). A multi-stakeholder view on director remuneration guidance in South Africa. *South African Journal of Accounting Research, 36*(3), 195–212.

Verster, J. P. (2022). *The finance ghost: Make the ED work for you.* Financial Mail. https://www.businesslive.co.za/

Vieito, J. P., Khan, W., Cerqueira, A., & Brandão, E. (2008). Is executive compensation different across S&P listed firms? *Quarterly Journal of Finance and Accounting*, 171–192.

Weisbach, M. S. (2007). Optimal executive compensation versus managerial power: A review of Lucian Bebchuk and Jesse Fried's pay without performance: The unfulfilled promise of executive compensation. *Journal of Economic Literature, 45*(2), 419–428.

Wooldridge, J. M. (2009). *Introductory econometrics: A modern approach* (4th ed.). South-Western Cengage Learning.

Yusuf, I., & Abubakar, Z. (2017). Executive compensations and earnings management in deposit money banks in Nigeria. *Sahel Analyst: Journal of Management Sciences, 15*(1), 91102.

Zikhali, P. T. (2021). Social grants and poverty alleviation in South Africa: Addressing dependency attitude and behaviour. *Gender and Behaviour, 19*(2), 17854–17863.

Suzanne Kieviet is a lecturer and subject co-ordinator at the School of Accountancy, Stellenbosch University, South Africa. She graduated from Stellenbosch University with a master's degree in accounting (taxation) and is also a Chartered Accountant (SA). Her work and academic experience over the past three decades have been largely linked to business and business education. Her previous practical experience includes manager positions at an auditors' firm, followed by manager positions in the finance divisions of two JSE listed companies. She has published in accredited journals and presented papers at various international conferences, focussing on issues related to corporate governance and sustainability. She has acted as supervisor of postgraduate students, and she regularly reviews articles written for publication related to corporate governance and entrepreneurship development. Correspondence concerning this chapter should be sent to kieviets@sun.ac.za.

Henriette Elsabè Scholtz is a senior lecturer at the School of Accountancy at Stellenbosch University. Graduated from Stellenbosch University with a Ph.D. in Accounting, her research interests lie in the realms of corporate governance sustainability and executive directors' remuneration. She is also a Chartered Accountant (SA). Prior to joining academia, she garnered valuable industry experience, having served as a general manager at a large banking group. She has published numerous papers in accredited journals and presented at several international conferences. She has supervised postgraduate students, acted as an examiner for master's students, and regularly reviews articles written for publication on corporate governance, ethics, internal auditing and executive directors' remuneration.

Towards Internationalisation of African Banks: Strategy, Legitimacy and Sustainability

Mesiet W. Kamihanda and Miguel M. Torres

1 Introduction

In September 2015, world leaders gathered in New York to approve the Doc. A/RES/70/1, also known as The 2030 Agenda for Sustainable Development. Ban Ki-Moon, the eighth secretary-general of the United Nations, discourses after the approval asserting that this initiative signals a 'paradigm shift for people and the planet' (Nilsson et al., 2018; United Nations, 2015). If it is for the planet, it is also for people, but, indeed, Agenda 2030s significant expectation is that governments, businesses, and civil society actors would get involved and responsible in progressing a sustainable world and achieving the SDGs (Sustainable Development Goals) (Zhan & Santos-Paulino, 2021). The introduction of SDGs and the call for a decade of action (Sachs & Sachs, 2021; Van Tulder et al., 2021) demand an understanding of how the private sector, specifically involved in adopting and implementing the SDGs (Ike et al., 2019).

Many uphold that the private sector, specifically MNEs, has exceptional capability to bring to bear in delivering on the SDGs through innovation, efficiency, responsiveness, and specific skills or resources' provision (Berning, 2019) as well as financer, community capacity builder, product and service provider, partner, and innovator (Eang et al., 2023). Indeed, the business sector, i.e., non-public or non-governmental firms, has a decisive role in influencing the implementation of the SDGs. MNEs play a critical role in achieving SDGs adoptions in host countries

M. W. Kamihanda (✉)
UEMF, Business School, Fès, Morocco
e-mail: w.mesiet@ueuromed.or

M. M. Torres
Kent Business School, University of Kent, Canterbury, UK
e-mail: m.torres@kent.ac.uk

J. Bento, M. M. Torres (eds.), *Global Perspectives on Climate Change, Inequality, and Multinational Corporations*, CSR, Sustainability, Ethics & Governance,
https://doi.org/10.1007/978-3-031-80797-8_6

(Liou & Rao-Nicholson, 2021; Van Zanten & Van Tulder, 2018; Zhan, 2021) specified that the SDGs cannot be achieved without the contributions of MNEs. Likewise, the role of international businesses as a critical agent in attaining SDGs can always be noticed (Kano, 2018; Mio et al., 2020; Van Zanten & Van Tulder, 2021; Verbeke & Kano, 2016).

This movement surrounding SDG implementation is an exciting opportunity (Pedersen, 2018) for global actors to work together to achieve SDG gains, reduce poverty and ensure a sustainable future (Van Tulder et al., 2021). However, corporate involvement in the SDGs is still limited and mainly associated with commitment to other sustainability-related themes (van der Waal & Thijssens, 2020). Precisely, what influences MNEs' adoption and implementation of SDGs (and how?) is not (yet) fully explored in the extant literature. While some studies have unveiled the legitimisation (Colleoni, 2013; King & Whetten, 2008; Schmidtke & Lenz, 2023; Zhang et al., 2022) of MNEs concerning sustainability, little is discussed about the quest for the link between SDG implementation and businesses' strategies for internationalisation performance (Boehe, 2016; Persson et al., 2016; Scheyvens et al., 2016; Siltaloppi et al., 2021; Strand, 2014; Verbeke & Kano, 2015; Yang et al., 2019). More evidence is needed about the role of MNEs in the financial sphere, particularly Multinational Banks (MNBs), in contributing to this global challenge.

Hence, MNBs' contribution to adopting and attaining SDGs still needs to be clarified. Studies before 2014 had little emphasis on sustainability. For instance, Amran et al. (2016) highlighted that few businesses integrated climate change into their business strategy due to factors such as the board of directors' international experience, country effects and organisation slack. Attention to sustainability has increased among international businesses since the announcement of the SDGs by the United Nations (Gomez-Trujillo & Gonzalez-Perez, 2020; Kolk, 2016). Nevertheless, International Business Research (IBR) studies tend to (almost) ignore the MNBs block when researching the multinational enterprise and sustainability (Kili, 2016; Mohieldin et al., 2022a; Sarfraz et al., 2018). Due to the sensitive nature of the activity (in essence, it is a relational activity), MNBs are subject to high scrutiny from host countries' local socio-political actors, especially the regulator and the media (Hong et al., 2015; Kili, 2016).

It is, therefore, crucial for IBR to turn the spotlight on the MNBs, especially from emerging market economies that have a greater chance of influencing the attainment of the SDGs. Like other developing regions, Africa is lagging in closing a sustainable development gap (Linnerud et al., 2021). Thus, this paper aims to contribute to the analysis of Emerging Market Multinational Banks (EMMNBs) and try to cover one or two of the several research gaps presented by the SDGs in developing countries (Castro et al., 2021; Mohieldin et al., 2022b). Specifically, we explore how EMMNBs' estimation through communication and discursive strategies and internalisation performance tend to influence the adoption of SDGs by these banks. We use three MNBs from Morocco present in African countries, especially West Africa, North Africa, and Central Africa. We deem this important because Africa is under-researched, as are MNBs from emerging markets (Bose & Khan, 2022). As Bose and Khan (2022) highlighted, the south-south MNEs' investment presents an opportunity worth exploring. This can be reinforced because corporations tend to behave and influence SDGs differently depending on the countries in which they are present. Their SDGs reporting and adoption differ for MNEs in the global North and the Global South (Arita, 2013; Uddini, 2017).

The contribution of this paper comes from different angles, the main one being the context: even though there is less IBR on Emerging Country Multinational Enterprises (EMNEs) (Aguilera et al., 2017; Cuervo-Cazurra et al., 2020; Marano et al., 2017) than there are (MNEs) in developed economies context (Bose & Khan, 2022; Nishitani & Kokubu, 2020). A large number of available studies are based on EMNEs from Latin America, Asia and Eastern Europe (Aguilera et al., 2017; Duque-Grisales & Aguilera-Caracuel, 2021; Duque-Grisales et al., 2020; Gomez-Trujillo & Gonzalez-Perez, 2020; Kili, 2016; Lin et al., 2019; Wei & Nguyen, 2020). IBR lacks the flavour of MNEs from the African continent (Barnard, 2019; Barnard et al., 2023). Furthermore, there needs to be more focus on MNBs. Only a few studies have considered MNBs concerning sustainability and SDGs (Barbaglia et al., 2023; Kili, 2016; Talbot & Ordonez-Ponce, 2022) and are not focused on the African continent. Therefore, the focus on EMNBs is a considerable gap that demands IBR attention. Thus, we contribute to the literature by studying one specific actor central to IBR yet often overlooked: MNBs. We assess the influence of legitimisation, internationalisation performance, and strategy on adopting SDGs from the African context, now termed AMBs.

The unique characteristics of the AMNBs can offer new insight into the internalisolation performance, legitimisation seeking, and strategy and how these tend to impact their adoption and, hence, implementation of SDGs. Integration and implementation of SDGs in the MNE business model, and particularly in the MNB business model, is cumbersome to assess, especially on the part of the motives and strategies aside from contributions. Schaltegger (2018) indicated that the interlinkage between the 17 SDGs and the difficulty in identifying a direct, explicit connection between firms and some goals negatively impact the development of any theoretical framework (Pizzi et al., 2021). The United Nations considers the 17 SDGs to be of equal dignity and should be implemented equally (Van Zanten & Van Tulder, 2018; Weitz et al., 2018). At the same time, each company has the legitimacy to decide which goals have priority according to its structure and objectives. These priorities can be conditioned by the country, the economic sector of activities, the business model, the regulatory framework, the firm culture, and the interests of the relevant stakeholders (Ike et al., 2019; Kolk et al., 2017; Van Zanten & Van Tulder, 2018; Weitz et al., 2018). However, the literature on developing countries reveals scarcity in this regard. Despite its structural relevance, the banking sector represents one of the research areas characterised by very few studies (Clayton & Schaab, 2022).

Nevertheless, emphasising the SDGs for MNBs remains a stiff topic of discussion. Several studies stressed that MNBs engage in SDGs to gain dominance, legitimate and protect their reputation (Borda et al., 2017; Chiu & Sharfman, 2011; Forcadell et al., 2020; Nguyen et al., 2021;Vaara & Tienar, 2008 ; Vaara & Tienari,

2011), while others claimed that the act of investing in SDGs and sustainability practices have significant implications on the MNBs profitability performance through gaining more customers and increasing shareholders' value (Murshudli et al., 2020; Wu & Shen, 2013). The SDGs are pressing as only 7 years are left before realising the goals. That is, 2030 is the deadline as specified by the United Nations. As MNEs are actively engaging in SDGs by incorporating sustainability, setting up sustainability departments, and doing sustainability reporting, it is paramount to assess the motives and strategies of MNEs towards SDGs and develop a solid consensus conclusion and recommendation for urgent policy implications.

This study tries to fill a gap by investigating why (motives) and how (strategies) MNBs approach SDGs using data from Moroccan Banks' expansion in West Africa. We will further assess the significance of these banks' strategies towards achieving SDGs in the region from both empirical and theoretical angles. Therefore, we will analyse the strategies employed by considering internal and external business characteristics. We will use sustainability reports (non-financial disclosure), interviews and questionnaires to analyse the pattern and consistency of typology used by CEOs and sustainability managers to assess the motives for incorporating SDGs in their business models, the SDGs that carry more weight and if all these MNBs' investment in SDGs is having a significant influence toward attaining SDGs in West African countries whose economies are still less advanced and like other SSA countries the theme of Sustainability is still lagging (Vaara & Tienar, 2008; Vaara & Tienari, 2011). Furthermore, it is essential to note that the previous literature weighed more on analysing SDGs and MNBs from developed economies such as Europe and Japan and MNBs from mainly Europe present in different developing countries, that is, North-South financial flow. There is a severe scarcity of studies investigating South-South financial flows of multinational banks about SDGs in Africa. Therefore, we intend to answer the following main research questions:

RQ1. What sustainability strategies are Multinational Moroccan banks implementing?
RQ2. What are the main motives for undertaking those strategies?
RQ3. How are these strategies adding (fastening) or weakening (delaying) the attainment of SDGs in the WAEMU region?
RQ4. What is the role of legitimisation and internationalisation performance in achieving these objectives?

These findings are relevant because knowledge about MNBs' sustainability strategies allows for identifying potential for improvement with a clear and robust social impact. Specifically, thoroughly comprehending such strategies can benefit stakeholders concerned with environmental and social issues, including sustainability managers, bank product managers, policymakers, and environmental organisations. Such knowledge clarifies where strategic action can be taken to move banks towards contributing to SDGs, especially in Africa, where the demand is still high.

2 Literature Review

2.1 Legitimisation and Communication and SDGs Adoption

Legitimacy Theory (LT) refers to the perceived or actual justification and validity of something, such as a government, authority, or action. It is the quality of being accepted as rightful or lawful (Bitektine & Haack, 2015). In the context of government, legitimisation is associated with the consent of the governed and adherence to established laws and principles. There are a variety of frameworks for analysing the way discourses construct legitimacy for social practices in public communication and the everyday interaction of firms (Vaara & Tienar, 2008). Vaara and Tienar (2008) identified the main categories of legitimisation as authorisation, that is, legitimisation by referencing the authority, law/tradition and people in whom institutional authority is vested. 'Moral evaluation' is legitimisation through referencing to discourses of value, 'rationalisation' is legitimisation through referencing to goals and the use of standardised social actions, and to the social knowledge that awards them (firms) with cognitive validity and mythopoesis, which is legitimisation conveyed via narrative whose results/outcomes award legitimate actions and punish un-legitimate ones (Vaara & Tienari, 2011; Van Leeuwen, 2007).

MNEs are constantly looking to gain legitimacy in the host countries by creating confidence and trust among stakeholders (García-Álvarez de Perea et al., 2019) due to the liability of newness and foreignness faced, especially by EMNEs (Kim & Mudambi, 2020; Klossek et al., 2012; Marano et al., 2017). This way, CSR is also used by MNEs that wish to establish legitimacy and gain a public perception of being 'a good corporate citizen' (Arvidsson, 2010; Khan et al., 2013), improve competitiveness and boost brand awareness (Scharf & Fernandes, 2013), balance corporate profits with society's prosperity and build an excellent corporate image in the host countries (Ebo Hinson, 2011). As a result, social responsibility is reported to constitute the third largest budgetary item of business entities' corporate communications (Hutton et al., 2001). This is not a surprise since maintaining legitimacy is critical for the survival and success of organisations. Despite digital transformation and sustainable business models, firms still need to gain legitimacy to get accepted, not only socially but also by political legitimacy. MNEs are constantly moving to gain legitimacy. At the same time, governments, specifically in less developed and emerging markets economies, are becoming more dependent on MNEs (Gomez-Trujillo & Gonzalez-Perez, 2020). However, it is unclear if MNEs can leverage the political legitimacy to adopt more SDGs and hence lead to sustainable development in emerging and developing countries by solving problems such as poverty reduction, reducing inequality, and curbing climate issues (Ghauri, 2022; Park & Ghauri, 2015).

2.2 Managing Legitimacy in the Context of SDGs

Legitimacy can be defined as a generalised perception or assumption that the actions of a firm/corporation or entity are proper, desirable, and appropriate within a specific social context and system of beliefs, values, norms and definitions (Schmidtke & Lenz, 2023; Suchman, 1995; Zhang et al., 2022) Companies communicate with stakeholders to manage their legitimacy (Deegan, 2019). Reporting offers a one-way communication channel from a corporation to its stakeholders. The research found that legitimacy concerns drive sustainability and embed companies into social and environmental reporting (Deegan, 2019; Hahn & Kuhnen, 2013). Legitimacy theory has been used for about a decade to underline why firms report environmental and social (Deegan, 2002). According to Fiaschi et al. (2017), speech and press freedom in the host countries have a significant influence on emerging markets MNEs' corporate social irresponsibility (CSIR) in a sense that firms are less likely to harm when investing in countries with solid speech and press freedoms (Fiaschi et al., 2017). Therefore, due to the sensitive nature of the activity, MNBs are subject to a high level of scrutiny from host countries' local socio-political actors, especially the regulators and the media (Fiaschi et al., 2017). The pressure for companies to contribute to the SDGs has been built up by persistent stakeholders, for instance, media, governments, and shareholders, and the pressure to comply with host countries' regulations (Deegan, 2019) and such pressure and expectations are likely to threat the firms' legitimacy if they are not addressed (Van Zanten & Van Tulder, 2018). Using a sample of 86 listed MNEs, Duque-Grisales et al. (2020) found that adopting sustainability initiatives is conditioned by institutional pressures to meet stakeholder expectations.

Sarfraz et al. (2018) state that internationalised banks, i.e., MNBs, are more motivated to undertake CSR initiatives than domestic banks. Whether out of symbolic legitimacy seeking or substantive motive, the SDGs adoption and disclosure remain a question that is not (yet) thoroughly answered. Companies' commitment to SDGs, especially with SDG 12, was a mixture of symbolic and substantive legitimisation strategies in companies that frequently adopt sustainability practices to gain legitimacy and a positive reputation (Fiaschi et al., 2017).

EMNEs are more likely to face the liabilities of origin, that is, the negative perceptions in host countries about these firms' willingness and ability to build legitimate businesses (Kim & Mudambi, 2020; Marano et al., 2017). Several studies have identified the negative MNEs' liability of foreignness consequences, including the loss of first-mover advantage (Edman, 2016) and reputational damage (Bianchi & Ostale, 2006; Zeng & Glaister, 2016) in the host country. Hence, from a legitimacy perspective, companies can overcome the liability of foreignness by improving their legitimacy among stakeholders in host countries (Luo, 2002). This legitimacy can be enhanced by being responsive and giving back to society (Wood, 1991). By implementing CSR and sustainability initiatives, companies can strengthen their acceptance among stakeholders in host countries and overcome the liability of foreignness (Caussat et al., 2019). Maruyama and Wu (2015) consider that firms try to gain a reputation and overcome the liability of foreignness by improving moral legitimacy attained by conducting CSR activities. In the same line, Silva (2021) indicated that two-thirds of companies tend to address the SDGs by mapping the SDGs to existing activities to legitimise their contributions. On the other hand, MNEs are keener in considering their negative externalities in host countries and tend to actively address SDGs in their business strategies contrary to addressing SDGs following stakeholders' pressure and liability of foreignness.

Firms that observed regulations, which have a higher employee relations level and are more social and community involvement, were more likely to have higher levels of environmental practice, including observing SDGs. CSR strategy can be linked with the SDGs when stakeholder theory and creating a shared value are two observed goals. Sustainability is an antecedent of corporate reputation and a tool to enhance stakeholders' acceptance and perceptions of companies' activities (Gomez-Trujillo & Gonzalez-Perez, 2020). Numerous studies have indicated that SDG involvement by MNEs is inspired by a mixture of legitimacy and institutional motives (Van der Waal & Thijssens, 2020).

Calabrese et al. (2022) highlighted that MNEs' adoption and general contribution to the SDGs is still symbolic and mostly a matter of legitimacy-seeking. They further indicated that firms are more likely to contribute to the SDGs related to environmental and economic targets. Therefore, MNEs with different business strategies implemented SDGs, focusing more on SDGs 5, 8, 9, 12 and 13. Moreover, MNEs are more engaged with SDG targets that are actionable and relevant to their business model (value chain) and more engaged on SDG targets that 'avoid harm' rather than those that 'do good' (Hult et al., 2018; Van Zanten & Van Tulder, 2018) the MNEs' international legitimacy is at risk if they harm the conduct of their business (Fiaschi et al., 2017). Therefore, although there are diverse studies on corporate strategy legitimisation, sustainability legitimisation and SDGs implementation, it is crucial to assess whether legitimisation by MNEs, specifically MNBs from emerging market economies, has a positive influence on the adoption and implementation of SDGs in the host country or not.

Hypothesis 1. *Legitimisation in discourse and communication positively affects the adoption and implementation of SDGs by AMBs.*

2.3 Internationalisation Performance and SDG Adoption

According to Boehe (2016), EMNEs can effectively internalise in the presence of firm and country-specific disadvantages by domestic-seeking internalisation, that is, by capturing and servicing foreign domestic markets (curbing market demand, adopting and local partnering). MNEs are constantly sensing and seizing internalisation opportunities and developing organisational flexibility by legitimising, leveraging, and launching strategies (Grøgaard et al., 2022). The motivations for the internationalisation of enterprises depend on their scales, industry type and ownership (Dunning & Lundan, 2008). According to Park and Ungson (2019), overcoming institutional voids, deepening localisation, creating inclusive markets, and fostering government support and a higher international network were the strategies for building high internalisation performance and sustained legacy. Gomez-Trujillo and Gonzalez-Perez (2020) highlighted common characteristics in studying internationalisation and sustainability in emerging markets. Therefore, it is the impetus to suggest a framework of research that considers internationalisation as a driver for the pursuit of SDGs and sustainability initiatives. Although MNEs are essential in attaining the SDGs, the roles and goals of MNEs' subsidiaries are linked to the headquarters or parent companies. No single accountability framework enables internalised companies to measure, assess and report their contribution to the SDGs (Ghosh & Rajan, 2019).

The internationalisation performance of MNEs and implementation of SDGs nexus still needs to be clarified in the extant body of literature. According to Shen et al. (2023), internationalisation experience and performance are significantly related to (and had a positive impact on) SDG activities and corporate social responsibility, and firms that employ a higher tier of environmental strategy portray a higher degree of internationalisation. Chen et al. (2016) indicated that internalisation performance is associated with the MNEs' higher engagement (adoption of SDGs). This study highlighted that MNEs' engagement with SDGs is due to stakeholders' pressure and the need to build legitimacy and overcome the liability of foreignness. The positive relationship between MNEs' internalisation and SDG engagement was more robust in counties with higher SDG index scores and poor institutional quality. At the same time, the relationship was weaker in firms facing high competition and more externalities. Gomez-Trujillo and Gonzalez-Perez (2020) highlighted a positive linkage between SDG adoption and internalisation in the sense that the adoption of SDGs and sustainability strategies enables EMNEs subsidiaries to develop internalisation standards and competitive advantages, address liabilities in foreignness and gain legitimacy among stakeholders in the face of internal and external pressure.

Duque-Grisales et al. (2020) revealed that MNEs with more proactive environmental strategy exhibit higher levels of internationalisation; hence, firms that implement advanced environmental and sustainability strategies portray companies' performance in sustainability matters and the implementation of SDGs demonstrate more significant sustainability commitment become more recognised and visible in foreign markets, leading to their easy expansion in countries with different institutional environments. However, Galeazzo et al. (2023) find that contrary to these results, institutional pressures associated with the country of origin and industry type positively and significantly impacted companies' engagement with the SDGs. Still, internationalisation does not considerably relate to companies' performance in sustainability matters and the implementation of SDGs. This aligns with prior research, and CSR initiatives are unrelated to competitiveness and internalisation performance (Marin et al., 2017). Thus, globalising subsidiaries and return on assets are essential in defining internalisation performance. However, the connectivity between internalisation performance and the adoption of SDGs by MNEs was not disclosed. Therefore, the second hypothesis is based on internalisation performance and adoption of SDGs by African multinational banks (AMBs).

Hypothesis 2. *Internationalisation Performance positively affects the adoption of SDGs by AMBs.*

2.4 Strategy and SDG Adoption

According to Miles et al. (1978), businesses fall into four strategies: defender, prospector, analyser and reactor. Where the defender, prospector, and analyser are categorised under a proactive strategy. Defender strategy is rooted in creating stable products/services and customers, extensive research, cautiousness about new opportunities, observing efficiency and cost, organisation efficiency, single niche robust technology, less flexibility, and routinisation. Prospectors are first movers and tend to take advantage of opportunities immediately as they arise. Therefore, it is rooted in exploiting new markets and product opportunities, growth through product and market development, aggressive investments, monitoring of wide environmental events and conditions, multiple technologies, and a more flexible, less routinisation degree.

On the other hand, analysts are concerned with locating and exploiting new market and product opportunities while maintaining a firm base of traditional products and customers. The strategy is a defender and prospector hybrid, maintaining some research and development, firm growth through product development and market penetration with stable and flexible technology. A reactive strategy exists in unstable organisations/industries where the former three strategies are improperly pursued. Tend to react or respond to patterns of the environment in sometimes unstable and inconsistent ways. Maintaining the current strategy-structure relationship for too long despite demanding and overwhelming changes in environmental conditions and lack of clear strategy and structure can cause companies to be reactors (Miles et al., 1978).

There has been an ongoing debate on how firms using these strategies approach sustainability. Luo (2002) indicated that the mechanisms that can overcome liabilities of foreignness might include a defensive strategy, with defensive mechanisms such as contract protection, parental control, output standardisation and parental service. According to Chen et al. (2016), firms that deploy proactive strategies correlate with greater internalisation. Several studies suggest that reactive strategies suit the needs of MNCs trying to learn from local contexts. Acquier et al. (2018), by combining global templates with locally mandated practices, find that firms using reactive strategies might succeed in achieving survival (Tsui-Auch & Chow, 2019), financial solid performance (Peng et al., 2008) and competitive advantage over local companies in host countries. Proactive firms are more engaged in innovative leadership, customer engagement, and market shaping in the countries of presence (Brege & Kindstrom, 2020). Proactive strategies significantly drive firms' performance

(Brege & Kindstrom, 2020). MNEs that employ advanced proactive environmental strategy will likely achieve higher internalisation performance in foreign markets (Duque-Grisales et al., 2020) as proactive sustainability strategy is positively associated with corporations' sustainability performance and control systems (Potrich et al., 2019). Although both reactive and proactive strategies minimise costs. Proactive strategies are crucial for survival during crises, as they foster dynamic capabilities that improve a firm's competitive advantage.

Talbot and Ordonez-Ponce (2022) indicated that banks tend to apply more proactive strategies and a solid commitment to stakeholders during times of crisis, such as the COVID-19 crisis. Furthermore, the proactive strategy positively impacted firms' green innovation practices, financial performance, and adoption of SDGs (Khan et al., 2021). While companies taking more proactive measures exhibit more socially responsible investments, reactive environmental practices negatively impact socially responsible investments. Similarly, proactive environmental strategies have often benefited MNEs and advanced their local aspirations. According to Carballo-Penela and Castroman-Diz (2015), firms that adopted proactive environmental strategies positively impacted managers' attitudes and initiatives toward sustainable development and firms' performance in the short run. A proactive environment strategy precedes green process innovation, mediates the positive relationship between green organisational culture (environmental capabilities) and green product innovation performance (competitive), and facilitates corporations' adoption of environmental sustainability initiatives that are different from the reactive strategy (Chang, 2015). A combination of CSR and innovative strategies if companies are to improve their financial performance while being socially responsible. On the other hand, proactive environmental strategy, a proactive environmental strategy, has a positive nonlinear moderating impact on stakeholders' pressure (legitimacy), including media, regulators and shareholders, and proactive environmental strategies do not directly influence or affect corporations' internalisation performance.

Although there is no apparent connection in the extant literature on the impact of corporations' strategies and Sustainability, specifically SDGs adoption, by MNEs, there is always a motive in which firms are motivated to SDGs implementation (Forcadell et al., 2020) extrinsic, self-driven or instrumental motives represent the potential improvement of legitimacy, competitiveness, efficiency gains, and innovativeness that a firm aims to realise through a specific sustainable behaviour and are connected with reactive strategies (Den Hond et al., 2015; Forcadell et al., 2020; Nguyen et al., 2021). In contrast, intrinsic motives are associated with ambiguous proactive strategies (Gabaldon & Groschl, 2015). Unlike other studies that argued that MNCs scale up self-legitimisation to defend their organisations' legitimacy considering public criticism, Schmidtke and Lenz (2023) used 28 regional MNCs from 1980 to 2019, demonstrated that MNEs strengthen self-legitimisation to mobilise additional support from relevant audiences and self-legitimisation aims primarily to achieve proactive legitimacy expansion instead of reactive legitimacy protection. Despite the strategic and operational dimensions, Linnerud et al. (2021) identified no formula for adopting the SDGs. Furthermore, motivations underlying corporate involvement with the 2030 agenda are not determined by the management level of focus on SDG integration.

Siltaloppi et al. (2021) analysed how management practices lead organisations to navigate CSR tensions in defensive and active ways. Results showed that, as opposed to a defensive strategy, reactive firms navigate into CSR easily. They added that the strategic sensemaking that integrates CSR into the corporation's long-term visions and the quest for competitive advantage is much considered. Businesses must incorporate SDGs in their business model rather than adopting them as an externality. This would involve research and rooted integration (proactive strategies).

Nevertheless, Jimenez et al. (2021) claimed that many businesses have not shifted from the 'business as usual' notion where sustainability is an externality (done for quick advantage reaping-reactive). Teh and Corbitt (2015) assessed what influences the extent of a large company's eco-sustainability and revealed the relevance of the company's 'hard' and 'soft' strategies (proactive and reactive, respectively) playing the primary role. However, the paper did not state which approach is likely to adopt eco/environmental sustainability initiatives actively.

Tsai and Liao (2017) revealed that corporations are more likely to adopt a proactive environmental strategy to improve eco-innovations but under high market demand. Heras-Saizarbitoria et al. (2022) indicated an SDGs-washing motive for the enormous multinational organisations adopting SDGs. Despite revealing the motive, the study was silent about the strategies behind it. Some studies suggested that diverse factors dictate the adoption of SDGs. For instance, corporate purpose drives MNEs' adaptation and implementation of SDGs, precisely the goal-based and duty-based purpose. These findings align with Yu et al. (2023), whose study identified that corporations tend to focus on or adopt certain aspects of the SDGs at the cost of others, depending on the firm's goals. Having observed the stick debate on corporate strategies (proactive and reactive strategies), the following hypothesis is depicted.

Hypothesis 3. *A more proactive strategy positively affects AMBs' adoption of SDGs.*

3 Materials and Methods

Preliminary data for this study was collected from CEO and presidents' discursive typology, as well as sustainability reports for three Moroccan Multinational Banks, namely the Attijariwafa Bank, the Banque Centrale Populaire and the Bank of Africa (MBCE group). These banks are AMNBs since they have FDI in other African countries, particularly North Africa, Central Africa, and West Africa. Corporate and sustainability disclosure, analysis of annual reports and narratives in the yearly reports, SCR reporting through annual reports, and sustainability reports have been a common way of assessing the firms' strategies, involvement, behaviours and contributions to sustainability and SDGs as well as their ways of legitimation seeking in the host countries and internalisation performance among others (Cosma et al., 2020; Deegan, 2002). Thus, three types of documents investigated were the annual reports, CEO and president letters, and sustainability reports. These sources of evidence were downloaded from the banks' official websites.

4 Results

Since the announcement of SDGs by the United Nations, MNEs' sustainability has caught great attention in IBR (Gomez-Trujillo & Gonzalez-Perez, 2020; Kolk, 2016; Sachs & Sachs, 2021; Van Tulder et al., 2021; Zhan & Santos-Paulino, 2021). Here, we join this discussion by proposing an explanatory model that considers the influence of AMBs' strategies, internalisation performance and legitimisation on SDGs adoption. Using a deep analysis of three case studies, our findings highlight the positive effect of legitimisation, internalisation performance, and strategy on the adoption of SDGs by these banks.

4.1 Legitimisation in Discourse and Communication Positively Affects the Adoption and Implementation of SDGs by AMBs

Table 1 summarises the results of our study, which strongly support all hypotheses. Regarding hypothesis 1, claiming that AMBs consistently integrate sustainability and CSR into their public communications as a legitimisation strategy, Attijariwafa Bank (AWB) positions CSR as a collective consciousness embedded in its corporate strategy, reinforced by its inclusion in the Casablanca ESG 10 CSR index and South-South cooperation activities. Similarly, Banque Centrale Populaire (BCP) highlights its alignment with international standards through its involvement with the International Finance Corporation (IFC) and the Environment, Social, and Governance (ESG) index, focusing on corporate citizenship and environmental stewardship. BMCE Bank of Africa further emphasises legitimacy by aligning its actions with internationally recognised frameworks such as the Equator Principles and ISO 14001 certification, with commitments to initiatives like the UN's Principles for Responsible Banking and Women's Empowerment Principles. Across all cases, the banks' discourses promote legitimacy by showcasing their alignment with the SDGs through sustainability reports, CSR commitments, and global partnerships.

4.2 Internationalisation Performance Positively Affects the Adoption of SDGs by AMBs

The findings also provide robust support for Hypothesis 2, indicating that internationalisation performance enhances AMBs' commitment to SDGs. AWB's pan-African expansion explicitly links its presence in multiple African markets with goals such as financial inclusion, support for small and medium-sized enterprises (SMEs), and infrastructure development, all key components of the SDGs. BCP's

Table 1 Evidence from annual reports, CEO (or president) letters and sustainability reports

Source of evidence	Impact of evidence	Attijariwafa Bank	Banque Centrale Populaire	BMCE Bank of Africa
Annual reports	**Hypothesis 1—Legitimisation in discourse and communication positively affects the adoption and implementation of SDGs by AMBs**	Evidence: AWB explicitly integrates CSR and sustainability into its communication, reinforcing its role as a socially responsible corporation. The bank consistently emphasises corporate citizenship, ethics, and sustainability, for example: The integration of CSR in the bank's strategy goes beyond mere managerial responsibilities and is positioned as a "collective consciousness" that guides AWB's practices. This suggests a legitimisation strategy that aligns with the values of the SDGs AWB's inclusion in the Casablanca ESG 10 CSR index (2018) and its recognition as a player in South-South cooperation reflect an attempt to gain legitimacy in the market by showcasing its alignment with SDGs The emphasis on education, environmental stewardship, and financial inclusion reinforces a discourse that legitimises its actions by aligning them with broader societal goals, including several SDGs	Evidence: The reports show that BCP and its leadership have significantly emphasised legitimising their activities through public statements, partnerships, and international cooperation, for example: The 2012 report mentions BCP's involvement with the International Finance Corporation (IFC), a World Bank subsidiary, highlighting an effort to legitimise its operations by aligning with well-established international financial institutions The 2018 discourse focuses on social and environmental responsibility, where BCP integrates into the "Environment, Social, and Governance (ESG)" index, thus signalling its alignment with internationally recognised sustainability frameworks Emphasis on corporate citizenship, with commitments to education, environmental protection, and inclusivity, is an apparent effort to gain legitimacy in the eyes of stakeholders by promoting values central to the SDGs	Evidence: BMCE Bank emphasises sustainable finance and a strong CSR commitment. For instance, the 2013 report highlights the bank's adherence to the Equator Principles and ISO 14001 certification for environmental standards. These initiatives reflect BMCE's effort to align its operations with internationally recognised frameworks, which bolster its legitimacy and align with the SDGs The African Entrepreneurship Award and the focus on financial inclusion, education, and environmental protection (as noted in 2015 and subsequent years) show BMCE's commitment to sustainable development as a core element of its legitimacy BMCE's participation in global sustainability initiatives, such as the UN's Principles for Responsible Banking (2019) and Women's Empowerment Principles (2020), further establishes its role as a legitimate advocate for SDG-related principles

Table 1 (continued)

Source of evidence	Impact of evidence	Attijariwafa Bank	Banque Centrale Populaire	BMCE Bank of Africa
	Hypothesis 2—Internationalisation performance positively affects the adoption of SDGs by AMBs	Evidence: AWB's pan-African expansion is a primary strategic focus, and its operations across multiple African countries align with sustainable economic development. The bank explicitly links its expansion in Africa with goals such as financial inclusion, support for SMEs, and infrastructure development, which are closely related to SDGs such as poverty reduction, decent work, and economic growth AWB's international presence, especially in African markets, positions it as a catalyst for regional integration and South-South cooperation. It further reinforces the bank's alignment with SDGs, such as partnerships for the goals and sustainable cities and communities The strategic focus on sectors like renewable energy (e.g., the financing of wind farms and clean energy projects) highlights how internationalisation is tied to environmentally sustainable goals. This supports the notion that international performance drives the adoption of SDG-related initiatives, particularly in emerging markets	Evidence: Many reports emphasise BCP's international expansion, particularly in Africa and partnerships with global financial institutions: In 2012 and 2015, there is a strong focus on BCP's African expansion (e.g., the acquisition of Banque Atlantique), promoting economic growth and financial inclusion in Sub-Saharan Africa BCP's South-South cooperation initiatives, often mentioned across several years, reinforce their cross-border engagement and contribute to sustainable economic development in emerging markets, aligning with SDGs like poverty reduction and financial inclusion The internationalisation strategy also supports microfinance efforts (e.g., Atlantic Microfinance for Africa) to promote inclusivity and economic growth, which are central to SDGs	Evidence: BMCE's international expansion, especially its operations in Sub-Saharan Africa through the Bank of Africa, has consistently been a core strategic focus. The reports from 2013 to 2020 emphasise the bank's growth across the African continent and into regions like Europe and Asia (e.g., the Shanghai branch in 2018) The bank's commitment to inclusive growth and sustainable finance is tightly linked with its African expansion, as seen in the 2014 and 2015 reports, in which BMCE extended its footprint in various African countries to promote entrepreneurship and financial inclusion BMCE's involvement in green bonds (2016) and renewable energy projects (2019) showcases how its international presence drives SDG-related initiatives, particularly in environmental sustainability

	Hypothesis 3—A more proactive strategy positively affects AMBs' adoption of SDGs	Evidence: AWB adopts a proactive strategy by committing to financial inclusion and offering innovative solutions, such as the "Hissab Bikhir" concept for low-income households. This proactive stance aligns with the SDGs, which focus on reducing inequalities and ensuring financial inclusion The bank's initiatives for youth entrepreneurship and small business support, mainly through programs like the "Entrepreneurship Hub" and Dar Al Moukawil centres, reflect a proactive approach to economic development, which aligns with SDG goals related to decent work and economic growth AWB's proactive participation in climate finance (e.g., involvement in COP22 and supporting the transition to a low-carbon economy) shows how its forward-thinking strategy is directly tied to environmental SDGs The South-South cooperation awards and participation in the International Africa Development Forum underscore AWB's proactive role in promoting regional cooperation and shared value creation, components of the SDGs	Evidence: There is a consistent reference to proactive strategies in the reports, for example: The 2016 and 2017 reports detail BCP's proactive approach to expanding financial services to underserved markets in Sub-Saharan Africa. They position the bank as a leader in financial inclusion and corporate social responsibility, directly related to several SDGs In 2016, BCP's focus on extending microfinance services and opening subsidiaries like AMIFA for financial inclusion demonstrated a proactive stance on adopting SDG-related goals, particularly those targeting poverty reduction and economic inclusivity Prominently featured in 2017 and 2018 reports, corporate citizenship and sustainability initiatives reflect a proactive approach toward social and environmental responsibility (key pillars of SDGs)	Evidence: BMCE adopts a proactive strategy in promoting SDG-related goals through innovative initiatives like the African Entrepreneurship Award (launched in 2014), which supports projects focused on education, the environment, and innovation. This proactive approach aligns with SDGs related to quality education, decent work, and economic growth The bank's proactive stance in sustainable finance is evident in its leadership role in Positive Impact Finance (2017) and its issuance of green bonds to finance eco-responsible projects (2016). These initiatives highlight BMCE's proactive measures toward environmental sustainability, a key component of SDGs BMCE's involvement in climate finance and its participation in initiatives like the Mainstreaming Climate Action Initiative (2018) demonstrate its proactive role in addressing climate change (SDG 13)

(continued)

Table 1 (continued)

Source of evidence	Impact of evidence	Attijariwafa Bank	Banque Centrale Populaire	BMCE Bank of Africa
CEO & president letters	Hypothesis 1—Legitimisation in discourse and communication positively affects the adoption and implementation of SDGs by AMBs	All letters emphasise CSR, climate action, and digital transformation, aligning with the bank's discourse on sustainability. This indicates a positive legitimisation effect on their efforts towards SDGs.	Several letters emphasise integrating sustainability efforts, including climate finance, CSR, and supporting local economies, suggesting that communication positively influences the SDGs	Many letters emphasise BMCE Bank's commitment to CSR, sustainable finance, and African development. These themes are repeated throughout their discourse, particularly concerning their contributions to COP22, their promotion of African entrepreneurship, and their CSR awards
	Hypothesis 2—Internationalisation performance positively affects the adoption of SDGs by AMBs	Their international expansion and focus on African growth, especially in Egypt, demonstrate that internationalisation supports broader SDG commitments	The reports highlight expanding international activities, particularly in sub-Saharan Africa and beyond, showcasing the role of global growth in facilitating SDG efforts	The bank's strategy focuses heavily on international expansion, especially in sub-Saharan Africa and Europe. Their global reach, mainly through acquisitions like the Bank of Africa, underscores their contribution to SDG-related initiatives across borders
	Hypothesis 3—A more proactive strategy positively affects AMBs' adoption of SDGs	Initiatives like Energies 2020 and the adoption of green finance highlight a proactive approach to supporting SDG adoption.	The proactive implementation of green finance initiatives and strategic growth plans, such as Energies 2020, supports the hypothesis that proactive strategies positively affect SDG adoption	BMCE Bank's proactive strategies, such as the "Energies 2020" plan and its investments in green finance, microfinance, and inclusive banking, demonstrate an approach to integrating SDG goals into its core operation

Sustainability reports	Hypothesis 1—Legitimisation in discourse and communication positively affects the adoption and implementation of SDGs by AMBs	The CSR reports emphasise sustainable finance, green growth, and social responsibility as core themes, suggesting that legitimisation through communication is a significant driver in promoting SDGs	The reports clearly emphasise sustainability and corporate responsibility in public communication, showcasing an alignment with SDG goals. Focusing on financial inclusion and green financing suggests legitimisation positively influences SDG adoption	The reports emphasise the communication of SDG commitments, especially in sustainability initiatives and partnerships with global organisations like UNEP
	Hypothesis 2—Internationalisation performance positively affects the adoption of SDGs by AMBs	Attijariwafa Bank's expansion into Africa and other international markets integrates SDG goals, particularly in regions where the bank supports renewable energy projects and entrepreneurship	Internationalisation Performance positively affects the adoption of SDGs by AMBs: The bank's international expansion in Africa and involvement in sustainable projects in these regions, including renewable energy, highlight how internationalisation supports SDG goals	The bank's international expansion, especially across Africa, aligns with its SDG efforts, particularly in financing green projects
	Hypothesis 3—A more proactive strategy positively affects AMBs' adoption of SDGs	The bank's proactive strategies, such as climate finance, renewable energy financing, and programs like the "Dar Al Moukawil" for SMEs, highlight its alignment with SDG adoption (AWB CSR Report 2019)	A more proactive strategy positively affects AMBs' adoption of SDGs: The proactive launch of green bonds and other sustainability initiatives supports this hypothesis, showing the bank's strategic efforts in integrating SDGs into its operations	The proactive strategies, such as green bond issuance and impact finance initiatives, directly contribute to SDG adoption

Source: Authors' elaboration with Banks' Annual Reports, CEO (or President) Letters and Sustainability Reports

international expansion, particularly in Africa through the acquisition of Banque Atlantique, demonstrates a strong focus on financial inclusion and poverty reduction, aligning with key SDGs. BMCE's international operations, especially in Sub-Saharan Africa, reinforce its commitment to SDGs by driving sustainable finance initiatives, such as green bonds and renewable energy projects. Across all banks, internationalisation is closely tied to broader SDG-related objectives, including sustainable economic growth, regional integration, and environmental sustainability.

4.3 A More Proactive Strategy Positively Affects AMBs' Adoption of SDGs

The results also support Hypothesis 3, highlighting how proactive strategies contribute to the integration of SDG-related goals within AMBs. AWB adopts a proactive stance by promoting financial inclusion through innovative solutions like the 'Hissab Bikhir' program for low-income households and youth entrepreneurship initiatives like the 'Entrepreneurship Hub'. Its leadership role in climate finance further demonstrates a proactive alignment with environmental SDGs. BCP shows similar proactive strategies, expanding microfinance services and opening subsidiaries like AMIFA to drive financial inclusion and poverty reduction. The bank's corporate citizenship and sustainability leadership, particularly in Sub-Saharan Africa, is another element of its proactive approach to SDG adoption. BMCE's proactive measures, such as the African Entrepreneurship Award and green finance initiatives, also align with the SDGs, promoting entrepreneurship, environmental sustainability, and inclusive growth. These proactive strategies demonstrate a forward-looking approach by the banks to align their operations with SDG goals.

5 Discussion

This study investigated the sustainability strategies of a set of AMNBs, their motives, and the influence of legitimisation and internationalisation performance on adopting SDGs. The findings from the analysis of the three Moroccan multinational banks—Attijariwafa Bank, Banque Centrale Populaire, and BMCE Bank of Africa—demonstrate clear connections between these banks' strategies and the SDGs. Our primary research objective (RO1) was to identify the sustainability strategies implemented by AMNBs and their impact on SDG adoption. Our findings show that these banks have incorporated primarily proactive strategies to align their operations with SDGs, focusing on financial inclusion, green finance, and regional economic integration. This confirms the hypothesis that AMNBs increasingly embed sustainability in their corporate strategies. Moreover, in line with our second objective (RO2), we examined the motives behind these strategies, revealing that legitimisation and internationalisation performance are significant drivers. As hypothesised (Hypothesis 1 and Hypothesis 2), legitimisation through discourse and communication positively influences SDG adoption, while internationalisation performance fosters the integration of SDG-related initiatives in foreign markets.

The novelty of this research lies in the African context of multinational banks. Previous studies have focused primarily on multinational enterprises (MNEs) from developed economies, leaving the financial sector in emerging markets underexplored (Barnard et al., 2023; Mohieldin et al., 2022a, b). Our results differ from earlier research on MNEs by emphasising the role of AMNBs in addressing sustainability challenges, particularly in underdeveloped regions such as Sub-Saharan Africa. For example, while Gomez-Trujillo and Gonzalez-Perez (2020) emphasised sustainability reporting as a legitimisation strategy in developed markets, our findings suggest that AMNBs employ proactive strategies like financial inclusion and green bonds to align with SDGs in host countries, contributing to regional development (Van Zanten & Van Tulder, 2018).

Theoretically, this research contributes to the discourse on sustainability and legitimacy within the framework of international business by highlighting how African banks integrate sustainability into their business models. Our findings support the legitimacy theory, demonstrating that AMNBs use SDGs to gain acceptance in host countries by aligning their strategies with social and environmental goals. This extends previous work on MNEs and legitimacy (Borda et al., 2017; Kim & Mudambi, 2020) by focusing on African banks' role in the global sustainability agenda.

This research has important practical implications for policymakers and business leaders. AMNBs should consider intensifying their efforts toward green finance and financial inclusion to strengthen their contribution to the SDGs. The findings also suggest that proactive strategies, such as developing SME-support programs and climate finance initiatives, are essential for achieving long-term sustainability objectives. Policymakers can foster regional collaboration by encouraging cross-border cooperation between AMBs.

This study has limitations that should be acknowledged. First, the focus on three Moroccan banks limits the generalizability of the findings to other regions in Africa or other sectors. Second, the reliance on publicly available sustainability reports and corporate disclosures may lead to an incomplete understanding of the strategies employed by these banks, particularly those not explicitly mentioned in public documentation. Additionally, while the study provides insights into the relationship between internationalisation performance and SDG adoption, the causality remains to be determined due to the cross-sectional nature of the data.

Future research should expand the sample to include a broader range of AMBs and other financial institutions operating in different regions. A longitudinal approach could provide a deeper understanding of how sustainability strategies evolve and their long-term impact on SDG attainment. Additionally, qualitative research, including interviews with key decision-makers in AMNBs, could offer further insights into the motives behind sustainability strategies. Future studies should also explore the role of digital transformation in enhancing sustainability performance within African banks, particularly concerning SDG 9 (Industry, Innovation, and Infrastructure).

6 Conclusion

This study advances our understanding of the strategies employed by AMNBs to align with the SDGs and highlights the significance of legitimisation and internationalisation performance in driving sustainability efforts. The findings underscore how MNBs' legitimisation strategies, proactive environmental and financial inclusion initiatives, and their internationalisation performance positively influence their alignment with and adoption of SDG-related goals. The banks examined, such as Attijariwafa Bank, Banque Centrale Populaire, and BMCE Bank of Africa, demonstrate an increasing commitment to sustainability through innovative programs that target financial inclusion, climate finance, and entrepreneurship. This indicates a growing trend among AMNBs to embed sustainability within their core operations.

However, the study acknowledges several limitations. One notable limitation is the focus on only three banks from a single country, which may limit the generalizability of the findings to other regions or financial institutions. Additionally, the reliance on secondary data from sustainability reports and CEO communications presents potential biases, as these reports often emphasise positive aspects of corporate performance. The study also does not explore the actual outcomes of these banks' SDG efforts, leaving room for further investigation into the tangible impacts of these strategies on the broader social and environmental landscape.

For future research, there is a clear need to explore the sustainability strategies of MNBs from other African countries and emerging markets to assess the long-term impact of their SDG efforts. This will provide a more comprehensive understanding of the role of MNBs in achieving SDGs. Policy implications include the necessity for governments and regulatory bodies to incentivize banks to prioritise sustainable finance, offering frameworks that facilitate the integration of SDGs. For practitioners, particularly sustainability managers and financial institutions, the study suggests the importance of developing proactive strategies that not only comply with international sustainability standards but also foster innovation in areas such as financial inclusion and green finance, crucial for addressing pressing global challenges.

References

Acquier, A., Carbone, V., & Moatti, V. (2018). "Teaching the sushi chef": Hybridization work and CSR integration in a Japanese multinational company. *Journal of Business Ethics, 148*(3), 625–645.

Aguilera, R. V., Ciravegna, L., Cuervo-Cazurra, A., & Gonzalez-Perez, M. A. (2017). Multilatinas and the internationalization of Latin American firms. *Journal of World Business, 52*(4), 447–460.

Amran, A., Ooi, S. K., Wong, C. Y., & Hashim, F. (2016). Business strategy for climate change: An ASEAN perspective. *Corporate Social Responsibility and Environmental Management, 23*(4), 213–227.
Arita, S. (2013). Do emerging multinational enterprises possess South-South FDI advantages? *International Journal of Emerging Markets, 8*(4), 329–353.
Arvidsson, S. (2010). Communication of corporate social responsibility: A study of the views of management teams in large companies. *Journal of Business Ethics, 96*(3), 339–354.
Barbaglia, M., Bianchini, R., Buttice, V., Elia, S., & Mariani, M. M. (2023). The role of environmental sustainability in the relocation choices of MNEs: Back to the home country or welcome in a new host country? *Journal of International Management, 29*(5), 101059.
Barnard, H. (2019). From the editor: The social side of international business policy – Mapping social entrepreneurship in South Africa. *Journal of International Business Policy, 2*(1), 1–8.
Barnard, H., Amaeshi, K., & Vaaler, P. M. (2023). Theorizing international business in Africa: A roadmap. *Journal of International Business Policy, 6*(4), 389–407.
Berning, S. C. (2019). The role of multinational enterprises in achieving sustainable development: The case of Huawei. *European Journal of Sustainable Development, 8*(3), 194–194.
Bianchi, C., & Ostale, E. (2006). Lessons learned from unsuccessful internationalization attempts: Examples of multinational retailers in Chile. *Journal of Business Research, 59*(1), 140–147.
Bitektine, A., & Haack, P. (2015). The "macro" and the "micro" of legitimacy: Toward a multilevel theory of the legitimacy process. *The Academy of Management Review, 40*(1), 49–75.
Boehe, D. M. (2016). The internationalization of service firms from emerging economies: An internalization perspective. *Long Range Planning, 49*(5), 559–569.
Borda, A., Newburry, W., Teegen, H., Montero, A., Najera-Sanchez, J. J., Forcadell, F., Lama, N., & Quispe, Z. (2017). Looking for a service opening: Building reputation by leveraging international activities and host country context. *Journal of World Business, 52*, 503.
Bose, S., & Khan, H. Z. (2022). Sustainable development goals (SDGs) reporting and the role of country-level institutional factors: An international evidence. *Journal of Cleaner Production, 335*, 130290.
Brege, H., & Kindstrom, D. (2020). Exploring proactive market strategies. *Industrial Marketing Management, 84*, 75–88.
Calabrese, A., Costa, R., Levialdi Ghiron, N., Tiburzi, L., & Villazon Montalvan, R. A. (2022). Is the private sector becoming cleaner? Assessing the firms' contribution to the 2030 Agenda. *Journal of Cleaner Production, 363*, 132324.
Carballo-Penela, A., & Castroman-Diz, J. L. (2015). Environmental policies for sustainable development: An analysis of the drivers of proactive environmental strategies in the service sector. *Business Strategy and the Environment, 24*(8), 802–818.
Castro, G. D. R., Fernandez, M. C., & Colsa, A. U. (2021). Unleashing the convergence amid digitalization and sustainability towards pursuing the Sustainable Development Goals (SDGs): A holistic review. *Journal of Cleaner Production, 280*, 122204.
Caussat, P., Prime, N., & Wilken, R. (2019). How multinational banks in India gain legitimacy: Organisational practices and resources required for implementation. *Management International Review, 59*(4), 561–591.
Chang, C.-H. (2015). Proactive and reactive corporate social responsibility: Antecedent and consequence. *Management Decision, 53*(2), 451–468.
Chen, D. L., Moskowitz, T. J., & Shue, K. (2016). Decision making under the gambler's fallacy: Evidence from asylum judges, loan officers, and baseball umpires. *The Quarterly Journal of Economics, 131*(3), 1181–1243.
Chiu, S.-C., & Sharfman, M. (2011). Legitimacy, visibility, and the antecedents of corporate social performance: An investigation of the instrumental perspective. *Journal of Management, 37*(6), 1558–1585.
Clayton, C., & Schaab, A. (2022). Multinational banks and financial stability. *The Quarterly Journal of Economics, 137*(3), 1681–1736.
Colleoni, E. (2013). CSR communication strategies for organizational legitimacy in social media. *Corporate Communications, 18*(2), 228–248.
Cosma, S., Venturelli, A., Schwizer, P., & Boscia, V. (2020). Sustainable development and European banks: A non-financial disclosure analysis. *Sustainability, 12*(15), 6146.

Cuervo-Cazurra, A., Newburry, W. E., & Park, S. (2020). *Building strategic capabilities in emerging economies* (1st ed.). Cambridge University Press.

Deegan, C. (2002). Introduction: The legitimising effect of social and environmental disclosures–a theoretical foundation. *Accounting, Auditing & Accountability Journal, 15*(3), 282–311.

Deegan, C. M. (2019). Legitimacy theory: Despite its enduring popularity and contribution, time is right for a necessary makeover. *Accounting, Auditing & Accountability Journal, 32*(8), 2307–2329.

Den Hond, F., De Bakker, F., & Doh, J. (2015). What prompts companies to collaboration with NGOs? Recent evidence from The Netherlands. *Business & Society, 54*(2), 187–228.

Dunning, J. H., & Lundan, S. M. (2008). *Multinational enterprises and the global economy* (2nd ed.). Edward Elgar Publishing Limited.

Duque-Grisales, E., & Aguilera-Caracuel, J. (2021). Environmental, social and governance (ESG) scores and financial performance of multilatinas: Moderating effects of geographic international diversification and financial slack. *Journal of Business Ethics, 168*(2), 315–334.

Duque-Grisales, E., Aguilera-Caracuel, J., Guerrero-Villegas, J., & Garcıa-Sanchez, E. (2020). Can proactive environmental strategy improve Multilatinas' level of internationalization? The moderating role of board independence. *Business Strategy and the Environment, 29*(1), 291–305.

Eang, M., Clarke, A., & Ordonez-Ponce, E. (2023). The roles of multinational enterprises in implementing the United Nations Sustainable Development Goals at the local level. *Business Research Quarterly, 26*(1), 79–97.

Ebo Hinson, R. (2011). Online CSR reportage of award-winning versus non award-winning banks in Ghana. *Journal of Information, Communication and Ethics in Society, 9*(2), 102–115.

Edman, J. (2016). Cultivating foreignness: How organizations maintain and leverage minority identities. *Journal of Management Studies, 53*(1), 55–88.

Fiaschi, D., Giuliani, E., & Nieri, F. (2017). Overcoming the liability of origin by doing no-harm: Emerging country firms' social irresponsibility as they go global. *Journal of World Business, 52*(4), 546–563.

Forcadell, F. J., Aracil, E., & Ubeda, F. (2020). Using reputation for corporate sustainability to tackle banks digitalization challenges. *Business Strategy and the Environment, 29*(6), 2181–2193.

Gabaldon, P., & Groschl, S. (2015). A few good companies: Rethinking firms' responsibilities toward common pool resources. *Journal of Business Ethics, 132*(3), 579–588.

Galeazzo, A., Miandar, T., & Carraro, M. (2023). SDGs in corporate responsibility reporting: A longitudinal investigation of institutional determinants and financial performance. *Journal of Management and Governance*, Forthcoming.

García-Álvarez de Perea, J., Ramírez-García, C., & Del Cubo-Molina, A. (2019). Internationalization business models and patterns of SMEs and MNEs: A qualitative multi-case study in the agrifood sector. *Sustainability, 11*(10), 2755.

Ghauri, P. N. (2022). The role of multinational enterprises in achieving sustainable development goals. *AIB Insights, 22*(1), 1–5.

Ghosh, S., & Rajan, J. (2019). The business case for SDGs: An analysis of inclusive business models in emerging economies. *International Journal of Sustainable Development & World Ecology, 26*(4), 344–353.

Gomez-Trujillo, A. M., & Gonzalez-Perez, M. A. (2020). What do we know about organizational sustainability and international business? *Management of Environmental Quality, 31*(2), 292–305.

Grøgaard, B., Colman, H. L., & Stensaker, I. G. (2022). Legitimizing, leveraging, and launching: Developing dynamic capabilities in the MNE. *Journal of International Business Studies, 53*(4), 636–656.

Hahn, R., & Kuhnen, M. (2013). Determinants of sustainability reporting: A review of results, trends, theory, and opportunities in an expanding field of research. *Journal of Cleaner Production, 59*, 5–21.

Heras-Saizarbitoria, I., Urbieta, L., & Boiral, O. (2022). Organizations' engagement with sustainable development goals: From cherry-picking to SDG-washing? *Corporate Social Responsibility and Environmental Management, 29*(2), 316–328.

Hong, J., Wang, C., & Kafouros, M. (2015). The role of the state in explaining the internationalization of emerging market enterprises. *British Journal of Management, 26*(1), 45–62.

Hult, G. T. M., Mena, J. A., Gonzalez-Perez, M. A., Lagerstrom, K., & Hult, D. T. (2018). A ten country-company study of sustainability and product-market performance: Influences of doing good, warm glow, and price fairness. *Journal of Macromarketing, 38*(3), 242–261.

Hutton, J. G., Goodman, M. B., Alexander, J. B., & Genest, C. M. (2001). Reputation management: The new face of corporate public relations? *Public Relations Review, 27*(3), 247–261.

Ike, M., Donovan, J. D., Topple, C., & Masli, E. K. (2019). The process of selecting and prioritising corporate sustainability issues: Insights for achieving the sustainable development goals. *Journal of Cleaner Production, 236*, 117661.

Jimenez, D., Franco, I. B., & Smith, T. (2021). A review of corporate purpose: An approach to actioning the sustainable development goals (SDGs). *Sustainability, 13*(7), 3899.

Kano, L. (2018). Global value chain governance: A relational perspective. *Journal of International Business Studies, 49*(6), 684–705.

Khan, A., Muttakin, M., & Siddiqui, J. (2013). Corporate governance and corporate social responsibility disclosures: Evidence from an emerging economy. *Journal of Business Ethics, 114*(2), 207–223.

Khan, P. A., Johl, S. K., & Johl, S. K. (2021). Does adoption of ISO 56002-2019 and green innovation reporting enhance the firm sustainable development goal performance? An emerging paradigm. *Business Strategy and the Environment, 30*(7), 2922–2936.

Kili, M. (2016). Online corporate social responsibility (CSR) disclosure in the banking industry: Evidence from Turkey. *International Journal of Bank Marketing, 34*(4), 550–569.

Kim, J. K., & Mudambi, R. (2020). An ecosystem-based analysis of design innovation infringements: South Korea and China in the global tire industry. *Journal of International Business Policy, 3*(1), 38–57.

King, B. G., & Whetten, D. A. (2008). Rethinking the relationship between reputation and legitimacy: A social actor conceptualization. *Corporate Reputation Review, 11*(3), 192–207.

Klossek, A., Linke, B., & Nippa, M. (2012). Chinese enterprises in Germany: Establishment modes and strategies to mitigate the liability of foreignness. *Journal of World Business, 47*(1), 35–44.

Kolk, A. (2016). The social responsibility of international business: From ethics and the environment to CSR and sustainable development. *Journal of World Business, 51*(1), 23–34.

Kolk, A., Kourula, A., & Pisani, N. (2017). Multinational enterprises and the sustainable development goals: What do we know and how to proceed? *Transnational Corporations, 24*(3), 9–32.

Lin, R., Gui, Y., Xie, Z., & Liu, L. (2019). Green governance and international business strategies of emerging economies' multinational enterprises: A multiple-case study of Chinese firms in pollution-intensive industries. *Sustainability, 11*(4), 1013.

Linnerud, K., Holden, E., & Simonsen, M. (2021). Closing the sustainable development gap: A global study of goal interactions. *Sustainable Development, 29*(4), 738–753.

Liou, R.-S., & Rao-Nicholson, R. (2021). Multinational enterprises and sustainable development goals: A foreign subsidiary perspective on tackling wicked problems. *Journal of International Business Policy, 4*(1), 136–151.

Luo, Y. (2002). Capability exploitation and building in a foreign market: Implications for multinational enterprises. *Organization Science, 13*(1), 48–63.

Marano, V., Tashman, P., & Kostova, T. (2017). Escaping the iron cage: Liabilities of origin and CSR reporting of emerging market multinational enterprises. *Journal of International Business Studies, 48*(3), 386–408.

Marin, L., Martın, P. J., & Rubio, A. (2017). Doing good and different! The mediation effect of innovation and investment on the influence of CSR on competitiveness. *Corporate Social Responsibility and Environmental Management, 24*(2), 159–171.

Maruyama, M., & Wu, L. (2015). Overcoming the liability of foreignness in international retailing: A consumer perspective. *Journal of International Management, 21*(3), 200–210.
Miles, R. E., Snow, C. C., Meyer, A. D., & Coleman, H. J. (1978). Organizational strategy, structure, and process. *The Academy of Management Review, 3*(3), 546–562.
Mio, C., Panfilo, S., & Blundo, B. (2020). Sustainable development goals and the strategic role of business: A systematic literature review. *Business Strategy and the Environment, 29*(8), 3220–3245.
Mohieldin, M., Wahba, S., Gonzalez-Perez, M. A., & Shehata, M. (2022a). Chapter 3: How businesses can accelerate and scale-up SDG implementation by incorporating ESG into their strategies. In M. Mohieldin, S. Wahba, M. A. Gonzalez-Perez, & M. Shehata (Eds.), *Business, government and the SDGs* (1st ed., pp. 65–104). Palgrave Macmillan.
Mohieldin, M., Wahba, S., Gonzalez-Perez, M. A., & Shehata, M. (2022b). Chapter 1: SDGs and the 2030 agenda: On crisis and opportunity. In M. Mohieldin, S. Wahba, M. A. Gonzalez-Perez, & M. Shehata (Eds.), *Business, government and the SDGs* (1st ed., pp. 1–17). Palgrave Macmillan.
Murshudli, F., Zapotichna, R., & Dilbazi, E. (2020). International banking business and bank strategy: Global trends and benchmarks for Post-Soviet states. *Economic Annals-XXI, 185*(9–10), 27–38.
Nguyen, D. K., Huynh, T. L. D., & Nasir, M. A. (2021). Carbon emissions determinants and forecasting: Evidence from G6 countries. *Journal of Environmental Management, 285*, 111988.
Nilsson, M., Chisholm, E., Griggs, D., Howden-Chapman, P., McCollum, D., Messerli, P., Neumann, B., Stevance, A.-S., Visbeck, M., & Stafford-Smith, M. (2018). Mapping interactions between the sustainable development goals: Lessons learned and ways forward. *Sustainability Science, 13*(6), 1489–1503.
Nishitani, K., & Kokubu, K. (2020). Can firms enhance economic performance by contributing to sustainable consumption and production? Analyzing the patterns of influence of environmental performance in Japanese manufacturing firms. *Sustainable Production and Consumption, 21*, 156–169.
Park, B. I., & Ghauri, P. N. (2015). Determinants influencing CSR practices in small and medium sized MNE subsidiaries: A stakeholder perspective. *Journal of World Business, 50*(1), 192–204.
Park, S. H., & Ungson, G. R. (2019). Rough diamonds in emerging markets: Legacy, competitiveness, and sustained high performance. *Cross Cultural Strategic Management, 26*(3), 363–386.
Pedersen, C. S. (2018). The UN sustainable development goals (SDGs) are a great gift to business! *Procedia CIRP, 69*, 21–24.
Peng, M., Wang, D., & Jiang, Y. (2008). An institution-based view of international business strategy: A focus on emerging economies. *Journal of International Business Studies, 39*(5), 920–936.
Persson, A., Weitz, N., & Nilsson, M. (2016). Follow-up and review of the sustainable development goals: Alignment vs. internalization. *Review of European, Comparative & International Environmental Law, 25*(1), 59–68.
Pizzi, S., Venturelli, A., & Caputo, F. (2021). The "comply-or-explain" principle in directive 95/2014/EU. A rhetorical analysis of Italian PIEs. *Sustainability Accounting, Management and Policy Journal, 12*(1), 30–50.
Potrich, L., Cortimiglia, M. N., & de Medeiros, J. F. (2019). A systematic literature review on firm-level proactive environmental management. *Journal of Environmental Management, 243*, 273–286.
Sachs, J. D., & Sachs, L. E. (2021). Business alignment for the "decade of action". *Journal of International Business Policy, 4*(1), 22–27.
Sarfraz, M., Qun, W., Hui, L., & Abdullah, M. I. (2018). Environmental risk management strategies and the moderating role of corporate social responsibility in project financing decisions. *Sustainability, 10*(8), 2771.

Schaltegger, S. (2018). Linking environmental management accounting: A reflection on (missing) links to sustainability and planetary boundaries. *Social and Environmental Accountability Journal, 38*(1), 19–29.

Scharf, E. R., & Fernandes, J. (2013). The advertising of corporate social responsibility in a Brazilian bank. *International Journal of Bank Marketing, 31*(1), 24–37.

Scheyvens, R., Banks, G., & Hughes, E. (2016). The private sector and the SDGs: The need to move beyond 'business as usual'. *Sustainable Development, 24*(6), 371–382.

Schmidtke, H., & Lenz, T. (2023). Expanding or defending legitimacy? Why international organizations intensify self-legitimation. *The Review of International Organizations*, Forthcoming.

Shen, S., Venaik, S., & Liesch, P. (2023). A novel model linking UN SDGs with international experience and firm performance. *International Business Review, 33*, 102170.

Siltaloppi, J., Rajala, R., & Hietala, H. (2021). Integrating CSR with business strategy: A tension management perspective. *Journal of Business Ethics, 174*(3), 507–527.

Silva, S. (2021). Corporate contributions to the sustainable development goals: An empirical analysis informed by legitimacy theory. *Journal of Cleaner Production, 292*, 125962.

Strand, R. (2014). Strategic leadership of corporate sustainability. *Journal of Business Ethics, 123*(4), 687–706.

Suchman, M. C. (1995). Managing legitimacy: Strategic and institutional approaches. *The Academy of Management Review, 20*, 571–610.

Talbot, D., & Ordonez-Ponce, E. (2022). Canadian banks' responses to COVID-19: A strategic positioning analysis. *Journal of Sustainable Finance & Investment, 12*(2), 423–430.

Teh, D., & Corbitt, B. (2015). Building sustainability strategy in business. *Journal of Business Strategy, 36*(6), 39–46.

Tsai, K.-H., & Liao, Y.-C. (2017). Sustainability strategy and eco-innovation: A moderation model. *Business Strategy and the Environment, 26*(4), 426–437.

Tsui-Auch, L. S., & Chow, D. (2019). MNEs' agency within institutional contexts: A study of Walmart's post-acquisition practices in Mexico, Germany, and Japan. *Journal of International Management, 25*(2), 100655.

Uddini, M. K. (2017). Climate change and global environmental politics: North-South divide. *Environmental Policy and Law, 47*, 3–4.

United Nations. (2015). *Transforming our world: The 2030 Agenda for sustainable development* (UN Doc. a/rcs/70/1 (September 25, 2015)). Technical report, United Nations Department of Economic and Social Affairs – Sustainable Development.

Vaara, E., & Tienar, J. (2008). A discursive perspective on legitimation strategies in multinational corporations. *The Academy of Management Review, 33*(4), 985–993.

Vaara, E., & Tienari, J. (2011). On the narrative construction of multinational corporations: An antenarrative analysis of legitimation and resistance in a cross-border merger. *Organization Science, 22*(2), 370–390.

Van der Waal, J. W. H., & Thijssens, T. (2020). Corporate involvement in sustainable development goals: Exploring the territory. *Journal of Cleaner Production, 252*, 119625.

Van Leeuwen, T. (2007). Legitimation in discourse and communication. *Discourse & Communication, 1*(1), 91–112.

Van Tulder, R., Rodrigues, S. B., Mirza, H., & Sexsmith, K. (2021). The UN's sustainable development goals: Can multinational enterprises lead the Decade of Action? *Journal of International Business Policy, 4*(1), 1–21.

Van Zanten, J. A., & Van Tulder, R. (2018). Multinational enterprises and the sustainable development goals: An institutional approach to corporate engagement. *Journal of International Business Policy, 1*(3–4), 208–233.

Van Zanten, J. A., & Van Tulder, R. (2021). Improving companies' impacts on sustainable development: A nexus approach to the SDGs. *Business Strategy and the Environment, 30*(8), 3703–3720.

Verbeke, A., & Kano, L. (2015). The new internalization theory and multinational enterprises from emerging economies: A business history perspective. *Business History Review, 89*(3), 415–445.

Verbeke, A., & Kano, L. (2016). An internalization theory perspective on the global and regional strategies of multinational enterprises. *Journal of World Business, 51*(1), 83–92.

Wei, Z., & Nguyen, Q. T. K. (2020). Chinese service multinationals: The degree of internationalization and performance. *Management International Review, 60*(6), 869–908.

Weitz, N., Carlsen, H., Nilsson, M., & Skanberg, K. (2018). Towards systemic and contextual priority setting for implementing the 2030 agenda. *Sustainability Science, 13*(2), 531–548.

Wood, D. J. (1991). Corporate social performance revisited. *The Academy of Management Review, 16*(4), 691–718.

Wu, M.-W., & Shen, C.-H. (2013). Corporate social responsibility in the banking industry: Motives and financial performance. *Journal of Banking & Finance, 37*(9), 3529–3547.

Yang, D., Jiang, W., & Zhao, W. (2019). Proactive environmental strategy, innovation capability, and stakeholder integration capability: A mediation analysis. *Business Strategy and the Environment, 28*(8), 1534–1547.

Yu, H., Bansal, P., & Arjalies, D.-L. (2023). International business is contributing to environmental crises. *Journal of International Business Studies, 54*(6), 1151–1169.

Zeng, J., & Glaister, K. W. (2016). Competitive dynamics between multinational enterprises and local internet platform companies in the virtual market in China. *British Journal of Management, 27*(3), 479–496.

Zhan, J. X. (2021). GVC transformation and a new investment landscape in the 2020s: Driving forces, directions, and a forward-looking research and policy agenda. *Journal of International Business Policy, 4*(2), 206–220.

Zhan, J. X., & Santos-Paulino, A. U. (2021). Investing in the sustainable development goals: Mobilization, channeling, and impact. *Journal of International Business Policy, 4*(1), 166–183.

Zhang, L., Ling, J., & Lin, M. (2022). Artificial intelligence in renewable energy: A comprehensive bibliometric analysis. *Energy Reports, 8*, 14072–14088.

Mesiet W. Kamihanda is an Assistant Lecturer in Economics at the Mwalimu Nyerere Memorial Academy in Dar es Salaam, Tanzania, and a PhD candidate in Management. Her research interests focus on the intersection of aid and sustainable development, addressing challenges such as poverty reduction, education, and gender equality while examining its implications for policy and governance in developing countries.

Miguel M. Torres, PhD in Economics, is a Lecturer in International Business at Kent Business School (KBS). His career includes consulting services to the European Commission and World Bank, applied research and teaching roles at Leeds, Milano, Aveiro, KU Leuven, and Uppsala universities, with appointments at the University Federal of Goiás, IF Goiano, and the University Catolica of Portugal. He has held multiple visiting professorships and is an INSOL International and Marie Skłodowska-Curie Fellow. He completed executive programs at Harvard, Carnegie Mellon University (USA), and University College London (UK). His research focuses on the intersection of international business, public policy, and strategic management, focusing on multinational management models, corporate political risk, and negotiation strategies, particularly emphasizing the complexities of cross-border restructuring and foreign direct investment drivers using cross-sectional, time-series, and panel data approaches. His work has been published in the *International Business Review, Journal of Business Ethics, Multinational Business Review, Journal of Economic Methodology, Arbitration International, International Insolvency Review, PLOS One, and AIB Insights*. He co-edited the books *Global Perspectives on Climate Change, Inequality, and Multinational Corporations* (Springer, 2025) and *Dead Firms: Causes and Effects of Cross-Border Corporate Insolvency* (Emerald, 2017).

Red Tape Roadblocks on M&As: Emerging Economies' Regulators as a Source of Inefficiency

Catherine Cai, Rifat Kamasak, and Aslihan Bozcuk

1 Introduction

Mergers and Acquisitions (M&As) constitute one of the most critical strategic decisions for multinational corporations (MNCs) and have become a primary method of internationalisation and growth (Aquaro et al., 2023; Battista et al., 2024; Ragozzino & Reuer, 2024). M&A transactions possess significant risks and require high investment costs and resource commitments. In the context of M&As, deal duration refers to the time (usually measured by the number of days) between deal announcement and completion (Aquaro et al., 2023). Deal duration is relatively different from deal completion, which indicates the successful finalisation of an M&A by fulfilling all legal, financial, and regulatory requirements and transferring ownership or control (DePamphilis, 2019). Although some deals are completed on the announcement day, many have been through the deal process and take time to complete. Both deal duration and completion are essential topics in understanding the M&A deal process. The time frame between announcing an M&A transaction and its finalisation poses substantial consequences for the involved parties. Delays during this period can lead to direct costs, including penalties for contract breaches (Gulapalli et al., 2017; Koirala et al., 2023), inefficient resource use (Sakhartov & Reuer, 2023), and

C. Cai
University of the West of England, Bristol, UK
e-mail: catherine.cai@uwe.ac.uk

R. Kamasak (✉)
University of Reading, Reading, UK
e-mail: r.kamasak@henley.ac.uk

A. Bozcuk
Akdeniz University, Antalya, Turkey
e-mail: aebozcuk@akdeniz.edu.tr

J. Bento, M. M. Torres (eds.), *Global Perspectives on Climate Change, Inequality, and Multinational Corporations*, CSR, Sustainability, Ethics & Governance,
https://doi.org/10.1007/978-3-031-80797-8_7

missed opportunities in international markets (Cataltepe et al., 2023; Simpson & Sariol, 2022). Moreover, indirect costs can include reduced legitimacy (Fuad & Venugopal, 2024), harm to reputation (Bauer & Friesl, 2024), and increased uncertainty among employees (Cartwright & Cooper, 2018). Therefore, the speed of identifying a target and announcing an M&A to the completion of the deal is an essential barometer of a deal's efficiency and prospective success.

The M&A literature (e.g., King et al., 2021; Zhou et al., 2024) mentions several factors predicting the duration of M&A deal completion. These include the involved parties' resource availability (Luypaert & De Maeseneire, 2015; Masulis & Simsir, 2018), physical (Cai et al., 2016) and institutional distance (Ermolaeva, 2019), institutional context (Yurdakul et al., 2022; Zakaria & Kiziler, 2023) and quality (Liou et al., 2016), managerial decisions and CEO roles (Kamasak et al., 2020; Lawrence et al., 2024), ownership (Lindner et al., 2023), deal size and characteristics (Singh et al., 2024), knowledge spillovers (Huang & Xie, 2023; Kamasak et al., 2017), market efficiency (Kim & Song, 2017), and environmental dynamism (Kamasak et al., 2016; Strobl et al., 2020) and other factors from the fields in political economy (Li et al., 2017) and public policy (Booji & Sahib, 2012).

While we do not repudiate the role of the factors mentioned above, the quality of institutions and regulatory bodies can play a more significant role, particularly in emerging economies. We can support this proposition through several arguments. First, the impact of institutions is more pronounced in emerging economies (Yurdakul & Kamasak, 2021; Duran et al., 2019; Peng et al., 2008), and mergers represent a business transaction where governments have the opportunity and the incentive to exert significant influence. Second, due to weak institutional quality in emerging economies (Kamasak & Yavuz, 2015; Paul & Jadhav, 2020), the structure of deals differs from that of advanced economies. Firms in emerging economies are generally less experienced than their developed economy counterparts and less likely to use complex deal structures to mitigate information asymmetry (Song et al., 2021). Finally, individuals' and firms' rent-seeking behaviours are more widespread in emerging economies (Kamasak et al., 2019; Liu et al., 2019). These situations necessitate stricter controls and open more room for government involvement in M&A deals (Col & Sen, 2019; Dinc & Erel, 2013). Thus, the focus on regulatory agencies is particularly relevant for acquirers from emerging economies, where institutions and regulatory agencies tend to be less efficient, and deal duration tends to be longer. This research defines regulatory bodies as government agencies or public authorities responsible for enforcing laws and regulations that govern M&A activities. During the deal, these entities comply with legal standards and protect the public interest (Gaughan, 2017). Regulatory agencies are essential in cross-border M&A deal completion and duration (Batista et al., 2023; Sales & Zanini, 2017). They can facilitate or delay the deal duration of the intended M&As, impacting operational efficiencies and the seller and acquirer's economic expectations (Cai et al., 2014). Therefore, research should cautiously explore the roles of institutions for acquirers in emerging economies. However, previous M&A literature has ignored the impact of regulators on M&A deal completion.

This paper explores how regulatory bodies in emerging and developed economies impact cross-border M&A transactions by causing delays. We argue that regulatory bodies in emerging economies will prolong the duration of the deal, whereas regulatory agencies in developed economies will not. There has been limited progress in empirical studies focusing on the pre-deal completion phase of M&As since Jemison and Sitkin's (1986) well-known study (Schijven et al., 2024; Welch et al., 2020). Most studies (e.g., Grant et al., 2022; Luypaert & de Maeseneire, 2015) focused on firm—or deal-level factors, neglecting pre-deal factors such as regulatory bodies' actions and their influence on M&A deal completion.

Drawing on institutional theory, we address this gap by empirically testing the effects of regulatory bodies on M&A deal durations in Brazil, a large emerging economy. Primarily, we aim to find an answer to the question: Does the involvement of antitrust regulatory agencies prolong the duration of M&A deals in international settings? We answer this question by analysing and evaluating a sample of cross-border M&As in Brazil between 2000 and 2015. In this period, Brazil has become a significant global player, has many MNEs, and is increasingly becoming a global player (Casanova & Kassum, 2013). Harle et al. (2012) documented that in 2010, Brazil conducted more than five times the number of M&A reviews compared to China. More recently, PricewaterhouseCoopers (2021) highlighted a historic record in M&A transactions in Brazil, with 1038 announced deals in 2020, marking a 14% increase from 2019 and a 48% rise compared to the five-year average from 2015 to 2019. Despite the adverse effects of the COVID-19 pandemic, this upward trend has persisted (Barros et al., 2024). Furthermore, Brazil represents a typical middle-class emerging economy in contrast to China and India, which have unique traditional and historical socioeconomic characteristics.

Brazil's main differences from China and India in regulatory involvement in M&As can be attributed to legal frameworks, market structures, and the level of state intervention in the economy (DePamphilis, 2019). The Brazilian economy is characterised by international alignment and a market-oriented approach, yet China has broader economic goals, such as fostering technological advancement or consolidating industries. India also has an economic policy that prioritises the protection of domestic industries and the prevention of market concentration. In China, the regulatory environment is heavily influenced by state intervention, with the government actively involved in shaping M&A activities to align with broader economic policies. Regulatory decisions in China often reflect state priorities, such as maintaining control over critical industries or fostering the growth of domestic champions. Similarly, the regulatory framework for M&As strongly emphasises protecting domestic industries and preventing foreign dominance, influenced by its historical experience with colonialism and economic liberalisation. However, the primary focus of regulatory bodies in Brazil is to prevent anti-competitive practices, ensuring that M&As do not create monopolies or harm consumer welfare. Brazil's legal framework is characterised by its alignment with international standards, promoting transparency and predictability in regulatory decisions.

Considering most prior M&A studies concentrate on China and India, our findings can offer more relevant and applicable implications to other emerging economies that share similarities with Brazil. Brazil's regulatory environment distinguishes it from China and India and makes it an ideal context for studying the role of regulatory bodies in M&A processes. The differences in legal frameworks, market structures, and the level of state intervention across these countries provide a rich basis for comparative analysis, offering insights into the diverse ways regulatory bodies influence M&A activities in different economic contexts. Thus, we contribute to the pre-deal M&A literature by showing that the effectiveness of antitrust regulatory agencies in home and host countries can affect the completion of M&A deals.

Finally, we utilise a unique dataset from the Thompson SDC Mergers and Acquisitions Database and Zephyr, encompassing public and private M&As in Brazil. Two-thirds of acquired firms worldwide are privately held or subsidiaries of divisions of other firms. By including private and public firms in the same sample, we can obtain more representative findings for real-world scenarios (Erel et al., 2015).

The rest of the paper is structured as follows: The next section reviews relevant studies on deal duration, constructs a theoretical framework, and develops hypotheses on the relationship between M&A deal duration and regulatory agency involvement. Then, we explain the methodology and empirical analyses. The following sections present the results and discussion. We conclude with policy implications and the ethical impact of regulatory delays.

2 Literature Review

2.1 M&A Deal Process

M&A deals include multi-stage processes. The M&A deal is complete when the acquirer pays the purchase price, and the seller transfers the shares or business. The initial pre-deal stage involves informal and formal negotiations between the potential target and the acquirer. During this stage, the acquirer and potential target express preliminary interest and negotiate the possibilities and contractual details of the deal. This private stage concludes when both parties sign a binding Letter of Intent (LOI) and publicly announce the deal. Subsequently, the public stage, the period between deal announcement and completion, starts. Figure 1 presents the private and public stages of an M&A deal (Boone & Mulherin, 2007).

The public stage necessitates both parties to file the documents within a specific timeframe under the pre-review system. Then, the competition authorities and other regulatory agencies evaluate the M&A's competitive, economic, and sectoral impact, and the transaction can only be finalised once the agencies approve it (ICLG, 2015). This decision may include approval with or without restrictions or outright prevention. The authorities may make other specific decisions, such as requiring the combined firm to divest assets in certain regions as part of the approval conditions. Operational synergies of the deal motivations can emerge after the transaction's

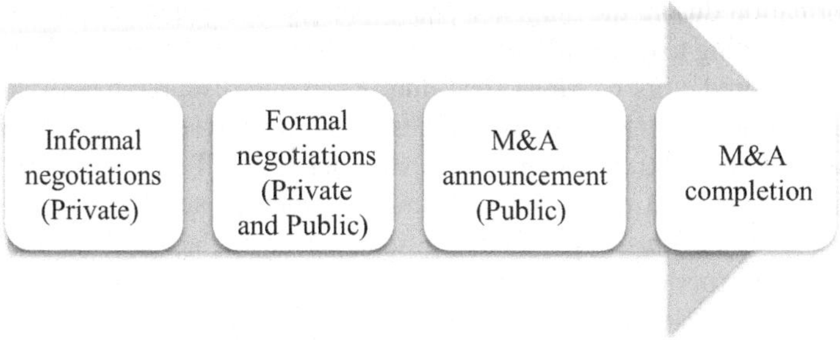

Fig. 1 M&A deal completion process. (Source: Elaborated by the authors)

approval, making the involvement of government agencies more critical. Therefore, regulatory institutions can influence the duration of M&A deal completions.

2.2 *Theoretical Background of Pre-merger Activities*

Few studies (i.e., Carlton, 2007; Chung et al., 2023) have analysed the process from the announcement of an M&A deal to completion, and most research focuses on the impact of internal factors. The literature explores pre-deal M&A activities from different perspectives. The upper echelons (Hambrick & Mason, 1984) and corporate governance (Short et al., 1999) literature examine the process at the managerial level (Lou et al., 2024). Both theories analyse a previously announced M&A deal process from shareholders' and top managers' perspectives and attribute the M&A delay or cancellations to their actions.

In particular, publicly listed firms and unsolicited or hostile bids require shareholder meetings, which are challenging in securing a target's shareholder approval and finalising the deal process. Takeover defences by target firms can also prolong deal duration, making deals involving publicly listed targets take longer than private ones. Managerial characteristics such as hubris can affect the likelihood of completion and the duration of deals. Research (e.g., Aktas et al., 2012) found that target and acquirer CEOs with high narcissism correlate with lower probabilities of M&A deal completions.

The Transaction Cost Economies (TCE) framework (Williamson, 1998) focuses on specific transaction structures in M&A deals to explain variations in deal durations. TCE can influence deal durations through comprehensive assessments, high due diligence costs, and complexity in bargaining, contracting, and negotiating, which stipulates terms for representations, warranties, indemnities, and earn-outs (Ahammad et al., 2017). Moreover, information asymmetry between buyers and sellers can extend the duration of the deal to ensure a fair valuation. Under the conditions of information asymmetry, acquirers often conduct extensive due diligence to examine the target's legal, financial, and operational aspects. The discovery of new information during the diligence phase can result in negative shareholder perceptions, extending to deal duration (Kau et al., 2008). Another research stream that examines the predictors of M&A deal duration is the resource-based view of the firm (RBV) (Barney & Arikan, 2005) and its offshoot knowledge-based view (KBV) (Eisenhardt et al., 2000) and dynamic capabilities (DC) (Teece et al., 1997). RBV emphasises the critical roles of firms' unique resources and capabilities, such as acquirer and target firms' reputation, legitimacy, learning ability, and international and contextual experience, to facilitate M&A deal completions. Similarly, resource dependency theory (RDT) (Salancik & Pfeffer, 1978) examines how the availability of internal resources influences a firm's reliance on external resources, determining firms to engage in M&A transactions and interact with regulatory bodies when acquiring diverse targets. A comprehensive body of strategy research (Bauer & Matzler, 2014; Harrison et al., 1991; Popli & Ahsan, 2024) has investigated the influence of a firm's internal resource base on M&A deal decisions. While internal factors are essential, the impact of institutions outside the firm is critical for understanding variations in M&A deal durations. The time required to complete an international deal cannot be examined in isolation from the context, as the effects of internal factors depend on external factors such as the institutional environment (Hoskisson et al., 2000). We have reviewed key studies that examine deal duration (i.e. the number of days between deal announcement and deal completion) as the dependent variable. Table 1 presents a review of previous studies that examine factors influencing deal duration. As presented in Table 1, most studies on factors impacting deal duration either focus on internal factors (such as deal characteristics and firm characteristics) or country factors (including home and host country factors, cultural distance, and institutional distance). We also observe that most studies are conducted in developed economies, indicating the necessity of more studies in emerging economies with unique environmental conditions. Studies have used institution-based views as theoretical background, considering the widespread influence of environmental factors. Given the unique environmental conditions, we also adopt the institutions-based view, which examines the social and regulatory contexts within which a firm's activities occur, offering a non-economic perspective on organisational behaviours and strategies (DiMaggio & Powell, 1991). North (1990) defines institutions as the rules of the game interconnected with dominant formal and informal institutions in their environment, which provide firms with resources and legitimacy. Therefore, the involved parties in M&A deals should understand, influence, negotiate, and even partially shape their institutional environments (Kostova et al., 2008). The acquirers generally overcome the liability of foreignness (Zaheer, 1995) by dealing with the regulatory structures, government agencies, legal systems, professional bodies, interest groups and public opinions in the host country (Oliver, 1991). One crucial aspect of the institutional context that directly influences the duration of M&A deals is the role of government institutions, particularly regulatory agencies.

Table 1 Literature review of factors influencing deal duration

Paper	Author	Year	Journal	Sample	Theory	Factors influencing deal duration	Findings
A. External factors (country-level)							
Cross-border acquisition abandonment and completion: The effect of institutional differences and organisational learning in the international business service industry, 1981–2001	Dikova, Sahib & van Witteloostuijn	2010	*Journal of International Business Studies*	2389 announced cross-border acquisition deals in the international business service industry	Institutional theory	Formal and informal institutions, experience	Differences in national formal and informal institutions explain part of the variation in the likelihood that an announced cross-border acquisition deal will be completed and in the duration of the deal-making Experience with completed cross-border acquisition deals shortens the deal duration in institutionally distant environments

(continued)

Table 1 (continued)

Paper	Author	Year	Journal	Sample	Theory	Factors influencing deal duration	Findings
Slow but certain: The pre-M&A process of European national champion deals	Booji & Sahib	2012	*Advances in Mergers and Acquisitions*	88 completed cross-border M&As within Europe between 1997 and 2010	Public policy	National champion, country power, M&A experience	National champion M&As face a longer pre-M&A process than other comparable European M&As. National champion M&As are almost always completed successfully, while comparable European M&As are often derailed
Cross-border acquisitions by state-owned firms: How do legitimacy concerns affect the completion and duration of their acquisitions?	Li, Xia & Lin	2016	*Strategic Management Journal*	Cross-border acquisition activities in the U.S. from 1990 to 2012	Political economy	Legitimacy perspective	State-owned foreign firms take longer to complete a deal than other foreign firms. Firm characteristics moderate this relationship
The effect of cultural and institutional factors on initiation, completion, and duration of cross-border acquisitions	Lawrence, Raithatha, & Rodriguez	2021	*Journal of Corporate Finance*	218.806 cross-border acquisitions from 1970 to 2016	Institutional theory	Cultural distance, institutional distance	The deal's completion and completion depend not on cultural and institutional factors

Overcoming institutional challenges in cross-border acquisitions: how home-and host-country acquisition experiences shorten the deal completion phase	Soleimani & Yang	2022	*Multinational Business Review*	A panel data of 8175 cross-border acquisitions from 2003 to 2009	Institutional theory	Institutional theory and organisational learning literature	Host-country acquisition experience has more impact on shortening deal completion duration The quality of business institutions in the host country and the institutional distance between the home and host countries amplify or attenuate the effect of past acquisition experiences, depending on their origin, the quality of business institutions, and the institutional distance of where they are used
B. Internal factors (deal and firm-level)							
Antecedents of time to completion in mergers and acquisitions	Luypaert & de Maeseneire	2015	*Applied Economics Letters*	1150 M&As between listed US companies during 1994–2011	Information asymmetry	Deal complexity, Information asymmetry and other deal characteristics	A lengthier acquisition duration characterises stock offers, deal hostility, mergers and larger deals Solid and transparent shareholder support accelerates deal completion, as does the likelihood of overpayment. Experienced bidders succeed more rapidly in completing transactions, implying learning effects

(continued)

Table 1 (continued)

Paper	Author	Year	Journal	Sample	Theory	Factors influencing deal duration	Findings
Stay close to me: What do ESG scores tell about the deal timing in M&A transactions?	Cardillo & Harasheh	2023	*Finance Research Letters*	415 M&A deals in the EU and UK from 2002 to 2020	ESG	Acquirer and target ESG rating, differences in ESG rating	Discrepancies in the pre-deal ESG ratings between target and acquirer firms lead to decreased closure speed
C. Both internal and external factors							
The causes and consequences of delayed/abandoned cross-border merger & acquisition transactions: a cross-case analysis in the dynamic industries	Reddi, Xie & Huang	2016	*Journal of Organizational Change Management*	Multiple case studies (3 cross-border M&As in India)	Multiple perspectives	Ownership factors, deal factors and country-level factors	Ownership structure, lack of international deal-making experience There is a legal dispute over international taxation, weak financial market laws, government and political intervention, and a proposed 'complex' post-merger ownership structure. Institutional determinants include market and competition regulations, higher levels of political intervention, and erratic government officials' behaviour

Source: Authors' elaboration

2.3 The Role of Regulatory Agencies

Compared to other market transactions, such as greenfield investments, M&As involve buying an asset and integrating the target into the organisational boundary (Cuypers et al., 2021; Geyskens et al., 2006). It provides a fast entry to foreign markets and offers learning opportunities from targets. M&As can generate cost-reducing synergies and introduce advanced M&As can generate cost-reducing synergies and introduce advanced technologies and superior managerial practices to the firms they merge with or acquire. They enhance efficiencies and innovation and improve product and service offerings. The positive impacts of M&As can be more pronounced in cross-border scenarios.

Despite their positive outcomes, M&As can also negatively affect markets and competition. M&As can potentially create large firms that are 'too big to fail' and lead to industry concentration that will harm consumer welfare. They can profit through rent-seeking activities and gain legal protections through licensing regulation. Some possible adverse outcomes are reduced competition, created price monopoly and diminished consumer choices if not appropriately monitored and regulated. Besides, M&As can increase their market power through economies of scale and scope rapidly compared to greenfield projects (Wang, 2009). Therefore, the role of institutions and regulatory bodies is pivotal in balancing these outcomes to promote healthy market competition and protect consumer interests through antitrust and competition laws (ACLs). ACLs aim to prevent monopolies, maintain competitive markets and ensure sufficient industry rivalry.

For example, in a market composed of small firms, the entry of a giant MNC through an M&A can negatively affect competition, alter industry structure, and deter new entrants. ACLs safeguard consumer rights by preventing monopolistic dominance and fostering a competitive environment. Some studies (Ekelund et al., 2001; Sirower, 1997) associate ACLs with the prolonged duration of M&A deals; however, the results are inconclusive. M&A control is significant. The law firm White & Case LLP (2023) states that competition authorities worldwide are increasingly cooperating, making it essential for merging parties to consider all relevant antitrust authorities.

Two policy regimes influence M&A deal completions: antitrust merger review policies and sectoral regulations (Brown et al., 2023). Antitrust authorities control and monitor M&A activities to ensure a healthy and well-functioning market, fair competition, and the prevention of monopoly and exclusive trading. Sectoral regulators, on the other hand, ensure the proper industry structure in sensitive sectors. They review M&A deals, both domestic and cross-border, within specific industries with high concentration, such as telecommunications, banking, and other critical sectors. Sometimes, a single deal potentially impacting national security may fall under multiple jurisdictions by sectoral regulators and multiple antitrust competition agencies. Thus, regulatory bodies can result in longer M&A deal durations, negatively impacting firms by delaying or preventing the intended merger (Bonaime et al., 2018; de Bodt et al., 2024). Based on the explanations above, we hypothesise:

Hypothesis H1 Regulatory body involvement in emerging economies is associated with longer M&A deal durations.

In developed economies, regulatory bodies may be minimised or consolidated under a single umbrella organisation to facilitate M&A transactions. However, this may be different for emerging economies. A transparent and consistent regulatory environment leads to a self-selection process, preventing M&As that do not comply with existing regulations from being pursued (Ahsan et al., 2024). Thus, in a perfect scenario, M&As should not be significantly impacted by antitrust regulations or regulatory agencies' scrutiny. However, the non-transparency of existing regulations means that antitrust regulatory agencies affect M&A processes. In emerging economies, regulatory agencies have fewer resources and less experience with merger reviews than in developed economies (He & Padrón-Hernández, 2024). Institutional consistency and transparency are even lower. A notable example is the acquisition of Companhia Vale do Rio Doce (VALE) by the Canadian mining company Inco Ltd. The deal was announced on 11 June 2006 and received approval from the Canadian Competition Bureau on 24 September 2006. Inco's board of directors and shareholders subsequently accepted the offer, and the transaction was completed on 5 January 2007. Due to the transparency and consistency of their decision-making processes, gaining legitimacy and obtaining approval from regulatory agencies in developed economies is relatively more straightforward. Therefore, we hypothesise:

Hypothesis H2 Regulatory body involvement in developed economies is not associated with M&A deal durations in developed economies.

3 Methods

3.1 Context and Sample

Brazil, a leading Latin American emerging economy, has undergone significant economic transformation over the last thirty years through various reform packages, including the Plano Real. The plan supports 'economic openness and regional integration, controlling inflation, and ensuring stability' (Kumar et al., 2023, p. 1594). The policies that the Brazilian Government adopted yielded some positive outcomes; however, the high and complicated tax burden, deficiencies in monetary and regulatory policies, high unemployment rate, populist government spending, and increased global competition drive Brazilian firms to pursue cross-border M&A activities (Battista et al., 2024). In Brazil, three agencies oversee antitrust regulations (IBA, 2022): The Conselho Administrativo de Defesa Econômica (CADE), an independent agency; the Secretaria de Direito Econômico (SDE) under the Ministry of Justice; and the Secretaria de Acompanhamento Econômico (SEAE) within the Ministry of Finance. The Brazilian New Competition Act (Law No. 12,529/2011), enacted in 2012, granted CADE enhanced authority to approve, block, or impose remedies on mergers and acquisitions unilaterally or through agreements with the involved parties. Our analysis will include examining the roles of these three regulatory agencies and the other regulatory agencies in emerging economies and developed economies in our sample.

The data collection proceeded in several steps. We extracted cross-border deals from Brazil between 2000 and 2015. We excluded internal transactions, rumoured deals, acquisitions of assets, or deals in which foreign firms acquired local firms through their subsidiaries. We have also removed zero-duration deals. We view duration as a non-zero continuous variable since if the announcement date is the same as the completion date, the results could be misleading, and we cannot observe factors that might delay the deal process (Dikova et al., 2010). Second, we merged the deal information with firm-level information from the Bureau van Dijk (BvD) database and other company sources. Third, we compiled country-level information.

Our sample consists of all cross-border M&A deals involving Brazilian acquirers in Brazil between 2000 and 2015, sourced from the Thomson SDC Mergers Database and Zephyr Database. We only included non-zero deal durations (Dikova et al., 2010) and selected 155 M&A deals with detailed information about announcements, completion dates, and firm characteristics. While 78 transactions occur in emerging economies, 77 M&A deals are from emerging economies. In our sample, Brazilian firms have predominantly conducted M&As in developed economies, including the USA, Canada, and Portugal, as well as in emerging economies such as Argentina, Uruguay and Colombia. The primary industries involved in these M&A deals are mining, oil and gas, construction, banking, steel, and food and beverage. The average deal completion time in the emerging economy sample is around 132 days, with the most extended M&A deal taking 1187 days. In contrast, the developed economy sample has an average completion time of 95 days, with the most extended deal duration being 400 days. Table 2 presents the average deal duration by host countries in our sample. Countries with high average durations are Chile (266 days), Ecuador (265 days), Russia (234 days), and Uruguay (175 days). The developed economy sample has an average completion time of 98 days, with the most extended deal duration being 400 days. The developing economy sample has an average of 132 days. Overall, the average non-zero deal duration in our sample of cross-border M&As from Brazil is 113 days.

3.2 Dependent and Explanatory Variables

The dependent variable is the deal duration, measured as the number of days between the announcement date and the completion date of the M&A. The completion date signals that both parties involved perceive that the most crucial issues for the acquisition are resolved (Dikova et al., 2010). For the independent variables, we use the regulatory agencies involved in transactions in developed and emerging economies as the independent variable. We obtained the regulatory agency data

Table 2 Average deal duration by host countries

Host country name	Host country code	Number of deals	Mean in days
United Arab Emirates	AE	1	103
Argentina	AR	25	119.64
Australia	AU	7	99.43
Bermuda	BM	2	48.5
Bolivia	BO	1	77
Bahamas	BS	1	123
Canada	CA	11	74.64
Chile	CL	6	265.83
China	CN	1	105
Colombia	CO	6	113.17
Czech Republic	CZ	1	87
Germany	DE	2	77.5
Denmark	DK	1	88
Dominican Republic	DO	1	25
Ecuador	EC	1	265
Spain	ES	3	109.33
France	FR	4	138.75
India	IN	2	46.5
Italy	IT	5	45.2
Luxembourg	LU	3	50.333
Mexico	MX	6	141.67
Netherlands	NL	4	114.5
Norway	NO	4	50.25
Peru	PE	3	33.33
Portugal	PT	10	145.4
Paraguay	PY	3	111.67
Russia	RU	1	234
Turkey	TR	1	21
United Kingdom	UK	6	117.83
United States	US	22	96.64
Uruguay	UY	9	175.11
South Africa	SA	2	115.5

Source: Authors' elaboration

from Zephyr and cross-verified it with company websites. We counted the number of regulatory agencies from emerging and developed economies involved with the focal deal. We assigned a code of (1) if the regulatory agency was from a host emerging economy and (0) otherwise. The classification of developed and emerging economies was based on the IMF's list.

We also employ control variables. We controlled for several variables to capture the most significant factors affecting deal durations. These factors include deal characteristics, industry, and firm-specific attributes. Cash payments can shorten M&A deal durations, as they are more straightforward and less administratively burdensome than stock offers. The need for stockholder approval can extend the deal duration, so we included dummy variables to control for the publicly listed status of both the acquirer and target firms, coded as (1) for listed and (0) otherwise. MNEs may opt for partial acquisitions or joint ventures to expedite market entry and avoid increased rivalry with local firms. A higher percentage of the acquired stake would require more negotiation, so we controlled for the percentage acquired in the deal. More experienced acquirers are likely to have developed an M&A routine, facilitating deal completion. Thus, we controlled for the acquirer's total experience.

Acquiring a subsidiary, a minor partner of a larger enterprise, can complicate transactions as the parent's influence often persists in the subsidiary's governance structure for an extended period. We used a dummy variable to indicate whether the transaction involved a subsidiary (1) or not (0). Since institutional distance, which refers to the cultural similarity or difference between the acquirer and target countries, can influence the duration of the deal (Donnelly et al., 2024), we controlled it. We employed the GLOBE project's uncertainty avoidance and power distance measures to assess institutional distance. Finally, industry characteristics defined by primary SIC code divisions and year dummies accounting for macroeconomic factors were also controlled.

4 Results

We used hierarchical Ordinary Least Squares (OLS) regression to analyse our dataset. First, we calculated inter-item correlations and variance inflation factor (VIF) scores. All were within acceptable levels, with inter-correlations below 0.80 and VIF scores under 5 (Gravetter & Forzano, 2008), indicating no multicollinearity issues. Table 3 shows the findings of our analyses.

We examined the impact of developed and emerging economy regulatory agencies on cross-border M&A deal duration, respectively. Initially, we included only the control variables in Model 1, which shows their effects on cross-border M&A deal duration and provides the model's benchmark specification with a constant. In Model 2, we added the involvement of emerging economy regulatory agencies as the independent variable. Model 3 includes the involvement of developed economy regulatory agencies. The models report coefficients, standard errors, R^2 values, and Wald chi-squared test results. Model 2 confirms a significant relationship between emerging economy regulatory agencies and M&A deal duration (β = 116.023, $p < 0.05$), indicating that the involvement of emerging economy regulatory agencies leads to prolonged M&A deal durations. Thus, we accept H1. In contrast, the analysis finds no relationship between the involvement of developed economy regulatory agencies and M&A deal duration in Model 3, leading us to accept H2. Therefore, the regulatory agencies of developed economies are not associated with the M&A deal durations in our sample.

Table 3 Regression results (Source: Authors' elaboration)

Dependent variable: Deal duration			
	Model 1	Model 2	Model 3
Controls			
Cash payment	13.21 (16.62)	7.814 (16.47)	11.48 (17.01)
Acquired stake	0.374 (0.255)	0.413 (0.251)	0.364 (0.256)
Acquirer listed	12.08 (20.51)	7.903 (20.42)	13.73 (19.39)
Acquirer experience	5.659 (15.44)	3.657 (15.77)	6.505 (15.65)
Target listed	15.44 (25.22)	4.373 (22.85)	15.19 (25.59)
Target subsidiary	2.729 (16.94)	4.278 (16.55)	3.594 (17.48)
Institutional distance	−62.17* (33.99)	−51.25 (33.30)	−68.77* (35.92)
Independent variables			
Emerging economy regulatory agencies		116.023** (45.41)	
Developed economy regulatory agencies			23.02 (33.15)
Constant	44.15 (54.62)	51.93 (58.14)	45.65 (55.86)
Year dummies Industry dummies	Yes Yes	Yes Yes	Yes Yes
Observations R-squared	155 0.294	155 0.359	155 0.297

Note: Robust standard errors in parentheses, *** $p < 0.01$, ** $p < 0.05$, * $p < 0.1$

5 Robustness Checks

We have conducted further studies to check the robustness of the results, as depicted in Table 4. Firstly, a bilateral country relationship would influence the attitude of foreign investors. Prior studies (i.e., Li et al., 2017) have shown that bilateral investment treaties facilitate outward foreign direct investment and increase regulatory agencies' acceptance of foreign acquisitions. Thus, we controlled whether the home-host country had a bilateral investment treaty, and the results proved the robustness. Secondly, we have included the trade balance of countries. Nations with a negative trade balance may exhibit stronger economic nationalism and are more likely to obstruct deals. Even after accounting for this variable, our results remain consistent. Thirdly, the industrial relatedness of deals could influence the relationship between regulatory agencies and deal duration (Rao-Nicholson & Cai, 2020). On the one hand, entering a new industry abroad might increase the complexity of the deals; on the other hand, entering the same industry is more likely to trigger

Table 4 Robustness checks

Dependent variable: Deal duration				
	Model 1	Model 2	Model 3	Model 4
Independent variables				
Emerging economy	0.461***	0.480***	0.481***	0.453***
Regulatory agencies	(0.0234)	(0.0238)	(0.0239)	(0.0275)
Developed economy	−0.0400	−0.00948	−0.00583	−0.0122
regulatory agencies	(0.0398)	(0.0397)	(0.0408)	(0.0409)
Additional explanatory variables				
Bilateral investment treaty	0.356***	0.378***	0.379***	0.364***
	(0.0254)	(0.0262)	(0.0264)	(0.0273)
Trade balance		−0.00979***	−0.00977***	−0.00979***
		(0.00157)	(0.00157)	(0.00157)
Deal relatedness			−0.00836	0.00468
			(0.0217)	(0.0225)
Brazil antitrust				0.0744**
				(0.0361)
Controls				
Cash payment	0.0452**	0.0538***	0.0535***	0.0521***
	(0.0200)	(0.0200)	(0.0200)	(0.0200)
Acquired stake	0.00273***	0.00294***	0.00293***	0.00289***
	(0.000335)	(0.000336)	(0.000337)	(0.000337)
Acquirer listed	0.152***	0.151***	0.150***	0.159***
	(0.0242)	(0.0243)	(0.0244)	(0.0248)
Acquirer experience	0.0203	0.00501	0.00783	0.00307
	(0.0242)	(0.0244)	(0.0255)	(0.0256)
Deal size	0.0903***	0.0809***	0.0805***	0.0811***
	(0.00610)	(0.00617)	(0.00628)	(0.00629)
Target listed	0.168***	0.132***	0.132***	0.123***
	(0.0315)	(0.0320)	(0.0320)	(0.0323)
Target subsidiary	−0.216***	−0.206***	−0.206***	−0.199***
	(0.0252)	(0.0251)	(0.0251)	(0.0253)
Institutional distance	−0.292***	−0.207***	−0.206***	−0.234***
	(0.0611)	(0.0630)	(0.0631)	(0.0644)
Mining	−0.130***	−0.139***	−0.142***	−0.122***
	(0.0373)	(0.0378)	(0.0387)	(0.0398)
Construction	0.101	0.0613	0.0564	0.0531
	(0.0651)	(0.0653)	(0.0665)	(0.0665)
Manufacturing	−0.0980***	−0.120***	−0.123***	−0.103***
	(0.0286)	(0.0293)	(0.0302)	(0.0318)
Transport	0.683***	0.680***	0.678***	0.707***
	(0.0466)	(0.0473)	(0.0476)	(0.0496)
Services	−0.578***	−0.639***	−0.640***	−0.616***

(continued)

Table 4 (continued)

Dependent variable: Deal duration				
	Model 1	Model 2	Model 3	Model 4
	(0.0647)	(0.0652)	(0.0653)	(0.0665)
Wholesale	−0.250***	−0.270***	−0.275***	−0.248***
	(0.0419)	(0.0422)	(0.0445)	(0.0464)
Retail	−0.0901	−0.193*	−0.187*	−0.174
	(0.108)	(0.109)	(0.110)	(0.111)
Year dummies	Yes	Yes	Yes	Yes
Constant	3.455***	3.489***	3.498***	3.458***
	(0.110)	(0.111)	(0.113)	(0.115)
Observations	132	131	131	131
R-squared	0.429	0.429	0.429	0.430
Chi2	4312	4242	4242	4246

Source: Authors' elaboration
Note: Standard errors in parentheses, *** $p < 0.01$, ** $p < 0.05$, * $p < 0.1$

antitrust regulatory reviews, potentially delaying the deals. We have accounted for deal industrial relatedness in Column 3 of Table 4. The findings indicate that deals within the same industry do indeed extend deal duration, but the impact of regulatory agencies on deal duration remains robust. Lastly, we have also considered the influence of home country regulatory agencies in Brazil. We gathered data on the number of Brazilian regulatory agencies involved in cross-border deals, and the results continue to hold strong. Overall, we are confident in the robustness of our results.

6 Discussion

We investigated whether regulatory agencies provide Brazilian acquirers advantages or disadvantages in reducing M&A deal durations across emerging and developed economies. We shed light on the fact that regulatory agency involvement yielded different outcomes in emerging versus developed host economies but did not serve as an advantage. In line with our expectations, regulatory agencies prolonged transactions in emerging economies. In contrast, they neither prolong nor facilitate M&A deal durations in developed economies. Agency neutrality can be attributed to several factors beyond the agencies merely performing their mandated duties. Our explanations for emerging economies encompass the reasons relevant to developed economies, and the opposite of one reason for emerging economies can effectively explain the findings of developed economies. Thus, we will concisely integrate the findings for both types of economies to avoid duplication.

Our findings may offer empirical evidence to the literature where emerging economy regulatory process inefficiencies are mentioned (Chen et al., 2024; Kamasak & Yavuz, 2016; Mishra & Kumar, 2023). Bureaucratic red tape requires multiple approvals with documentation sets from various agencies. Firms engaged in M&A deals must navigate through layers of documentation, compliance checks, and multiple rounds of approvals. The complexity of regulatory frameworks, limited financial and human resources, and a lack of clear guidelines and necessary expertise to evaluate complicated M&A deals can slow down the M&A review process in emerging economies. The agencies may seek additional information from other advisory firms or struggle to understand the nuances of the transactions, prolonging the process even more. In India, various regulatory bodies, such as the Competition Commission of India (CCI), the Securities and Exchange Board of India (SEBI), and the Reserve Bank of India (RBI), have their own set of requirements and timelines with overlapping and conflicting demands (Bhattacharjea et al., 2019). Moreover, the need for firms to adhere to sector-specific regulations adds further delays. For example, Walmart's acquisition of Flipkart faced prolonged scrutiny from CCI. Delays were attributed to extensive information requests and the need to address multiple regulatory concerns (Saraswathy, 2019). Since developed economies have well-resourced regulatory agencies staffed with experts and adequately funded, this capacity ensures thorough and timely reviews of M&A transactions.

Emerging economies' weak and underdeveloped institutional infrastructure, consisting of inconsistent enforcement and frequent changes in regulatory policies and frameworks, can result in unpredictable regulatory landscapes and delays. Unlike in developed economies, one of their primary institutional problems is the slow and inefficient legal and judicial system. When regulatory decisions are challenged in court, the resulting legal proceedings and jurisdictions can significantly extend the duration of M&A deals. Developed economies have reliable frameworks for resolving disputes, with specialised courts and arbitration processes that can effectively handle complex commercial cases and promptly implement regulatory decisions. So, we can link our findings with these institutional differences between emerging and developed economies.

In particular, if the deals involve political agendas, key industry M&As or foreign investors, government or political interference and the imposition of additional conditions can be overwhelming. Additionally, favouritism and nepotism toward certain businesses can stem from personal relationships, political connections, economic interests, or individuals. This favouritism can create an unfair and unethical environment and create challenges for other firms. Governmental and political interventions may include rent-seeking behaviours and corruption. Politicians, bureaucrats or regulatory officials may demand bribes or favours in exchange for expedited approvals. Through delays in processing applications, unwarranted rejections, or the imposition of additional, arbitrary requirements, firms may find themselves caught in a cycle of negotiations and payments. The rent-seeking and corruption environment increases business costs and introduces uncertainty and delays in M&A transactions in emerging economies.

Emerging economy firms' efforts to obtain more sophisticated knowledge and overcome their legitimacy and image problems (Zhang et al., 2018) can compel them to rely more on government-related bodies (Brown et al., 2023) when conducting M&A transactions in other markets. However, our findings indicate that

Brazilian acquirers do not benefit from regulatory agency involvement in M&A transactions. Theoretically, we suggest that regulatory agencies in emerging economies may generate institutional costs, such as corruption, excessive bureaucracy, and administrative inefficiencies. When the deal scenario includes an emerging economy acquirer and target, these costs may contribute to even longer M&A deal durations, where the acquirer's government bodies might exploit a more conducive environment to assert their illegal or unethical advantages.

7 Conclusion

7.1 Policy Implications

Our study offers several managerial and policy implications for mitigating M&A deal completion delays. Managers should engage with regulatory agencies early in the pre-deal stage to proactively identify and address potential issues. Networking and relationship management are critical in emerging economies (Bai et al., 2021; Kamasak, 2013, 2014). Thus, building decent relationships with authorities and demonstrating a commitment to compliance can facilitate approvals. Employing advisors with expertise in the local regulatory environment can be beneficial in navigating complex processes. Managers should organise firms with experienced staff who can provide comprehensive documentation and respond to information requests quickly and transparently. Contingency plans that include various scenarios for potential delays can help manage the impact on deal timelines and address unexpected regulatory challenges with sound strategies.

Policymakers should standardise and simplify regulatory processes by streamlining transactions. Digital solutions like online portals and automated review systems can expedite these processes. Limited skilled and qualified staff in regulatory agencies is more pronounced in emerging economies. More investment and adequate funding for training and development for regulatory staff enable them to build capacity and expertise and handle M&A transactions effectively. Strengthening anti-corruption frameworks and promoting transparency in regulatory institutions are vital steps to combat rent-seeking behaviours and corruption. Policymakers must ensure that clear guidelines and accountability mechanisms are established. They can do it in several ways. Firstly, streamlining regulatory processes can help improve the efficiency of M&A approvals. Inspired by the Hart-Scott-Rodino Antitrust Improvements Act in the U.S., which sets clear timelines for antitrust reviews, emerging economies could benefit from implementing standardised timelines to minimise delays (Federal Trade Commission, 2020). Secondly, enhancing the capacity of regulatory bodies through intellectual capital and technology investment can also have a positive impact. Drawing from Singapore's approach, investing in training for regulatory staff and adopting advanced technology could significantly enhance the processing efficiency of M&A applications (MAS, 2019).

Thirdly, fostering public-private partnerships through establishing advisory boards that include the private sector, academia, and civil society stakeholders can also provide diverse insights and facilitate more informed policymaking (European Commission, 2018).

For example, in the United Kingdom, the Competition and Markets Authority (CMA) was established under the Enterprise and Regulatory Reform Act 2013, replacing the Competition Commission and parts of the Office of Fair Trading. The CMA uses a system that effectively differentiates between straightforward and complex M&A cases, ensuring timely reviews (CMA, 2024). The system consists of two phases. Phase 1 includes a preliminary to determine if a detailed investigation is needed. This review is usually completed within 40 working days to provide swift initial assessments. If potential competition issues are found, the merger moves to Phase 2. Phase 2 is more in-depth review is conducted when there is a realistic prospect of a substantial lessening of competition (SLC). This review involves a detailed analysis of the merger's impact on competition. Since the investigations are more thorough, this review typically takes 24 weeks, with a possible extension of up to 8 weeks.

Another successful model is the Australian Competition and Consumer Commission (ACCC), which has a transparent and consultative process for reviewing M&A deals. Established in 1995, the ACCC enforces the Competition and Consumer Act 2010 and other legislations, focusing on promoting competition and fair trading. The ACCC publishes detailed merger assessments, which include rationales behind each decision and invites public comments. This approach fosters transparency and trust in the regulatory process. It also helps stakeholders understand and anticipate the requirements and outcomes of merger reviews. It aims to complete public reviews within 12 weeks, providing a timely response and predictability to reduce uncertainty for the businesses involved (Ergas et al., 2009). Finally, political interference in M&A transactions should be minimised. Independent regulatory agencies with clear mandates can ensure impartiality and consistency in the approval processes and reduce stakeholders' legal objections, leading to shorter deal durations.

7.2 The Ethical Impact of Regulatory Delays

Regulatory delays can significantly impact the ethical landscape of business operations, particularly in emerging markets. These delays often lead to uncertainties that can hinder business growth, investment, and innovation, disproportionately affecting stakeholders ranging from small businesses to large corporations and consumers. The ethical implications are profound, as delayed regulatory processes can erode trust in public institutions, foster corruption, and create an uneven playing field. Enhancing transparency and accountability within regulatory frameworks is crucial to mitigate these ethical concerns. Transparency ensures that regulatory bodies' actions are visible and understandable to all stakeholders, which helps maintain public trust. Accountability ensures that regulators are answerable for their actions, which helps maintain integrity and ethical standards in regulatory processes (OECD, 2016). For example, requiring regulators to declare conflicts of interest and document decision-making processes publicly can help mitigate perceptions of bias and corruption and help build a culture of transparency, accountability, and trust (Grundy et al., 2022; Kilic & Kamasak, 2009).

Strengthening the institutional frameworks that underpin regulatory bodies can also enhance accountability. This includes establishing clear lines of responsibility, robust oversight mechanisms, and consequences for ethical breaches. For example, the UK's experience regarding the evolution of the regulatory impact assessment (RIA) and its integration into policy processes to enhance regulatory quality and effectiveness is quite notable (Humpherson, 2007). The 'Better Regulation Task Force' that was introduced in 1997 and the incorporation of RIA into broader government initiatives aimed at enhancing regulatory efficiency, including the introduction of the 'Regulatory Reform Action Plan' in 2002, are important milestones that underline the importance of the evolving nature of this issue (Jacobs, 2007). Despite the unique obstacles encountered in each case, different countries within the East African region have also adopted RIA to improve regulatory governance and policymaking (Welch, 2007), further underlining the continuous improvement mindset required to alleviate regulatory delays and the associated ethical impact.

7.3 Limitations and Future Research Suggestions

Our study has several limitations, and further studies that address these limitations can contribute to a more comprehensive understanding of the regulatory efficiency and effectiveness in M&A deals. Our study relies on a small sample size and a period between 2000 and 2015 since specific private M&A deals might not be fully reported in our database. Given the dynamic regulatory environments in emerging economies, longitudinal studies with additional data sources might provide more profound insights covering more recent regulatory reforms or shifts in political climates that could affect M&A processes. The study centres on the experiences of Brazilian M&As, which may not fully represent the situation in other emerging economies. Future research could benefit from a broader geographic scope to capture a more diverse range of emerging economies. Moreover, comparative studies could identify unique regulatory challenges across different countries and provide more generalised findings. Our study does not focus on sector-specific variations in regulatory impact. However, different industries, i.e. technology, energy, healthcare, logistics and finance, might experience varying levels of regulatory scrutiny, resulting in different M&A deal durations. Future research that delves into sector-specific analyses could offer insightful findings into the regulatory impacts on M&A deal durations across different industries.

References

Ahammad, M. F., Leone, V., Tarba, S. Y., Glaister, K. W., & Arslan, A. (2017). Equity ownership in cross-border mergers and acquisitions by British firms: An analysis of real options and transaction cost factors. *British Journal of Management, 28*(2), 180–196.

Ahsan, F. M., Popli, M., & Kumar, V. (2024). Formal institutions and cross-border mergers and acquisitions: A systematic literature review and research agenda. *International Business Review, 102306*, 102306.

Aktas, N., De Bodt, E., Bollaert, H., & Roll, R. (2012). CEO narcissism and the takeover process: From private initiation to deal completion. *Working paper in AFA 2012 Chicago Meetings Paper.*

Aquaro, M., Damioli, G., & Lengyel, B. (2023). Innovative mergers and acquisitions and the broker regions of European integration. *Regional Studies, 57*(2), 287–299.

Bai, W., Johanson, M., Oliveira, L., & Ratajczak-Mrozek, M. (2021). The role of business and social networks in the effectual internationalisation: Insights from emerging market SMEs. *Journal of Business Research, 129*, 96–109.

Barney, J. B., & Arikan, A. M. (2005). The resource-based view: Origins and implications. In M. A. Hitt, R. E. Freeman, & J. S. Harrison (Eds.), *The Blackwell handbook of strategic management* (pp. 123–182). Blackwell Publishing Ltd.

Barros, T. D. S., Cárdenas, J., & Hott, A. R. (2024). Shortening the distance between firms for M&A: The influence of small-world network structure on mergers and acquisitions. *International Journal of Emerging Markets, 19*(6), 1521–1541.

Batista, A. T. N., Lamounier, W. M., & Mário, P. D. C. (2023). Does economic policy uncertainty affect M&A operations? Evidence from the Brazilian market. *BBR. Brazilian Business Review, 20*, 133–156.

Battista, E., Nirino, N., Leonidou, E., Thrassou, A., & Vrontis, D. (2024). Mergers and acquisitions and environmental conditions: A systematic literature review to investigate their linkage. *Management Research Review, 47*(2), 226–244.

Bauer, F., & Friesl, M. (2024). Synergy evaluation in mergers and acquisitions: An attention-based view. *Journal of Management Studies, 61*(1), 37–68.

Bauer, F., & Matzler, K. (2014). Antecedents of M&A success: The role of strategic complementarity, cultural fit, and degree and speed of integration. *Strategic Management Journal, 35*(2), 269–291.

Bhattacharjea, A., De, O., & Gouri, G. (2019). Competition law and competition policy in India: How the competition commission has dealt with anticompetitive restraints by government entities. *Review of Industrial Organization, 54*, 221–250.

Bonaime, A., Gulen, H., & Ion, M. (2018). Does policy uncertainty affect mergers and acquisitions? *Journal of Financial Economics, 129*(3), 531–558.

Booij, M., & Sahib, P. R. (2012). Slow but certain: The pre-M&A process of European national champion deals. *Advances in Mergers and Acquisitions, 11*, 55–68.

Boone, A. L., & Mulherin, J. H. (2007). Do termination provisions truncate the takeover bidding process? *Review of Financial Studies, 20*(2), 461–489.

Brown, L. W., Liou, R. S., & Hasija, D. (2023). Emerging market multinational corporations' cross-border mergers and acquisitions and political distance: Does corporate political activity matter? *Thunderbird International Business Review, 65*(5), 533–546.

Cai, Y., Van Veen, K., & Gubbi, S. (2014). Does government stake influence cross-border deal completion? Evidence from Brazil. *Academy of Management Proceedings, 2014*(1), 13556.

Cai, Y., Tian, X., & Xia, H. (2016). Location, proximity, and M&A transactions. *Journal of Economics & Management Strategy, 25*(3), 688–719.

Cardillo, G., & Harasheh, M. (2023). Stay close to me: What do ESG scores tell about the deal timing in M&A transactions? *Finance Research Letters, 51*, 103498.

Carlton, D. W. (2007). Does antitrust need to be modernized? *Journal of Economic Perspectives, 21*(3), 155–176.

Cartwright, S., & Cooper, C. L. (2018). The psychological impact of merger and acquisition on the individual: A study of building society managers. In D. C. Cooper & R. Payne (Eds.), *Managerial, Occupational and Organizational Stress Research* (pp. 429–450). Routledge.

Casanova, L., & Kassum, J. (2013). Brazilian emerging multinationals*: In search of a second wind.*

Cataltepe, V., Kamasak, R., Palalar, A. D., & Bulutlar, F. (2023). Dynamic and marketing capabilities as determinants of firm performance: Evidence from the automotive industry. *Journal of Asia Business Studies, 17*(3), 617–638.

Chen, G., Dong, J., Hu, J., & Zhang, F. (2024). Regulatory, institutional misalignment and cross-border acquisitions: Evidence from an emerging-market country. *Journal of International Business Studies, 55*(2), 172–193.

Chung, C. Y., Hasan, I., Hwang, J., & Kim, I. (2023). The effects of antitrust laws on horizontal mergers: International evidence. *Journal of Financial and Quantitative Analysis*, 1–52.

CMA. (2024). *Mergers: Guidance on the CMA's jurisdiction and procedure (2024—Revised guidance)*. UK Competition and Markets Authority.

Col, B., & Sen, K. (2019). The role of corporate governance for acquisitions by the emerging market multinationals: Evidence from India. *Journal of Corporate Finance, 59*, 239–254.

Cuypers, I. R., Hennart, J. F., Silverman, B. S., & Ertug, G. (2021). Transaction cost theory: Past progress, current challenges, and suggestions for the future. *Academy of Management Annals, 15*(1), 111–150.

de Bodt, E., Cousin, J. G., Officer, M. S., & Roll, R. (2024). The deterrence effect of M&A regulatory enforcements. *Journal of Corporate Finance, 85*, 102559.

DePamphilis, D. (2019). Mergers, acquisitions, and other restructuring activities: An integrated approach to process, tools, cases, and solutions. Academic Press.

Dikova, D., Sahib, P., & van Witteloostuijn, A. (2010). Cross-border acquisition abandonment and completion: The effects of institutional differences and organizational learning in the international business service industry, 1981–2001. *Journal of International Business Studies, 41*, 223–245.

DiMaggio, P. J., & Powell, W. W. (1991). *The new institutionalism in organizational analysis* (Vol. 17). University of Chicago Press.

Dinc, I., & Erel, I. (2013). Economic nationalism in mergers and acquisitions. *Journal of Finance, 68*(6), 2471–2514.

Donnelly, R., Purkayastha, S., Manolova, T. S., & Edelman, L. F. (2024). Institutional distance, slack resources, and foreign market entry. *Journal of International Business Studies, 55*(2), 194–211.

Duran, P., Van Essen, M., Heugens, P. P., Kostova, T., & Peng, M. W. (2019). The impact of institutions on the competitive advantage of publicly listed family firms in emerging markets. *Global Strategy Journal, 9*(2), 243–274.

Eisenhardt, K. M., Santos, F. M., Pettigrew, I. A., Thomas, H., & Whittington, R. (2000). Knowledge based view. In A. Pettigrew, H. Thomas, & R. Whittington (Eds.), *Handbook of Strategy and Management* (pp. 139–164). Sage Publications.

Ekelund, R. B., Jr., Ford, G. S., & Thornton, M. (2001). The measurement of merger delay in regulated and restructuring industries. *Applied Economics Letters, 8*(8), 535–537.

Erel, I., Jang, Y., & Weisbach, M. S. (2015). Do acquisitions relieve target firms' financial constraints? *The Journal of Finance, 70*(1), 289–328.

Ergas, H., Ralph, E. K., & Robson, A. R. W. (2009). The ACCC merger guidelines: A review. *SSRN Electronic Journal*. https://ssrn.com/abstract=1424721

Ermolaeva, L. (2019). M&A deals completion and abandonment by Russian MNE. *International Journal of Emerging Markets, 14*(3), 475–494.

European Commission. (2018). Public private partnerships in the EU: Widespread shortcomings and limited benefits. *Special Report No* 09/2018.

Federal Trade Commission. (2020). Hart-Scott-Rodino Annual Report. https://www.ftc.gov/system/files/documents/reports/hart-scott-rodino-annual-report-fiscal-year-2020/fy2020_-_hsr_annual_report_-_final.pdf

Fuad, M., & Venugopal, A. (2024). Deal completion in mergers and acquisitions: Past accomplishment and future direction. *Cross Cultural & Strategic Management, 31*(1), 1–25.
Gaughan, P. H. (2017). Facilitating Meaningful Change Within US Law Schools. *UNHL Review, 16*, 243.
Geyskens, I., Steenkamp, J. B. E., & Kumar, N. (2006). Make, buy, or ally: A transaction cost theory meta-analysis. *Academy of Management Journal, 49*(3), 519–543.
Grant, A., Johnstone, D., & Kwon, O. K. (2022). How an idiosyncratic (zero-beta) risk can greatly increase the firm's cost of capital. *Australian Journal of Management, 47*(4), 664–685.
Gravetter, F. J., & Forzano, L.-A. B. (2008). *Research methods for the behavioural sciences* (3rd ed.). Wadsworth.
Grundy, Q., Parker, L., Wong, A., Fusire, T., Dimancesco, D., Tisocki, K., Walkowiak, H., Vian, T., & Kohler, J. (2022). Disclosure, transparency, and accountability: A qualitative survey of public sector pharmaceutical committee conflict of interest policies in the World Health Organization South-East Asia Region. *Globalization and Health, 18*(33), 1–23.
Gulapalli, E. K., Morand, C. L., & Wolski, G. E. (2017). Merger and acquisition transaction disputes. In M. A. Hitt, R. E. Ireland, & R. D. Ireland (Eds.), *Litigation services handbook: The role of the financial expert* (pp. 1–35). Wiley.
Hambrick, D. C., & Mason, P. A. (1984). Upper echelons: The organization as a reflection of its top managers. *Academy of Management Review, 9*(2), 193–206.
Harle, N., Ombregt, P., & Cool, K. (2012). *Merger control and practice in the BRIC countries vs. the EU and the US: Review thresholds (No. 2012/100)*. INSEAD Working Paper.
Harrison, J. S., Hitt, M. A., Hoskisson, R. E., & Ireland, R. D. (1991). Synergies and post-acquisition performance: Differences versus similarities in resource allocations. *Journal of Management, 17*(1), 173–190.
He, G., & Padrón-Hernández, I. (2024). Home country influence in cross-border mergers and acquisitions by emerging market firms: A systematic review. *Multinational Business Review, 32*, 265.
Hoskisson, R. E., Eden, L., Lau, C. M., & Wright, M. (2000). Strategy in emerging economies. *Academy of Management Journal, 43*(3), 249–267.
Huang, J., & Xie, T. (2023). Technology centrality, bilateral knowledge spillovers and mergers and acquisitions. *Journal of Corporate Finance, 79*, 102366.
Humpherson, E. (2007). Auditing regulatory impact assessment: UK experience. In C. Kirkpatrick & D. Parker (Eds.), *Regulatory impact assessment: Towards better regulation?* (pp. 155–173). Edward Elgar Publishing.
International Bar Association. (2022). *Brazil Negotiated M&A Guide 2022*. IBA Corporate and M&A Law Committee. https://www.ibanet.org/document?id=Corporate-m-a-minority-Brazil-22
International Comparative Legal Guide. (2015). *Brazil-merger control 2015*. http://www.iclg.co.uk/practice-areas/merger-control/merger-control-2015/brazil
Jacobs, C. (2007). The evolution and development of regulatory impact assessment in the UK. In C. Kirkpatrick & D. Parker (Eds.), *Regulatory impact assessment: Towards better regulation?* (pp. 135–154). Edward Elgar Publishing.
Jemison, D. B., & Sitkin, S. B. (1986). Corporate acquisitions: A process perspective. *Academy of Management Review, 11*(1), 145–163.
Kamasak, R. (2013). *Resource-based view (RBV) in all its aspects*. Scholars' Press Publishing.
Kamasak, R. (2014). Another attempt to look inside the 'black box' of performance creation: A case study approach. *International Journal of Business and Social Research, 4*(7), 1–16.
Kamasak, R., & Yavuz, M. (2015). Economic development, market characteristics and current business conditions in Turkey: A guide for successful operations. In B. Christiansen & M. M. Erdogdu (Eds.), *Comparative economics and regional development in Turkey* (pp. 336–354). IGI Global Publications.
Kamasak, R., & Yavuz, M. (2016). The use of business processes as a source of competitive advantage: Evidence from a high-velocity market. *Marmara Journal of Pure and Applied Sciences, 28*(S1), 28–33.

Kamasak, R., Yavuz, M., & Altuntas, G. (2016). Is the relationship between innovation performance and knowledge management contingent on environmental dynamism and learning capability? Evidence from a high turbulent market. *Business Research, 9*(2), 229–253.

Kamasak, R., Yozgat, U., & Yavuz, M. (2017). Knowledge process capabilities and innovation: Testing the moderating effects of environmental dynamism and strategic flexibility. *Knowledge Management Research & Practice, 15*(3), 356–368.

Kamasak, R., James, S. R., & Yavuz, M. (2019). The interplay of corporate social responsibility and corporate political activity in emerging markets: The role of strategic flexibility in nonmarket strategies. *Business Ethics: A European Review, 28*(3), 305–320.

Kamasak, R., Özbilgin, M. F., Küçükaltan, B., & Yavuz, M. (2020). Regendering of dynamic managerial capabilities in the context of binary perspectives on gender diversity. *Gender in Management: An International Journal, 35*(1), 19–36.

Kau, J. B., Linck, J. S., & Rubin, P. H. (2008). Do managers listen to the market? *Journal of Corporate Finance, 14*(4), 347–362.

Kilic, B., & Kamasak, R. (2009). Factor pattern of Auckland Individualism-Collectivism Questionnaire: A study of Turkish sample. *İş Güç, The Journal of Industrial Relations & Human Resources, 11*(5), 25–34.

Kim, H., & Song, J. (2017). Filling institutional voids in emerging economies: The impact of capital market development and business groups on M&A deal abandonment. *Journal of International Business Studies, 48*, 308–323.

King, D. R., Wang, G., Samimi, M., & Cortes, A. F. (2021). A meta-analytic integration of acquisition performance prediction. *Journal of Management Studies, 58*(5), 1198–1236.

Koirala, S., Rao, S., Farag, H., & Marshall, A. (2023). The market for corporate control and risk-taking: Evidence from global merger and acquisition laws. *British Journal of Management, 34*(2), 997–1022.

Kostova, T., Roth, K., & Dacin, M. T. (2008). Institutional theory in the study of multinational corporations: A critique and new directions. *Academy of Management Review, 33*(4), 994–1006.

Kumar, D., Sengupta, K., & Bhattacharya, M. (2023). Macroeconomic influences on M&A deal outcomes: An analysis of domestic and cross-border M&As in developed and emerging economies. *Journal of Business Research, 161*, 113831.

Lawrence, E. R., Raithatha, M., & Rodriguez, I. (2021). The effect of cultural and institutional factors on initiation, completion, and duration of cross-border acquisitions. *Journal of Corporate Finance, 68*, 101950.

Lawrence, E. R., Nguyen, T. D., & Upadhyay, A. (2024). Independence of board leadership of acquirers and the success of mergers and acquisitions. *Journal of Corporate Finance, 86*, 102581.

Li, J., Xia, J., & Lin, Z. (2017). Cross-border acquisitions by state-owned firms: How do legitimacy concerns affect the completion and duration of their acquisitions? *Strategic Management Journal, 38*(9), 1915–1934.

Lindner, T., Müllner, J., & Puhr, H. (2023). Ownership, institutions, and the agency of M&A completion. *Global Strategy Journal.*

Liou, R. S., Chao, M. C. H., & Yang, M. (2016). Emerging economies and institutional quality: Assessing the differential effects of institutional distances on ownership strategy. *Journal of World Business, 51*(4), 600–611.

Liu, J., Hu, M., Zhang, H., & Carrick, J. (2019). Corruption and entrepreneurship in emerging markets. *Emerging Markets Finance and Trade, 55*(5), 1051–1068.

Lou, B., Shepherd, N. G., Strobl, A., & Bauer, F. (2024). Playing politics: An upper echelons perspective on political behaviour during acquisition decision making. *British Journal of Management.*

Luypaert, M., & De Maeseneire, W. (2015). Antecedents of time to completion in mergers and acquisitions. *Applied Economics Letters, 22*(4), 299–304.

MAS (Monetary Authority of Singapore). (2019). *Financial Sector Technology and Innovation Scheme (FSTI 3.0).* https://www.mas.gov.sg/schemes-and-initiatives/fsti-scheme

Masulis, R. W., & Simsir, S. A. (2018). Deal initiation in mergers and acquisitions. *Journal of Financial and Quantitative Analysis, 53*(6), 2389–2430.

Mishra, B., & Kumar, A. (2023). Contouring of the regulatory governance framework in emerging economies: A novel multidimensional approach. *International Journal of Emerging Markets.*

North, D. (1990). *Institutions, institutional change and economic performance*. Cambridge University Press.

OECD. (2016). Governance of regulators' practices: Accountability, transparency and co-ordination. *The Governance of Regulators*. https://doi.org/10.1787/9789264255388-en

Oliver, C. (1991). Strategic responses to institutional processes. *Academy of Management Review, 16*(1), 145–179.

Paul, J., & Jadhav, P. (2020). Institutional determinants of foreign direct investment inflows: Evidence from emerging markets. *International Journal of Emerging Markets, 15*(2), 245–261.

Peng, M., Wang, D., & Jiang, Y. (2008). An institution-based view of international business strategy: a focus on emerging economies. *Journal of International Business Studies, 39*, 920–936. https://doi.org/10.1057/palgrave.jibs.8400377

Popli, M., & Ahsan, F. M. (2024). Springboarding and activity load: Constraints on managerial attention and dependency on advisory firms in cross-border acquisitions. *Journal of International Management, 30*(2), 101121.

Ragozzino, R., & Reuer, J. J. (2024). Implications of mergers and acquisitions for information disclosures in earnings calls. *Long Range Planning, 57*(1), 102393.

Rao-Nicholson, R., & Cai, C. (2020). The effects of ownership identity on corporate diversification strategy of Chinese companies in foreign markets. *Asia Pacific Journal of Management, 37*(1), 91–126. https://doi.org/10.1007/s10490-018-9578-8

Sakhartov, A. V., & Reuer, J. J. (2023). Resource redeployment in corporate acquisitions: Going beyond horizontal acquisitions. *Long Range Planning, 56*(6), 102287.

Salancik, G. R., & Pfeffer, J. (1978). A social information processing approach to job attitudes and task design. *Administrative science quarterly*, 224–253.

Sales, A. C. S., & Zanini, M. T. F. (2017). Investigating the deal-making failure in M&A: Deal makers' perspective in Brazil. *Revista de Administração, 52*, 467–478.

Saraswathy, B. (2019). The Flipkart-Walmart deal in India: A look into competition and other related issues. *The Antitrust Bulletin, 64*(1), 136–147. https://doi.org/10.1177/0003603X18823619

Schijven, M., Heimeriks, K. H., Graebner, M. E., Haspeslagh, P., & Mitchell, W. (2024). Thirty-three years after "managing acquisitions": Reflections, insights, and research directions. *Strategic Management Review.*

Short, H., Keasey, K., Wright, M., & Hull, A. (1999). Corporate governance: From accountability to enterprise. *Accounting and Business Research, 29*(4), 337–352.

Simpson, J. J., & Sariol, M. (2022). Uncertainty, entrepreneurial orientation, and the pursuit of M&A: Managing the unpredictable. *Journal of Business Research, 142*, 423–434.

Singh, V., Maheshwari, G. C., & Yadav, R. (2024). Effect of deal size on M&A outcome on HR parameters: An empirical analysis. *International Journal of Economics and Business Research, 27*(3), 404–433.

Sirower, M. (1997). *The synergy trap: How companies lose the acquisition game*. Free Press.

Soleimani, A., & Yang, K. M. (2022). Overcoming institutional challenges in cross-border acquisitions: How home-and host-country acquisition experiences shorten the deal completion phase. *Multinational Business Review, 30*(3), 410–433.

Song, S., Zeng, Y., & Zhou, B. (2021). Information asymmetry, cross-listing, and post-M&A performance. *Journal of Business Research, 122*, 447–457.

Strobl, A., Bauer, F., & Matzler, K. (2020). The impact of industry-wide and target market environmental hostility on entrepreneurial leadership in mergers and acquisitions. *Journal of World Business, 55*(2), 100931.

Teece, D. J., Pisano, G., & Shuen, A. (1997). Dynamic capabilities and strategic management. *Strategic Management Journal, 18*(7), 509–533.

Wang, A. (2009). The choice of market entry mode: Cross-border M&A or greenfield investment. *International Journal of Business and Management, 4*(5), 239–245.

Welch, D. (2007). Regulatory impact assessment in East Africa. In C. Kirkpatrick & D. Parker (Eds.), *Regulatory impact assessment: Towards better regulation?* (pp. 245–263). Edward Elgar Publishing.

Welch, X., Pavićević, S., Keil, T., & Laamanen, T. (2020). The pre-deal phase of mergers and acquisitions: A review and research agenda. *Journal of Management, 46*(6), 843–878.

White & Case. (2023). *Shining a light on the massive global surge in merger control filings.* https://www.whitecase.com/insight-our-thinking/shining-light-massive-global-surge-merger-control-filings.

Williamson, O. E. (1998). Transaction cost economics: How it works; where it is headed. *De Economist, 146*, 23–58.

Yurdakul, H., & Kamasak, R. (2021). Public private partnership (PPP) as a mechanism to improve the infrastructure needs of countries. In E. Akman, C. Babaoglu, & O. Kulac (Eds.), *Global Challenges to Improving Public Services and Government Operations* (pp. 222–241). IGI Global Publications.

Yurdakul, H., Kamasak, R., & Ozturk, T. Y. (2022). Macroeconomic drivers of public-private partnership (PPP) projects in low-income and developing countries: A panel data analysis. *Borsa Istanbul Review, 22*(1), 37–46.

Zaheer, S. (1995). Overcoming the liability of foreignness. *Academy of Management Journal, 38*(2), 341–363.

Zakaria, R., & Kiziler, E. (2023). Sectoral and institutional characteristics of acquirers and targets: A dyadic approach to deal completion. *Review of International Business and Strategy, 33*(4), 649–668.

Zhang, H., Young, M. N., Tan, J., & Sun, W. (2018). How Chinese companies deal with a legitimacy imbalance when acquiring firms from developed economies. *Journal of World Business, 53*(5), 752–767.

Zhou, S., Lan, F., Hu, Z., & Liu, Y. (2024). Determinants of successful mergers and acquisitions in China: Evidence from machine learning. *Digital Economy and Sustainable Development, 2*(1), 4.

Catherine Cai is a Senior Lecturer in Strategy and International Management at the University of the West of England (UWE Bristol) in the United Kingdom. Her research interests include emerging market multinational enterprises, cross-border M&As and corporate sustainability.

Rifat Kamasak is a Professor of Strategy and Entrepreneurship at the University of Reading, Henley Business School, UK. He also holds board membership positions in several companies listed on the Istanbul Stock Exchange. He worked in the food, confectionery, carpet, textile, aluminium, metal, retailing, trading and consulting industries for nearly twenty years. He has done research, consultancy and training at a large number of organisations. His primary areas of interest are strategic management, knowledge, innovation, and diversity management.

Aslihan Bozcuk is a Professor of Finance at Akdeniz University in Antalya, Turkey. She has extensive experience in empirical finance and accounting in corporate and academic settings. Her primary areas of interest are accounting and auditing, financial markets, and investment management.

The Link Between Climate Change and Human Health

Shabnum Soomro, Jam Ghulam Murtaza Sahito, and Farahnaz Gilal

Abbreviations

IPCC	Intergovernmental Panel on Climate Change
GHG	Greenhouse Gases
CO_2	Carbon Dioxide
CH4	Methane
N_2O	Nitrous oxide
US EPA	United States Environmental Protection Agency
NASA	National Aeronautics and Space Administration
NASEM	National Academies of Science, Engineering, and Medicine
EHE	Extreme Heat Events
CCA	Climate Change Anxiety
WHO	World Health Organization
COVID-19	Coronavirus Disease of 2019
WPR	World Population Review
CAF	Clean Air Fund
PM	Particulate Matter
NO_2	Nitrogen Dioxide
O_3	Ozone
COPD	Chronic Obstructive Pulmonary Disease
PTSD	Post-Traumatic Stress Disorder

S. Soomro (✉) · J. G. M. Sahito · F. Gilal
Department of Agricultural Economics, Singh Agriculture University, Tando Jam, Sindh, Pakistan
e-mail: sahito@sau.edu.pk

J. Bento, M. M. Torres (eds.), *Global Perspectives on Climate Change, Inequality, and Multinational Corporations*, CSR, Sustainability, Ethics & Governance,
https://doi.org/10.1007/978-3-031-80797-8_8

1 Introduction

The word "climate" refers to the general environmental conditions that can be observed as changes in temperature, precipitation, pressure, and humidity in the atmosphere. "Climate change" refers to the alteration of these environmental conditions through natural phenomena or human activities. Climate change causes global warming, unpredictable weather patterns, melting of ice glaciers, rising sea levels, and many other events related to the atmosphere all over the world. Climate change can be interpreted as a change in the mean and/or the variability of climate attributes detected and measured, for instance, by statistical testing, over a period of time sufficiently prolonged to ensure an adequate signal-to-noise ratio, usually several decades or longer (Fawzy et al., 2020). Greenhouse gas emissions from natural systems and human activities primarily drive this change in climate. Human activities have caused approximately a 1.0 °C increase in global warming compared to pre-industrial levels, and this is expected to further increase to 1.5 °C between 2030 and 2052 if current emission rates continue. In 2018, there were 315 natural disasters globally, primarily linked to the climate, impacting around 68.5 million individuals and leading to economic losses of 131.7 billion US dollars. Storms, floods, wildfires, and droughts made up about 93% of the overall damage.

The increasing frequency of heatwaves, illnesses linked to the climate, and natural disasters are all consequences of climate change, which seriously threaten world health. Extreme weather occurrences like floods, droughts, earthquakes, cyclones, volcanic eruptions, tsunamis, hurricanes, rising sea levels, poor air quality, and degraded food and water security all harm people's health. It also causes population migration and has an impact on vulnerable shelters (Gulzar et al., 2021; Hong, 2008; Hong & Seo, 2011). One of the more common factors influencing human health has been the climate and weather. Extreme weather, in particular, can have an impact on the environment's ability to provide clean air, food, water, shelter, and security because it is a factor that relates to climate and weather variability. Along with naturally occurring and man-made health stressors, climate variations pose a wide range of dangers to human health and well-being.

The primary cause of worsening climatic conditions, which leads to global warming, and the influence of extreme weather occurrences is human activity. There is a lack of planning for expected climate change effects and a poor consideration of how climate change may affect extreme events. Human-induced climate change is a significant and real problem, supported by observational evidence and scientific understanding. Climate change already has significant and severe effects. The general public is not adequately aware of or alarmed about the problem. Climate change is a generational issue with equity issues. Society is not making enough efforts to lessen the issue or prepare for its effects (Trenberth, 2018).

Degradation of the environment may accelerate the spread of heart and lung problems, diarrhea, malnourishment, and infectious and vector-borne diseases. The less obvious consequences of these climatic variations include migration, stress-related mental health issues, homelessness, and unstable economies. The groups most susceptible to the negative effects are mostly the children, the old, and those who are mostly poor (Kim et al., 2014). According to Franchini and Mannucci (2015), the rapid changes in the climate are becoming more and more noticeable. Numerous negative impacts on human health are expected from changes, which are mostly the results of exceptional increases in greenhouse gas emissions caused by human activity. A surge in allergies and vector-borne infectious diseases are two of them, with their distribution changing globally.

The emissions of greenhouse gases (GHGs) are the main drivers of climate change, which is naturally occurring and has a large impact on the climate. Examples of GHGs are CO_2, CH_4, N_2O, and H_2O. The continuous increase in extreme atmospheric situations due to global warming is one of the adverse effects of climate change. These changes are not solely caused by human activities, but natural phenomena such as solar cycles, earthquakes, and volcanic eruptions also contribute to environmental degradation (Roy, 2018). Thus far, natural sinks have prevented greenhouse gas emissions from rising above a safe limit. Yet ongoing human activities like the expansion of vehicle usage, farming, deforestation (Huang et al., 2016), industrialization, and combustion of fossil fuels (Yousaf et al., 2017) have increased global GHG emissions more rapidly, except for water vapor (Bousquet et al., 2006). The rise of global warming due to the emission of greenhouse gases has degraded the climate and upset the natural atmospheric equilibrium, a condition that was not present before the Industrial Revolution. It's generally accepted that using fossil fuels raises the atmospheric concentrations of greenhouse gases, which is causing climate change. Numerous causes could contribute to the health implications, including greater heat wave frequency and severity, fewer deaths from the cold, more frequent floods and droughts, shifts in the distribution of vector-borne diseases, and effects on the likelihood of disasters and famine. People living in less developed countries are more likely to be impacted by the effects, leading to overall damage to health conditions. High-income nations may potentially be negatively impacted, as demonstrated by the heat wave that struck Europe in 2003.

Pakistan is becoming more and more susceptible to the rampant impacts of climate change, with frequent floods, droughts, and storms becoming the norm. Despite contributing very little to global warming, the country is facing severe consequences of climatic variations. Comprehensive policies to address these natural disasters are lacking, and as a developing country, there is a lack of study on environmental vulnerability and perceptions of local risks, and Pakistan is not making nearly enough effort to adapt. Due to people's increased susceptibility to heat waves and other intense weather events like earthquakes, volcanic eruptions, tsunamis, flash floods, droughts, and tropical cyclones, the risks and harmful impacts of climate change are anticipated to rise substantially (Arndt et al., 2010; Rahmstorf & Coumou, 2011). Among the various changes, the worsening of the atmospheric environment is particularly important, as these changes can ultimately lead to various illnesses and diseases, such as cardiovascular deaths, respiratory illnesses, diseases carried by insects, lack of food leading to malnutrition, the spread of infectious diseases, and diarrhea caused by poor sanitation (Confalonieri et al., 2007; Costello et al., 2009).

Climate change also increases air pollution, which significantly impacts human health, mainly for those with asthma and cardiac difficulties (Snow & Snow, 2015). It shows that the gastrointestinal tract and cardiovascular system are more fragile to the negative impacts of global warming. Furthermore, several infectious diseases and the animals that carry them are impacted by climate change; the likelihood of contracting cholera, dengue fever, typhus, malaria, and West Nile infection rises. In middle latitudes, warming may reduce the occurrence of cold-related illnesses such as pneumonia, bronchitis, and arthritis; nevertheless, the risks of warming are likely to exceed these advantages. The reports claim that climate alterations cause at least 150,000 deaths per year, with this number expected to double by 2030. These fatalities are attributed to shifts in temperature and precipitation patterns resulting from anthropogenic climate change over the past 30 years. Climate fluctuations are linked to various human diseases, including respiratory and cardiovascular illnesses, heatwaves, crop failures leading to malnutrition, and altered transmission of infectious diseases. However, due to a lack of sufficient, high-quality data over an extended period of time, as well as relationship-affecting factors like treatment resistance, changes in immunity, and socioeconomic conditions, it is challenging to link the spread of diseases to climate change. There is still evidence to suggest that there are substantial health concerns associated with climate change and that the trend of recent decades has already increased mortality in many parts of the world. These kinds of dangers could be more common in large cities and could increase the effects of extreme weather events by creating an urban heat island effect in sub-Saharan Africa and temperate latitudes with variable rainfall.

Mental health is negatively impacted by climate variability, particularly anxiety and climate change anxiety (CCA), which is linked to psychiatric symptoms and functional impairment (Schwartz et al., 2023). Extreme weather, subacute weather, and environmental changes all have a substantial impact on mental health. These events can cause economic losses, decreased production, health risks, displacement of population, disconnect from the natural environment, and a breakdown in society. The timing and form of these occurrences, as well as the relationships between events and psychological outcomes, are critical for developing and implementing strategies to prevent and treat climate-related mental health issues. These events have a profound impact on mental health, and understanding when and how they occur is critical for effective policy creation.

The motivation for this discussion of the connection between human health and climate change is to increase the public's awareness of the significant impact that shifting environmental circumstances have on public health. Events that may represent a serious threat to human health include those that involve more frequent and severe weather, opportunities for the spread of infectious diseases, and other health hazards. With awareness of this connection, preventative actions can be taken to lessen the negative impacts and shield vulnerable groups from the long-term health consequences of environmental deterioration. The questions to be answered are the following:

- How does climate change contribute to the increased occurrence and severity of extreme weather events, and what are the health implications of these events?
- What steps can be taken to reduce the health risks of climate change for children, the elderly, and poor populations?
- What measures can be implemented to mitigate the health risks associated with climate change, particularly for vulnerable populations such as children, the elderly, and impoverished communities?
- What are the direct and indirect health consequences of climate change, and how do they manifest in different demographic groups and regions?

This chapter focuses on how climate change has a significant impact on human health and the spread of infectious diseases. It covers the issues, including the connection between vector-borne illness and climate change, the impact of intensive weather events caused by climate change and air pollution, and the health and environmental risks caused by these variables. Additionally, the public is made aware of the implications of climate change on human health in this chapter, particularly as it relates to vulnerable populations such as the elderly, children, and those who are living in poverty.

2 Climate Change Adaptation and Mitigation

Climate change adaptation refers to the process of adjustment in response to actual and expected climate change impacts, including rising temperatures, changed precipitation patterns, and increased frequency of extreme weather events, among others. On the other hand, climate change mitigation pertains to the deliberate actions taken to diminish the severity of climate change by reducing greenhouse gas emissions and augmenting carbon sinks. Füssel (2008) investigated climate change adaptation assessments with the goal of assisting policymakers in mitigating the concerns related to health caused by changing climate dimensions of the world. This study focuses on important components of health, climate, and adaptations that must be considered in climate change adaptation evaluations. It then assesses if these issues are appropriately addressed in current guidelines for controlling climate impacts. The analysis found that three evaluation standards based on a generic risk management framework are the most effective for guiding health-related adaptation assessments. However, because none of these standards fully addresses all of the fundamental issues of the adaptation decision process, practical adaptation evaluations must include parts from multiple recommendations. Since many climate-sensitive health concerns are not covered by the prevalent toxicological model of environmental health, current conceptual frameworks of ecological epidemiology are found to be of limited usefulness for assessing and arranging adaptation to climate change.

Additionally, Patel et al. (2022) conducted a study that revealed that the warming climate of the earth is leading to more frequent, longer-lasting, and hotter heat waves in regions unaccustomed to such weather events. In recent years, extreme heat events (EHEs) in various parts of the world have led to severe consequences due to climate change. These events directly lead to increased illness and mortality from heat-related conditions, exacerbate existing health issues, and intensify the adverse health effects of social factors. Indirectly, they contribute to corollary events like wildfires and air pollution. It is projected that climate change-related EHEs will continue to worsen for at least the next three decades, necessitating the need for health systems to prepare for a growing burden of heat-related illnesses, build heat resilience, and minimize healthcare-related climate impacts. Moreover, the possible dangers and negative effects of climate change are expected to increase significantly due to more frequent heat waves and extreme weather events such as flash floods, droughts, earthquakes, volcanic eruptions, tsunamis, and tropical cyclones. Among the various changes, the worsening of the atmospheric environment is particularly important, as these changes can ultimately lead to various illnesses and diseases, such as cardiovascular deaths, respiratory illnesses, diseases carried by insects, lack of food leading to malnutrition, the spread of infectious diseases, and diarrhea caused by poor sanitation. Moreover, climate change contributes to air pollution, which has a significant impact on human health, especially for people with asthma and cardiovascular difficulties. The cardiovascular system and gastrointestinal tract are especially vulnerable to the harmful consequences of global warming. Additionally, climate change affects some infectious diseases and their animal vectors, raising the risk of typhus, cholera, malaria, dengue fever, and West Nile virus infection.

3 Climate Change and Infectious Diseases

Climate change is a significant factor influencing the spread of infectious diseases. The danger of infections increases due to the shifting climate, which facilitates the spread of animals, mosquitoes, ticks, and the diseases they carry into new regions. Pathogens and their carrier animals expand due to various factors, including changing land use, warmer temperatures, and severe weather conditions. For example, warmer summers and milder winters give disease-carrying insects such as ticks and mosquitoes more time to develop and transmit diseases like West Nile virus and Lyme disease. Additionally, hazardous blooms of algae in groundwater and marine systems may become more severe due to climate change, which could endanger human health by polluting water or seafood. Climate change worsens over half of the identified human pathogenic diseases, affecting 58% of infectious diseases globally at some time (Mora et al., 2022). One of the main causes of the climatic dangers that worsen infectious diseases in humans is the release of greenhouse gases. To reduce these hazards and safeguard human health, we must act quickly and decisively to limit greenhouse gas emissions. Climate change plays a crucial role in spreading infectious diseases by altering habitats, weather patterns, and environmental conditions that influence the transmission of pathogens from animals to humans. Understanding these interactions is essential for mitigating the risks associated with infectious diseases in a changing climate. Climate change significantly impacts the prevalence of vector-borne diseases, such as malaria, dengue fever, and Zika virus.

Changes in precipitation and temperature can result in the spread of illnesses. The current mortality rate due to vector-borne diseases stands at 700,000, but without preventive measures, the number of cases may increase in the future. Climate change can lead to long-term challenges, including anxiety and post-traumatic stress disorder, due to factors such as environmental breakdown and displacement. Vector-borne infections kill about 700,000 people annually, accounting for more than 17% of all infectious diseases. These diseases can be caused by parasites, bacteria, or viruses. Mosquitoes and ticks are vectors for viral infections such as chikungunya fever, Zika virus fever, yellow fever, West Nile fever, Japanese encephalitis, and tick-borne encephalitis. According to a World Health Organization report released in 2020, preventive measures and community involvement can help to prevent a significant percentage of vector-borne infections. Several infectious diseases are significantly affected by climate change (dengue fever, malaria, cholera, Lyme disease, West Nile virus, flesh-eating bacteria, Zika virus, and chikungunya).

3.1 Dengue Fever

Dengue fever is a mosquito-borne viral disease highly responsive to climate change. Warmer temperatures and fluctuations in precipitation can boost mosquito populations and increase disease transfer to new places. Dengue is the most common viral infection carried by Aedes mosquitoes. Dengue affects more than 3.9 billion individuals in 129 countries, resulting in an estimated 96 million symptomatic cases and 40,000 fatalities every year.

3.2 Malaria

Malaria is a parasitic disease transmitted by mosquitoes, and its spread is affected by temperature, rainfall, and humidity. Extreme weather events, such as heatwaves and flooding, have a direct impact on malaria transmission and illness prevalence. As temperatures rise, mosquitos that carry the malaria parasite proliferate into previously inhospitable places. Malaria is a parasite ailment spread by Anopheline mosquitoes. It causes an estimated 219 million illnesses worldwide and kills over 400,000 people each year. The majority of deaths involve children under the age of five. According to a World Health Organization report published in 2023, global malaria cases in 2022 topped the pre-pandemic level in 2019 by 16 million, reaching an estimated 249 million. The COVID-19 pandemic disrupted malaria services, causing a considerable increase in both incidence and mortality rates. Pakistan,

Ethiopia, Nigeria, Papua New Guinea, and Uganda were among the countries most severely affected. Pakistan saw the greatest increase, with approximately 2.6 million cases in 2022, up from 500,000 in 2021.

3.3 Cholera

Cholera is an acute diarrheal infection caused by the bacterium Vibrio cholera, which is found in contaminated water. Extreme weather events such as floods can cause the right conditions for cholera outbreaks, especially in regions with inadequate water, sanitation, and hygiene infrastructure. In 2022, an unprecedented surge of larger and more lethal cholera outbreaks occurred globally, with approximately 30 countries reporting infections, significantly exceeding the average of fewer than 20 countries in the preceding 5 years. These outbreaks not only increased in frequency but also intensified in severity compared to previous occurrences. Factors contributing to this escalation in cholera cases include the influence of conflict and mass displacement on a global scale, along with the visible impact of climate change. Extreme climate events such as floods, cyclones, and droughts have led to reduced access to clean water, thereby creating an optimal environment for the transmission of cholera. This cholera crisis has unfolded across diverse regions, encompassing the Horn of Africa, the Sahel, Pakistan, Haiti, Lebanon, Malawi, and Syria. For example, Pakistan witnessed an excess of 500,000 reported cases of watery diarrhea following devastating summer floods, with a smaller number of laboratory-confirmed cholera cases.

3.4 Lyme Disease

Lyme disease is a bacterial infection transmitted by tick bites. Warmer temperatures and longer summers can lead to increased tick populations and the spread of the disease to new areas. Between 1991 and 2018, the number of recorded cases of Lyme disease in the United States saw a rapid surge, from 3.74 to 7.21 cases per 100,000 people. Climate change is one of the causes contributing to this increase, as it is increasing the area of ticks that carry the Lyme disease virus. Studies indicate that as a result of the rising temperature, there will likely be more suitable tick habitats across a wider area, which will raise the risk of Lyme disease. For instance, by the middle of the century, an increase in temperature is probably going to increase the number of Lyme disease cases by more than 20% (Kizielewicz, 2018).

3.5 *West Nile Virus*

West Nile virus is a mosquito-borne disease that can cause fever, encephalitis, and meningitis. Milder winters and warmer summers can lead to increased mosquito populations and the spread of the disease.

3.6 *Flesh-Eating Bacteria (Vibrio vulnificus)*

It lives in sea or brackish water when the temperature rises above 20 °C. It can lead to serious illness, such as necrotizing fasciitis, which can result in amputation. The rising water temperature has been linked to a rise in disease caused by this bacterium.

3.7 *Zika Virus*

Zika is a mosquito-borne viral disease that can cause severe birth defects in babies born to infected mothers. Warmer temperatures and changes in precipitation can lead to increased mosquito populations and the spread of the disease to new areas. There are no current outbreaks of Zika virus worldwide. However, health experts caution that new outbreaks could occur at any time, potentially due to a single mutation leading to a new virus variant. The World Health Organization (WHO) reports that while cases of Zika virus disease declined globally from 2017 onwards, transmission persists at low levels in several countries in the Americas and other endemic regions. Furthermore, in Europe, the first cases of the Zika virus spread locally by mosquitoes were reported in 2019, while in India, outbreak activity was noticed in 2021. Surveillance remains limited globally, but 89 countries and territories have reported evidence of mosquito-transmitted Zika virus infection to date.

3.8 *Chikungunya*

Chikungunya is a mosquito-borne viral disease that can cause fever, joint pain, and rash. Warmer temperatures and changes in precipitation can lead to increased mosquito populations and the spread of the disease to new areas. These diseases are affected by climate change through various mechanisms, such as changes in temperature, precipitation, and extreme weather events, which can influence the distribution, reproduction, and survival of the vectors (e.g., mosquitoes and ticks) and the pathogens they transmit.

4 Spread of Vector-Borne Diseases

The spread of vector-borne diseases (VBDs) is a multifaceted issue influenced by an array of factors such as climate change, human migration and travel, urbanization and land use changes, global trade and commerce, and inadequate public health infrastructure. With the ongoing climate warming altering ecosystems, vectors, such as mosquitoes, ticks, and fleas, are expanding their ranges, leading to an increased risk of VBD transmission. Additionally, human migration and travel can introduce vectors and pathogens to previously unaffected regions, while urbanization and land use changes can create new habitats for vectors and increase human-vector interactions. Global trade and commerce can also unintentionally introduce vectors and pathogens to new areas, underscoring the necessity for robust surveillance and monitoring systems.

Understanding the biology and ecology of vectors is imperative for predicting and preventing VBD transmission. Effective prevention and control measures encompass vector control, personal protection, vaccination, surveillance and monitoring, and public awareness and education. Addressing these factors and implementing effective strategies is crucial for mitigating the spread of VBDs and safeguarding public health. The geographic range in which vectors (such as ticks and mosquitoes) can live and reproduce can be increased by warmer temperatures. On the other hand, droughts can also facilitate breeding by generating pools of standing water from formerly flowing sources. Higher rainfall increases the number of breeding places for many vectors. By changing temperature patterns, global warming may increase the season during which diseases are transmitted. Pathogens, vectors, and hosts linked to vector-borne illnesses become more active as temperatures rise. Nevertheless, it can be difficult to accurately attribute these effects to climate change alone because other factors (such as changes in land use and human migration) also come into play. Due to climate change, an additional 4.7 billion people may be vulnerable to vector-borne illnesses, including dengue and malaria, by 2070. Globally, the incidence and geographic distribution of major vector-borne diseases is anticipated to rise. Certain illnesses might reappear in places where they were previously eradicated or spread to other areas (Ren et al., 2021).

5 Health-Related Illnesses and Risk Factors

Heatwaves, defined as prolonged periods of excessively hot weather, pose significant risks to human health. These extreme weather events are becoming more frequent and severe due to climate change, affecting populations globally. The health impacts of heatwaves are profound, particularly for vulnerable groups such as the elderly, children, those with preexisting health conditions, and low-income communities.

5.1 Health Effects of Heatwaves

Heatwaves have a devastating impact on human health, causing heat exhaustion, heat stroke, and cardiovascular strain. They also exacerbate respiratory issues, lead to dehydration and electrolyte imbalances, and can cause kidney damage and even death. The mental health effects, including anxiety, depression, and PTSD, can linger long after the heat subsides. Vulnerable populations, such as the elderly and those with pre-existing conditions, are disproportionately affected.

Heat-related illnesses can pose significant risks to individuals, especially during periods of extreme heat. These conditions include heat exhaustion, heat stroke, and heat cramps. Heat exhaustion is characterized by heavy sweating, weakness, dizziness, nausea, and fainting. It occurs when the body loses excessive water and salt, often due to prolonged exposure to high temperatures. Heat stroke is a severe condition in which the body's temperature regulation fails, leading to a rapid increase in core body temperature. Symptoms of heat stroke include confusion, seizures, and loss of consciousness and can be fatal if not treated promptly. It is crucial to seek immediate medical attention if someone is suspected of having heat stroke. Additionally, heat cramps can occur due to the excessive loss of electrolytes through sweating, resulting in painful muscle cramps. These cramps are a sign that the body is depleted of essential minerals and salts, and it's important to rest and rehydrate to alleviate this condition. Increased heat-related illnesses can range from mild conditions like heat cramps to severe life-threatening situations such as heat stroke. Heat exhaustion, a more severe form than heat cramps, results from a loss of water and salt in the body due to extreme heat and excessive sweating. Heat stroke, the most serious form, occurs when the body's heat-regulating system is overwhelmed by excessive heat, leading to a rapid rise in body temperature and a life-threatening emergency that requires immediate medical attention. Recent research attributes 37% of heat-related deaths to human-induced climate change. Heat-related deaths among those over 65 have risen by 70% in two decades.

5.1.1 Cardiovascular and Respiratory Problems

The impact of heat waves on human health, particularly concerning cardiovascular and respiratory issues, is a topic of great concern. Firstly, it is well-documented that heat waves can worsen cardiovascular and respiratory diseases, ultimately contributing to higher mortality rates. This is especially concerning for vulnerable populations such as the elderly and individuals with chronic conditions. The exacerbation of these health issues during heatwaves underscores the need for proactive measures to protect these at-risk groups. Secondly, heatwaves place a significant strain on the heart. As the body attempts to regulate its temperature, the heart must work harder to pump blood to the skin for heat dissipation. This increased workload on the heart elevates the risk of heart attacks and strokes, further emphasizing the serious health implications of heat waves. In conclusion, the impact of heat waves on cardiovascular and respiratory health is a critical area of concern. It is imperative to raise awareness about these issues and implement strategies to mitigate the adverse effects of extreme heat on human health, particularly for those who are most vulnerable.

5.1.2 Dehydration and Kidney Problems

Excessive heat can lead to various health issues, including dehydration and kidney problems. When the body is exposed to high temperatures, it responds by increasing sweating as a way to cool down. However, if fluid intake is not sufficient to compensate for the loss of fluids through sweating, dehydration can occur. Dehydration can have serious consequences, such as dizziness and confusion, and in severe cases, it can lead to heatstroke. Moreover, chronic dehydration and heat stress can also negatively impact kidney function. Filtering the waste and extra fluid from the circulation to create urine is a critical function of the kidneys. When the body is dehydrated, the kidneys may struggle to perform this function effectively, potentially leading to kidney stones and other renal problems. Therefore, staying well-hydrated, especially during hot weather, is essential to prevent these health issues and support overall kidney function.

5.1.3 Quality of Life and Mental Health

The impact of extreme heat on mental health is a significant concern that should not be overlooked. Firstly, extreme heat can lead to psychological stress, causing individuals to experience heightened levels of stress and anxiety. Furthermore, it can exacerbate existing mental health conditions, making it particularly challenging for those already dealing with such issues. Additionally, extreme heat has impacted cognitive function and decision-making, potentially leading to impaired judgment and decision-making abilities. Moreover, high nighttime temperatures can disrupt sleep patterns, leading to fatigue, irritability, and reduced overall well-being. Sleep disruption caused by higher temperatures can significantly affect a person's mental and emotional well-being, affecting their ability to function optimally and maintain a positive outlook. In conclusion, the impact of extreme heat on mental health is multifaceted, affecting psychological well-being, cognitive function, and overall quality of life. It is essential to recognize and address these concerns to ensure the holistic well-being of individuals, especially in the face of rising global temperatures.

5.1.4 Risk Factors for Heat-Related Illness

The risk factors for heat-related illnesses include both environmental and personal factors. Environmental risk factors include high temperatures, especially with high humidity, direct sun exposure, lack of wind or breeze to cool the body, proximity to engines or other hot equipment, and air temperature. Personal risk factors include medications, health conditions such as diabetes and heart disease, being overweight or obese, poor physical fitness, age over 60, non-breathable clothing or personal protective equipment, alcohol use within 24 h of working in heat, and prior heat illness. Additionally, dehydration, lack of acclimatization, and certain medical conditions like diabetes, infections, skin disorders, sickle cell trait, and cardiovascular disease can increase the risk of heat-related illnesses. Symptoms of heat-related illnesses include heavy sweating, nausea, muscle cramps, tiredness, confusion, and more. Preventive measures include staying hydrated, wearing light clothing, scheduling activities for cooler times of the day, taking rest periods in shady areas, and gradually increasing time spent outdoors to acclimate to the heat. It's crucial to be aware of warning signs like high body temperature, hot skin, headache, dizziness, and nausea for prompt action in case of heat-related illnesses.

6 Air Quality Degradation and Human Health

Climate change-related deterioration of air quality is a serious problem with broad effects on both the environment and human health. The problems that people around the world confront are made worse by the connection between climate change and air pollution. Through a number of mechanisms, including heat waves and wildfires that intensify as temperatures rise and release particulate matter and greenhouse gases into the atmosphere, climate change affects the quality of the air (Gomez et al., 2023). Additionally, drier and hotter conditions can elevate air pollution levels, particularly high ozone, by accelerating photochemical production processes. The degradation of air quality poses significant risks to public health, contributing to chronic disorders like cardiovascular and respiratory diseases and resulting in a substantial number of premature deaths annually.

In fact, air pollution is linked to between 28,000 and 36,000 deaths annually in the United Kingdom, where disadvantaged groups are disproportionately exposed to pollutants because of things like lack of access to green spaces and socioeconomic inequality. Reducing emissions of pollutants such as methane, black carbon, and ground-level ozone in addition to carbon dioxide is crucial for achieving sustainable development and improving health outcomes. Coordinated efforts that consider climate change and air pollution are required to address the degradation of air quality. Better air quality, less global warming, ecosystem protection, and eventually a healthier planet for all can be achieved by civilizations through concerted efforts to tackle air pollution and climate change.

Asthma, respiratory allergies, airway disorders, hypertension, coronary artery disease, heart attacks, and strokes are a few respiratory and cardiovascular conditions that are negatively impacted by air pollution related to climate change. Breathing problems such as coughing, throat discomfort, congestion, decreased lung function, and chest pain can be caused by exposure to pollutants such as particulate matter and ground-level ozone. Additionally, as temperatures rise, increased concentrations of pollen can exacerbate sensitivity to allergens and trigger severe allergic reactions, such as asthma attacks brought on by allergies. A larger risk of acute respiratory consequences exists for people who already have pollen allergies, as the combination of increased air pollution and allergens can worsen these medical conditions.

The impact of poor air quality on human health is a matter of significant concern, with a range of adverse effects that can manifest in individuals exposed to polluted air. Poor air quality can lead to a variety of respiratory issues, including an increased incidence and severity of asthma and allergic reactions. Additionally, exposure to harmful air pollutants can significantly contribute to the development and worsening of chronic respiratory diseases, such as bronchitis and emphysema. These respiratory issues can significantly impair an individual's quality of life and overall health.

The impact of poor air quality extends beyond respiratory issues to affect cardiovascular health as well. Individuals exposed to air pollutants face a higher risk of acute cardiovascular events, including heart attacks and strokes. Furthermore, poor air quality can lead to increased blood pressure, which in turn can contribute to chronic health conditions related to hypertension. Long-term exposure to certain pollutants, such as PM2.5 and benzene, has been linked to an elevated risk of developing lung cancer. This underscores the serious and potentially life-threatening consequences of prolonged exposure to poor air quality. The effects of poor air quality are not limited to physical health; they can also impact neurodevelopment and cognitive function. Evidence suggests that exposure to air pollution can affect cognitive development in children and contribute to cognitive decline in the elderly. Additionally, potential links have been identified between prenatal exposure to pollutants and conditions such as autism spectrum disorders, highlighting the wide-ranging impact of poor air quality on neurological health. In conclusion, the health effects of poor air quality are diverse and far-reaching. It is imperative to raise awareness about these effects and take proactive measures to address air quality concerns. By implementing policies and practices to reduce air pollution and protect vulnerable populations, we can work toward safeguarding public health and well-being. It is essential for individuals, communities, and governing bodies to prioritize initiatives that promote cleaner air and mitigate the adverse health effects associated with poor air quality.

Poor air quality has significant impacts on vulnerable populations, including children, the elderly, individuals with preexisting health conditions, and low-income communities. Children are particularly susceptible to air pollution due to their developing respiratory systems and higher breathing rates relative to their body weight. Exposure to pollutants such as particulate matter, nitrogen dioxide, and ozone can lead to chronic respiratory issues, impaired lung development, and an increased risk of asthma. Poor air quality can also affect cognitive development and academic performance, leading to more school absences due to respiratory illnesses among children.

The elderly are more likely to suffer from preexisting health conditions such as chronic obstructive pulmonary disease (COPD), asthma, and cardiovascular diseases. Air pollution exacerbates these conditions, leading to increased hospital admissions, reduced quality of life, and higher mortality rates. Aging weakens the immune system, making the elderly more susceptible to infections and other health complications triggered by air pollution. Individuals with preexisting health conditions such as asthma, heart disease, and diabetes are particularly vulnerable to the effects of air pollution. Air pollutants can trigger asthma attacks, worsen heart disease, and complicate the management of diabetes. These individuals often require more frequent medical care and interventions, placing additional strain on healthcare systems. Low-income communities often reside in areas with higher levels of pollution, such as near industrial sites, highways, and waste treatment plants. These areas typically have poorer air quality and fewer resources to mitigate exposure, leading to environmental injustice for these communities. In conclusion, addressing poor air quality is crucial to protecting the health and well-being of these vulnerable populations. Implementing policies to reduce air pollution and providing support for at-risk communities can help mitigate the adverse effects of poor air quality on these groups. Furthermore, raising awareness and promoting environmental justice are essential steps toward creating a healthier and more equitable environment for all.

7 Water Contamination and Waterborne Diseases

Water contamination and waterborne diseases remain critical public health challenges, particularly in developing countries like Pakistan. The sources of contamination are diverse, ranging from industrial waste and agricultural runoff to human waste due to inadequate sanitation. These contaminants, whether biological, chemical, or physical, pose significant risks to human health, necessitating a comprehensive and multifaceted approach to address these issues effectively.

Water contamination stems from several sources. The discharge of untreated waste from industrial enterprises and factories directly into water bodies introduces hazardous chemicals and heavy metals, significantly contributing to water pollution. Likewise, the widespread use of pesticides and fertilizers in agriculture results in the leaching of these substances into groundwater and surface water through agricultural runoff, posing a substantial risk to overall water quality and environmental health. In regions lacking adequate sanitation infrastructure, water sources are vulnerable to contamination by human waste, introducing waterborne disease-causing pathogens. Consequently, polluted rivers and lakes, stemming from the cumulative impacts of domestic, industrial, and agricultural waste, exacerbate water quality issues, necessitating comprehensive remedial action. The contaminants found in polluted water can be classified into three main types such as *biological contaminants* (bacteria, viruses, and parasites that can cause diseases such as *cholera, dysentery, typhoid fever, hepatitis A*, and *giardiasis)*; *chemical contaminants*, including heavy metals like lead and mercury, pesticides, and nitrates, which can have long-term health impacts; and *physical contaminants*, such as sediments and organic materials suspended in the water, which can affect water clarity and quality.

Contaminated water may cause the spread of waterborne diseases. Some common waterborne diseases include *waterborne diseases*, which are a significant health concern, particularly in areas with poor sanitation and limited access to clean water. These illnesses are transmitted through the consumption or contact with contaminated water. Among the most common waterborne diseases is *cholera*, caused by the bacteria Vibrio cholerae, which leads to severe *diarrhea* and dehydration that can be life-threatening if not treated promptly. *Dysentery*, another prevalent waterborne disease, is often caused by Shigella bacteria or amoebae, resulting in painful diarrhea with blood and mucus. *Typhoid fever*, caused by *Salmonella typhi*, is characterized by prolonged fever, weakness, *abdominal pain*, and sometimes a rash. This disease can lead to severe complications if untreated. *Hepatitis A*, a viral infection affecting the liver, is spread through the ingestion of contaminated water or food. Symptoms include *jaundice*, *fatigue*, and *nausea*, and although it is typically self-limiting, it can cause severe liver damage in rare cases. *Giardiasis*, caused by the protozoan Giardia lamblia, leads to gastrointestinal issues such as *diarrhea, gas,* and *stomach cramps*. This infection can persist for weeks, causing significant discomfort and dehydration. Preventing these diseases requires ensuring access to clean drinking water, practicing good hygiene, and implementing proper sanitation measures.

8 Storms, Floods, and Public Health

Severe storms and floods are natural disasters that have devastating impacts on public health. These events can cause immediate physical injuries, disrupt healthcare services, and create long-term health issues. The aftermath of storms and floods often exacerbates existing health disparities, particularly affecting vulnerable populations. Severe storms and floods have a significant impact on public health, affecting individuals across various dimensions. These events result in immediate physical injuries and contribute to long-term health issues, particularly among vulnerable populations. The acute health ramifications encompass trauma, including cuts, fractures, and blunt force trauma due to swift floodwaters and debris from storms. Moreover, there is a notable risk of drowning in floodwaters and the hazard of electrocution from downed power lines and electrical equipment within waterlogged areas. Exposure to hazardous materials carried by floodwaters, such as chemical contaminants from industrial sources and sewage, leads to potential poisoning and chemical burns. Additionally, water damage in residential and commercial structures fosters the growth of mold and mildew, precipitating respiratory issues and allergic reactions.

The protracted health impact involves potential contamination of drinking water, culminating in outbreaks of waterborne diseases such as *cholera*, *dysentery*, and *hepatitis A*. Furthermore, lingering standing water serves as a breeding ground for mosquitoes, elevating the risk of vector-borne diseases like *malaria* and *dengue fever*. Long-term exposure to mold and damp conditions post severe storms and floods contributes to the onset of chronic respiratory ailments, including *asthma* and *bronchitis*. The psychological toll of natural disasters may induce *stress, anxiety,* and *depression*, especially regarding property loss and relocation. Survivors may develop post-traumatic stress disorder (PTSD), particularly those who have faced significant loss or life-threatening situations. Additionally, damage to healthcare facilities and transportation disruptions impede access to medical care, which may adversely affect chronic disease management due to compromised access to required medications and consistent care.

Floods and storms account for a significant percentage of natural disasters worldwide. The occurrences of floods and storms combined accounted for approximately 67% of all natural disasters (Padli et al., 2013). Furthermore, floods alone contributed to 16% of weather-, climate-, and water-related deaths documented between 1970 and 2021 (Alves, 2024). Over the past 20 years, a total of 6457 reported floods, hurricanes, heatwaves, droughts, and other weather-related incidents have been held responsible for 90% of catastrophic disasters. During the period from 1995 to 2015, floods emerged as the cause of 47% of all weather-related disasters, impacting a staggering 2.3 billion individuals and claiming 157,000 lives. In addition to floods, storms became the deadliest type of weather-related disaster, resulting in 242,000 deaths worldwide, which accounted for 40% of all weather-related deaths. In summary, floods and storms make up a substantial portion, around 67–90%, of natural disasters globally. Floods alone account for nearly half of all weather-related disasters and affect many people. Storms are also a major contributor to weather-related deaths worldwide.

Based on the analysis of existing literature, it is evident that a significant portion of natural disasters in Pakistan can be attributed to floods and storms. Pakistan has experienced approximately 25 cyclonic storms/tornadoes and around 38 floods, including 20 major floods, within the last 100 years (Hamidi et al., 2020). The impact of floods in Pakistan has been extensive, affecting approximately 69.1 million people, surpassing the impact of earthquakes (which affected 6.6 million people) and storms (which affected 2.6 million people) (Sajjad et al., 2014). Over its 75-year history, Pakistan has encountered 28 super riverine floods, collectively affecting 616,558 km^2 of land. The region that constitutes present-day Sindh province has a notable history of recurrent riverine floods, experiencing at least 18 major flood events in the nineteenth and twentieth centuries (Raza, 2022). The findings suggest that floods and storms are the most prevalent and impactful natural disasters in Pakistan, affecting millions of individuals and vast areas of land over the past century.

9 Droughts and Water Scarcity-Related Health Challenges

Droughts and water scarcity pose significant and growing threats to public health globally. As climate change exacerbates water stress in many regions, it is crucial to understand and address the health impacts of limited water access and quality. Human health and well-being can suffer greatly from water scarcity. Lack of safe drinking water and inadequate sanitation services facilitate the transmission of waterborne diseases like cholera, typhoid, polio, hepatitis A, and diarrhea. These deprivations impact over two billion people worldwide.

Water scarcity also threatens food security, as it makes agriculture much more difficult. Food-insecure communities face acute and chronic hunger, putting children at risk of malnutrition-related conditions like stunting and wasting. Poor diet can also contribute to chronic illnesses such as diabetes. Beyond direct health impacts, water scarcity takes an economic toll. Time spent procuring water cannot be devoted to work or school, hindering productivity and development. The COVID-19 pandemic highlighted the importance of handwashing, but 2.3 billion people globally could not wash their hands at home at the pandemic's onset due to water access issues. Climate change will likely exacerbate water stress worldwide, as rising temperatures lead to more unpredictable weather and extreme events like droughts and floods. Regions like South Asia face the largest global concentration of water-related risks, while Sub-Saharan Africa exhibits the greatest variability in crop production and economic impacts of inadequate water supply. Addressing these health challenges requires sustainable water management practices, improved water infrastructure, and international cooperation. Governments, water authorities, and healthcare providers must urgently coordinate efforts to prevent and reduce the health impacts of droughts and water scarcity. Targeted, equity-based approaches are needed to protect vulnerable populations. In conclusion, droughts and water scarcity pose grave and growing threats to public health globally. Urgent action is required to implement existing solutions, integrate climate change into health policies, and invest in long-term resilience to safeguard human health and well-being in the face of water stress.

10 Climate Change, Agricultural Productivity, and Food Availability

Climate change profoundly impacts agricultural productivity and food availability, presenting significant challenges for global food security. Some key ways in which climate change influences these aspects include *shifts in growing conditions*: climate change varies in temperature and rainfall patterns, leading to shifts in growing conditions. Some regions may experience longer growing seasons, while others face increased droughts or floods. These changes can disrupt crop cycles, reduce yields, and impact overall food production. *Water stress:* Variation in the patterns of rainfall and higher temperature-related evaporation rates contribute to water stress in agricultural regions. Water scarcity can limit irrigation capabilities, affecting crop growth and leading to decreased agricultural productivity. *Pests and Diseases:* Higher temperatures and altered weather patterns foster an environment that is conducive to the growth of illness and pests. This may lead to heightened pest infestations, crop damage, and the transmission of plant diseases, further reducing crop yields and food availability. *Extreme weather events*: Hurricanes, heat waves, and heavy rains have become more intense due to climate changes. These events can cause widespread harm to crops, livestock, and infrastructure, disrupting food production and supply chains. *Sea level rise and coastal agriculture:* Rising sea levels impact coastal agricultural areas, leading to saltwater intrusion, soil salinity, and loss of arable land. Coastal farming communities face challenges in adapting to these changes, affecting food production and livelihoods. *Shifts in crop suitability:* climate change alters the suitability of regions for specific crops. New places may become appropriate for cultivation while traditional crop-growing areas will become less feasible. Farmers need to adapt by selecting resilient crop varieties and adopting sustainable farming practices. *Food security risks:* the combination of these factors poses significant food security risks, especially for vulnerable populations. Reduced agricultural productivity, increased food prices, and disruptions in food supply chains can exacerbate hunger, malnutrition, and food insecurity.

The changing climate, including increased frequency of catastrophic weather events, changing patterns of precipitation, and rising temperature, is causing disruptions to yields, quality, and distribution of crops. These changes have far-reaching effects on human health, economic stability, and environmental sustainability. The evidence is clear: climate change is already affecting agricultural productivity, with negative impacts on crop yields and quality observed globally. The cumulative effect of climate change on key crop yields between 1961 and 2021 has been a 21% reduction in global agricultural productivity, with low-latitude regions experiencing the most severe declines. Projections suggest that warming exceeding 3 °C will lead to global declines in yields, with tropical regions being disproportionately affected.

The concerns of these changes are far-reaching, with climate change disrupting food availability through production declines, local limitations, and price increases. Extreme meteorological conditions, such as droughts, floods, and tropical cyclones, exacerbate food insecurity by disrupting food production and supply chains. The most vulnerable populations, including those already suffering from hunger and malnutrition, are at greatest risk from these impacts. The imperative for action is clear. Adapting agricultural practices to the changing climate, investing in climate-resilient agriculture, and promoting sustainable food systems are essential to maintaining productivity, ensuring food availability, and protecting global food security. The international community must come together to address the climate crisis, support climate-resilient agriculture, and safeguard the health, well-being, and economic stability of current and upcoming generations. Addressing the impact of climate change on agriculture requires a coordinated effort involving adaptation strategies, sustainable agricultural practices, investment in resilient crop varieties, improved water management, and international cooperation. By enhancing climate resilience in agriculture, we can mitigate the adverse effects on food availability and work toward a more sustainable and food-secure future.

11 Malnutrition, Foodborne Illnesses, and Human Health

Climate change poses a grave threat to global health and nutrition, exacerbating the burden of malnutrition and foodborne illnesses worldwide. Increasing temperatures, erratic rain patterns, and extreme climatic conditions are disturbing food systems, contaminating water sources, and facilitating the transmission of pathogens, with profound impacts on human health, particularly among vulnerable populations. Climate change's effects on food diversity, quality, and availability are widely discussed, with decreased food production and nutritional value contributing to the global malnutrition crisis. Warming temperatures and elevated atmospheric carbon dioxide concentrations lead to reduced yields and declining nutritional quality in staple crops, increasing the risk of micronutrient deficiencies and diet-related non-communicable diseases. Extreme weather events further disrupt food supply chains, disproportionately affecting developing countries and heightening the risk of food scarcity and nutritional deficiency.

Shifts in climate also facilitate the spread of foodborne illnesses through multiple pathways. Altered environmental conditions favor the transmission of waterborne, airborne, foodborne, and vector-borne pathogens, increasing the incidence of infectious diseases globally. Extreme events like floods and sea-level rise contaminate water sources with pathogens, leading to outbreaks of waterborne and foodborne diseases. The burden of infectious diseases with epidemic potential is expected to rise, particularly in developed countries, while underdeveloped countries remain more vulnerable to these sensitive health risks. Vulnerable populations are disproportionately affected by climate-related health problems. The aftermath of climatic changes on human health is extensive, encompassing direct effects from harsh weather events and indirect impacts through disruptions in food systems and foodborne, waterborne, and vector-borne diseases. Vulnerable populations are disproportionately affected by climate-related health problems, such as women, children, ethnic minorities, poor communities, migrants, the elderly, and those with underlying health conditions. The burden on health systems caused by the increased occurrence of infectious diseases can lead to disruptions and inequality in access to care, as witnessed during events such as the COVID-19 pandemic.

The symptoms of malnutrition caused by changes in climate can vary based on the particular nutritional deficiencies and health impacts experienced by individuals. Some common symptoms of malnutrition resulting from climate change-induced food insecurity and reduced nutritional quality include undernutrition, obesity, and nutrient deficiencies. Undernutrition encompasses various health impacts, including stunting, wasting, being underweight, fatigue and weakness, delayed healing, and cognitive impairment. Stunting refers to impaired growth and development in children, resulting in a shorter height for their age, while wasting involves rapid weight loss and muscle wasting, often associated with acute malnutrition. Underweight individuals exhibit a low body weight for their age, indicating inadequate nutrition intake. Fatigue and weakness manifest as a lack of energy and strength due to insufficient nutrient intake, and delayed healing involves sluggish wound healing and recovery from illnesses due to weakened immune function. Additionally, undernutrition can lead to cognitive impairment, resulting in reduced cognitive function and learning difficulties, particularly in children.

Obesity is a condition that can lead to various health issues. Excessive weight gain due to inadequate diet quality and limited access to nutritious foods is a significant concern. This can result in an increased risk of metabolic disorders such as diabetes and cardiovascular diseases. Moreover, being overweight may lead to joint pain, causing mobility difficulties. Additionally, obesity can contribute to fatigue and reduced physical activity levels, impacting overall well-being. Individuals affected by obesity also have a higher likelihood of developing chronic conditions associated with this health issue.

Nutrient deficiencies play a significant role in impacting human health. *Iron deficiency* presents symptoms such as *fatigue, weakness, pale skin*, and impaired cognitive function. Similarly, a *zinc deficiency* can result in impaired immune function, delayed wound healing, and skin problems. *Vitamin A deficiency* is characterized by symptoms including night blindness, increased vulnerability to infections, and skin issues. Lastly, inadequate protein intake may lead to muscle wasting, fatigue, and a weakened immune system. Maintaining a balanced diet is imperative to ensure the body receives essential nutrients for optimal functioning. These symptoms highlight the diverse health impacts of malnutrition, emphasizing the importance of addressing food insecurity, promoting access to nutritious foods, and implementing sustainable agricultural practices to minimize the consequences of changes in climate on global nutrition and health.

The most common types of malnutrition caused by climate change include undernutrition and obesity. It also impacts the availability, quality, and diversity of food, leading to increased food and financial insecurities that contribute to malnutrition. Changes in environmental conditions are affecting crop yields and their nutritional quality, which, as a result, is declining the availability of affordable, nutritious food, leading to a higher risk of undernutrition and obesity globally. The rise in temperature, erratic rainfall, and extreme weather events exacerbate food insecurity, particularly in developing countries. The widespread nutritional deficiencies and diet-related diseases like diabetes mellitus and cardiovascular diseases are expected to further increase. The burden of undernutrition and obesity is also expected to increase, leading to reductions in food availability and quality, posing significant health risks to populations worldwide.

12 Changing Habitats, Vectors of Diseases, and Human Health

The impact of changing habitats on disease vectors and human health is a critical aspect influenced by climate change. As habitats evolve due to climate change, disease vectors, such as mosquitoes and ticks, may experience shifts in their geographic distribution and abundance. These changes can have profound implications for the transmission dynamics of vector-borne diseases and human health. Climate change is severely changing the environmental conditions, affecting disease vectors' survival, behavior, and reproduction. Warmer temperatures can expand the geographic range where vectors can thrive, while increased rainfall and drought can create more breeding sites for vectors. Changes in habitats can lead to extended disease transmission seasons and influence the behavior of vectors, such as altering their biting patterns. Understanding the impact of changing habitats on disease vectors is crucial for predicting and mitigating the spread of vector-borne diseases. By studying how climate change influences vector habitats, researchers and public health authorities can develop strategies to prevent and control the transmission of diseases like malaria, dengue, Zika virus, and Lyme disease. Effective surveillance, early warning systems, and targeted interventions are essential to protect human health, considering that vector habitats are changing as a result of climate change.

13 Mental Health Challenges and Climate Change

The increasing consequences of changes in climate are causing serious and complex psychological health issues for humans, such as anxiety, depression, and many other mental disorders. Globally, there is a growing influence on mental health and emotional well-being due to the warming of the planet, extreme weather, and environmental degradation brought on by climate change. Individuals affected by climate change, including extreme weather, food and water insecurity, and community disintegration, may suffer from a variety of mental health problems such as *anxiety, depression, post-traumatic stress disorder* (PTSD), and substance use disorders.

The growth of climate anxiety, sadness, and eco-anxiety is indicative of the psychological impact that climate change is taking on people, especially young people who experience everyday stress and worry due to their high levels of concern about climate change. Climate variability is expected to have a bigger effect on mental health as temperatures rise. This will result in more negative moods, disturbed sleep patterns, and a higher likelihood of acute hospital admissions for psychiatric disorders because of poor air quality and pollution exposure. Severe weather-related catastrophes and other climate-related occurrences can be traumatic, resulting in anxious, depressed, or hopeless sentiments as well as PTSD. Individuals suffering trauma akin to PTSD and elevated psychological distress may have long-lasting consequences on their mental health from direct exposure to catastrophes such as hurricanes and wildfires. The impact of global warming on psychological well-being extends beyond those who are immediately impacted by natural disasters; a warmer globe is having a widespread psychological impact as people who are not affected by climate-related events might also report feeling anxious and distressed about it.

It's expected that the effects on mental health will worsen as global warming continues to produce more extreme weather, pollutants in the air, and economic upheavals. The occurrence of violent incidents, suicide attempts, and hospitalizations due to mental issues has all increased with rising temperatures. A rise in suicides and higher anxiety levels have been linked to prolonged exposure to atmospheric pollution, demonstrating the complex connection among environmental variables and psychological outcomes. Addressing the mental health challenges caused by climate change requires a comprehensive approach that integrates mental health services, public health interventions, and climate change mitigation strategies. To improve mental health, resilience, and adaptability to climate change. Therefore, it is essential to acknowledge and tackle the psychological consequences of global warming.

14 Conclusion

There are serious health hazards associated with climate change worldwide, especially for low-income nations like Pakistan. Extreme weather events and environmental degradation are caused by human activity, especially the release of greenhouse gases, which increases disease, injury, and population relocation. The rising frequency and intensity of heat waves, floods, and droughts underscore the urgent need for proactive measures to mitigate these health risks. As previously discussed, numerous disorders, including respiratory and cardiovascular conditions, vector-borne infections, and mental health problems, are made worse by climate change. Over the past 30 years, it has been linked to over 150,000 deaths annually, with its adverse effects on the cardiopulmonary system, gastrointestinal tract, and mental health posing significant risks. Effective policy development and implementation are crucial for addressing these health challenges.

Climate change has a substantial influence on the occurrence as well as dissemination of infectious ailments, in particular, those carried by vectors, including *dengue fever*, *malaria*, and the *Zika virus*. Warmer temperatures and altered precipitation patterns increase disease-carrying vectors' populations and geographic range, heightening transmission risks. Immediate action is necessary to mitigate these effects and protect global health. Additionally, there are serious health hazards associated with heat waves expanding in frequency and intensity, especially for vulnerable populations. Preventive measures are crucial to mitigate the adverse effects of extreme heat, which exacerbates cardiovascular and respiratory issues and leads to higher mortality rates. Climate change exacerbates poor air quality, which also causes respiratory disorders, cardiovascular diseases, and other health conditions that mostly affect underprivileged communities.

Water contamination from industrial waste, agricultural runoff, and inadequate sanitation, worsened by severe storms and floods, remains a significant public health challenge. Climate change's impact on agricultural productivity and food availability leads to reduced food security and increased malnutrition. Addressing these challenges requires coordinated global action, including adapting agricultural practices and investing in climate-resilient agriculture. Overall, climate change is causing a global health crisis, affecting food systems, water sources, and pathogen transmission pathways. This results in reduced food availability and increased food insecurity, malnutrition, and the spread of foodborne illnesses. Ensuring a sustainable and resilient future while addressing the complex health concerns caused by climate change requires concerted efforts, global cooperation, and integrated approaches that incorporate public health treatments, surveillance, and mental health services.

Addressing the health challenges induced by climate change necessitates a multifaceted strategy integrating policy measures, public health interventions, and climate change adaptation strategies. Policymakers should prioritize the implementation of mitigating measures meant to lower greenhouse gas emissions and promote the adoption of sustainable energy sources. Concurrently, public health interventions targeting vulnerable communities, enhanced disease surveillance systems, and improved access to healthcare services should be developed and deployed. Furthermore, the formulation of adaptation strategies to bolster the resilience of infrastructure against extreme weather events, the enhancement of early warning systems, and the establishment of comprehensive response plans for heat waves are imperative. Educational campaigns should be orchestrated to raise awareness regarding the hazards to health that are caused by climate change, empowering individuals to take proactive measures. International collaboration and partnerships should be fostered to deal with global difficulties with health attributed to climate variability. This entails sharing best practices, resources, and expertise to orchestrate comprehensive and coordinated responses. Moreover, investments in research and development are crucial to understanding more about the relationship between health and climate change and to cultivating innovative solutions to mitigate health risks. By conscientiously implementing these solutions, policymakers can work toward safeguarding public health amidst the challenges posed by climate change.

References

Alves, B. (2024). *Floods-statistics and facts*. https://www.statista.com/topics/11261/floods/#topicOverview

Arndt, D. S., Baringer, M. O., & Johnson, M. R. (2010). State of the climate in 2009. *Bulletin of the American Meteorological Society, 91*(7), 1–222. https://doi.org/10.1175/BAMS-91-7-StateoftheClimate

Bousquet, P., Ciais, P., Miller, J. B., Dlugokencky, E. J., Hauglustaine, D. A., Prigent, C., & White, J. (2006). Contribution of anthropogenic and natural sources to atmospheric methane variability. *Nature, 443*(7110), 439–443. https://doi.org/10.1038/nature05132

Confalonieri, U., Menne, B., Akhtar, R., Ebi, K. L., Hauengue, M., Kovats, R. S., & Woodward, A. (2007). Human health. In *Climate change 2007: Impacts, adaptation, and vulnerability: Contribution of working group II to the fourth assessment report of the intergovernmental panel on climate change*. https://doi.org/10.1175/BAMS-91-7-StateoftheClimate

Costello, A., Abbas, M., Allen, A., Ball, S., Bell, S., Bellamy, R., & Patterson, C. (2009). Managing the health effects of climate change: Lancet and University College London Institute for Global Health Commission. *The Lancet, 373*(9676), 1693–1733. https://doi.org/10.1016/s0140-6736(09)60935-1

Fawzy, S., Osman, A. I., Doran, J., & Rooney, D. W. (2020). Strategies for mitigation of climate change: A review. *Environmental Chemistry Letters, 18*, 2069–2094. https://doi.org/10.1007/s10311-020-01059-w

Franchini, M., & Mannucci, P. M. (2015). Impact on human health of climate changes. *European Journal of Internal Medicine, 26*(1), 1–5. https://doi.org/10.1016/j.ejim.2014.12.008

Füssel, H. M. (2008). Assessing adaptation to the health risks of climate change: What guidance can existing frameworks provide? *International Journal of Environmental Health Research, 18*(1), 37–63. https://doi.org/10.1080/09603120701358416

Gomez, J., Allen, R. J., Turnock, S. T., Horowitz, L. W., Tsigaridis, K., Bauer, S. E., & Ginoux, P. (2023). The projected future degradation in air quality is caused by more abundant natural aerosols in a warmer world. *Communications Earth & Environment, 4*(1), 22. https://doi.org/10.1038/s43247-023-00688-7

Gulzar, A., Islam, T., Gulzar, R., & Hassan, T. (2021). Climate change and impacts of extreme events on human health: An overview. *Indonesian Journal of Social and Environmental Issues (IJSEI), 2*(1), 68–77.

Hamidi, A. R., Zeng, Z., & Khan, M. A. (2020). Household vulnerability to floods and cyclones in Khyber Pakhtunkhwa, Pakistan. *International Journal of Disaster Risk Reduction, 46*, 101496. https://doi.org/10.1016/j.ijdrr.2020.101496

Hong, Y. C. (2008). Climate change and human health. *Journal of Korean Medical Association, 51*(8), 764–769. https://doi.org/10.5124/jkma.2008.51.8.764

Hong, S. J., & Seo, J. H. (2011). Climate change and human health. *Journal of the Korean Medical Association/Taehan Uisa Hyophoe Chi, 54*(2), 149.

Huang, W., Gao, Q. X., Cao, G. L., Ma, Z. Y., Zhang, W. D., & Chao, Q. C. (2016). Effect of urban symbiosis development in China on GHG emissions reduction. *Advances in Climate Change Research, 7*(4), 247–252.

Kim, K. H., Kabir, E., & Ara Jahan, S. (2014). A review of the consequences of global climate change on human health. *Journal of Environmental Science and Health, Part C, 32*(3), 299–318.

Kizielewicz, C. (2018). Lyme disease Predicted To Rise in U.S. as Climate Warms.

Mora, C., McKenzie, T., Gaw, I. M., Dean, J. M., von Hammerstein, H., Knudson, T. A., & Franklin, E. C. (2022). Over half of known human pathogenic diseases can be aggravated by climate change. *Nature Climate Change, 12*(9), 869–875.

Padli, J., Habibullah, M. S., & Baharom, A. H. (2013). Determinants of flood fatalities: Evidence from a panel data of 79 countries. *Social Sciences and Humanities, 21*, 81–98.

Patel, L., Conlon, K. C., Sorensen, C., McEachin, S., Nadeau, K., Kakkad, K., & Kizer, K. W. (2022). Climate change and extreme heat events: How health systems should prepare. *NEJM Catalyst Innovations in Care Delivery, 3*(7), CAT-21.

Rahmstorf, S., & Coumou, D. (2011). Increase of extreme events in a warming world. *Proceedings of the National Academy of Sciences, 108*(44), 17905–17909.

Raza, M. (2022). *Pakistan's history of disasters and the lessons we fail to learn*. Dawn.

Ren, Z., Bi, P., & Li, S. (2021). Climate change, extreme weather, and vector-borne diseases: Spatial and temporal risks. *Frontiers in Environmental Science*. www.frontiersin.org/

Roy, I. (2018). Role of "natural factors" on recent climate change underestimated, research shows. ScienceDaily, University of Exeter.

Sajjad, S. H., Waheed, S. A., Khan, T., Qadri, S. T., & Gilani, N. (2014). Natural hazards and related contents in the curriculum of geography in Pakistan. *Asian Journal of Natural and Applied Science, 3*, 40–48.
Schwartz, S. E., Benoit, L., Clayton, S., Parnes, M. F., Swenson, L., & Lowe, S. R. (2023). Climate change anxiety and mental health: Environmental activism as a buffer. *Current Psychology, 42*(20), 16708–16721.
Snow, M., & Snow, R. (2015). The impact of climate change on human health. *Journal of Climatology & Weather Forecasting, 3*(1), e109.
Trenberth, K. E. (2018). Climate change caused by human activities is happening and it already has major consequences. *Journal of Energy & Natural Resources Law, 36*(4), 463–481.
Yousaf, B., Liu, G., Abbas, Q., Wang, R., Ali, M. U., Ullah, H., & Zhou, C. (2017). Systematic investigation on combustion characteristics and emission-reduction mechanism of potentially toxic elements in biomass-and biochar-coal co-combustion systems. *Applied Energy, 208*, 142–115.

Shabnum Soomro is an M.Sc. scholar in the Department of Agricultural Economics at Sindh Agriculture University, Tando Jam. She earned the degree of M.Sc. Agri. (Hons.) in Agricultural Economics from the Sindh Agriculture University Tando Jam.

Jam Ghulam Murtaza Sahito is an associate professor of agricultural economics at Sindh Agriculture University, Tando Jam. He has a Ph.D. in Economics from Germany. His research interests include econometrics and applied economics empirical research.

Farahnaz Gilal is an M.Sc. scholar in the Department of Agricultural Economics at Sindh Agriculture University, Tando Jam. She holds a degree of M.Sc. Agri. (Hons.) in Agricultural Economics from the Sindh Agriculture University Tando Jam.

Towards Understanding the Impact of a Refugee Crisis on MNEs: A Theoretical Essay in the Light of Hannah Arendt's Political Theory

Jenerton Arlan Schütz, Marcos de Moraes Sousa, and Miguel M. Torres

1 Introduction

This study is a theoretical essay that draws upon the political theories of Hannah Arendt to examine the phenomenon of statelessness and its implications for human rights and multinational enterprises (MNEs) corporate social responsibility (CSR) practices. The research integrates Arendt's philosophical perspectives with contemporary discussions on CSR and the role of MNEs in addressing statelessness. Hannah Arendt's experiences as a refugee and exile during World War II and the Holocaust profoundly shaped her political thought and critique of totalitarianism (Horst & Lysaker, 2021). On fleeing Nazi Germany in 1933, Hannah Arendt became a stateless person and remained in this vulnerable state for 18 years, when she was granted US citizenship (Kerber, 2007). The experience that Arendt went through significantly shaped her trajectory as a thinker in political theory, provoking her to search for the very meaning of politics and the foundation of her perspective (Canovan, 1992).

Arendt's provoking work has significantly impacted contemporary political theory, influencing thinkers such as Jürgen Habermas, Michel Foucault, and Judith Butler, whose ideas have considerably shaped modern thinking in advancing CSR

J. A. Schütz
Catholic University of Brasília, Brasília, Brazil
e-mail: jenerton.schutz@ifgoiano.edu.br

M. de Moraes Sousa
Federal University of Goiás, Goiânia, Brazil
e-mail: marcos.moraes@ifgoiano.edu.br

M. M. Torres (✉)
Kent Business School, University of Kent, Canterbury, UK
e-mail: m.torres@kent.ac.uk

J. Bento, M. M. Torres (eds.), *Global Perspectives on Climate Change, Inequality, and Multinational Corporations*, CSR, Sustainability, Ethics & Governance,
https://doi.org/10.1007/978-3-031-80797-8_9

initiatives by MNEs. Habermas emphasised communicative action and the public sphere, promoting transparency and stakeholder engagement (Habermas, 1984, 1989). Foucault's insights on power dynamics and governmentality encourage self-regulation and ethical governance (Foucault, 1975, 1977, 1991). Butler's theories on performativity and vulnerability guide MNEs towards building socially responsible identities and addressing the needs of marginalised groups (Butler, 1990, 2004).

Jürgen Habermas, Michel Foucault, and Judith Butler provide a robust theoretical foundation for root thinking about CSR practices that balance corporate interests with social and environmental responsibilities. This study explores a new frontier for CSR in MNEs, who have historically played a significant role in advancing CSR initiatives, focusing on sustainable development, ethical labour practices, and community engagement. Recently, there has been a growing recognition of MNEs' role in addressing the plight of stateless individuals, similar to how they have tackled other social issues through CSR (Kingston, 2019). We focus on Arendt's discussion of statelessness and human rights. It initially questions the meaning of statelessness in that losing civil rights also implies losing human rights. Arendt (1958) argues that stateless people face a fundamental condition of lack of rights resulting from losing a political community that recognises their legal, political, and human position. The prospect of these individuals being reduced to their most basic humanity led Arendt to critique human rights (Arendt, 1968). The second part of this chapter deals with Arendt's expression of the "right to have rights", outlining her specific criticisms. Her analysis suggests that the eighteenth-century declarations of rights emphasised the assumption that every human being is a citizen. This is a problematic assumption in the face of the large refugee populations generated by the mass denationalisation and denaturalization movements since the Second World War (WWII). However, Arendt does not see her analysis of statelessness as a renunciation of rights but as a call to rethink them in terms of the "right to have rights", as we will discuss in the third part of this article (by way of conclusion), elucidating the contemporary relevance of Arendt's revelations about statelessness and human rights and their significance for MNEs' practices that balance corporate interests with social and environmental responsibilities and, more broadly, for society.

Considering these reflections, it is necessary to emphasise that this chapter does not aim to exhaust the subject but rather to present and clarify the problem whose damaging effects of these practices have been deeply felt since WWII. Therefore, the main focus of this chapter is to shed light on the issue of statelessness and the fragile protection of human rights in this condition, as well as possible solutions forged in the light of Arendtian thought. Thus, by bringing up Hannah Arendt's ideas as a theoretical reference, this writing highlights the most relevant aspects of this discussion.

2 Without Rights as a Condition

Hannah Arendt endeavoured to understand statelessness as one of contemporary history's most recent mass phenomena. This phenomenon generated political instability that led to the unravelling of empires that were considered to be multi-ethnic after the First World War. As diagnosed by Arendt (1958), with the collapse of the nation-state system and totalitarian systems' disregard for human life, many people have been left stateless. This resulted in the denial of the "right to have rights"—being without a state or losing one's nationality implies the loss of all rights, making these people stateless in the world shared by human beings.

In this sense, Arendt starts to use the term stateless people (SP) to refer not only to those who formally lost their nationality but also to those who were no longer entitled to benefit from their citizenship rights, namely refugees,[1] asylum seekers, economic migrants, and even naturalised citizens who faced the threat of denaturalization in times of emergency. Thus, SPs are individuals not recognised as citizens by any country, resulting from conflicting nationality laws, discrimination, and administrative issues. What brought together these people, who possessed different legal *status*, is that they were all ejected from the "ancient trinity of state-people-territory", and Arendt argued that this exclusion left them in a condition of *rightlessness* (Arendt, 1958).

Stateless people are people deprived of their legal subjectivity and right to discourse, speech, and action (Blitz & Lynch, 2009; Siegelberg, 2020). These are people expelled from their countries, states, cities, homes, and humanity because they have lost their rights as citizens and as humans belonging to humanity and the common human world (Namondo, 2020). In other words, the stateless person does not have a legal intermediary to look after him or her or even other higher bodies that can recognise him or her. So, statelessness has to do with the non-existence of the legal link of nationality, i.e., the absence of a formal relationship/connection between a person and any state.

However, the drama of statelessness extends beyond the loss of their homes and the social fabric that makes up their place in the world; it is about non-acceptance and non-membership of a nation-state, without which they cannot regain importance in political organisation. In her essay, Arendt (1943, p. 69) writes: "[...] we refugees have lost our home, which means the familiarity of everyday life. We have lost our occupation, which means the confidence that we had some use in this world. We have lost our language, which means the naturalness of our reactions, the simplicity of our gestures, the impassive expression of our feelings".

In this context, what should be an opportunity to promote fraternity and solidarity between people often becomes a reason for hostility and various prejudices. Refugees are frequently perceived as expendable bodies, abandoned to their fate

[1] For clarification purposes, stateless people and refugees, from Arendt's perspective, are the same. According to the author, the heart of statelessness is identical to that of the refugee question (Arendt, 1958).

and their complex realities relegated to oblivion, forcing them into a radical struggle for survival. In the context of Arendt's (1943) lived experiences, she writes that the meaning of refugee has changed with what she has experienced. For Arendt, "refugees" have come to the misfortune of arriving in a new country without means and have had to be helped by refugee committees. For Lafer (1988, p.148), stateless people are "[...] condemned innocents, deprived of a place in the world—a place that makes their opinions meaningful and their actions effective". According to Chauí (2019), the contemporary state tends to take the form of a company, an institution incorporated by the market. Socially and economically, introducing structural unemployment and the Toyotist outsourcing of labour gives rise to a new working class that some scholars have called the precariat to indicate a new worker without a stable job, an employment contract, unionisation, and social security. They are not even in the fringes of society because they are not full citizens. They have a mind fuelled and motivated by fear, lack of self-esteem, dignity, and insecurity. It is also important to remember that the reasons why people migrate from their countries of origin are not restricted to extreme situations such as wars or disasters. They also include cases where a country can no longer guarantee minimum rights or meet basic needs such as income and employment. Even when there is no forced displacement, this does not mean that immigrants should be marginalised because of their unprotected living conditions. So, we have the paradox revealed by the precarious condition of stateless people. When one appears as nothing but human, stripped of all social and political attributes, it is challenging to claim and exercise the rights to which one is entitled by being born human (Table 1).

According to Arendt (1958), stateless people were unimportant and just a legal anomaly. They received late attention and consideration when, after WWII, their legal position was also applied to refugees expelled from their countries by the social revolution and denationalised by the victorious governments. After WWII, the issue of statelessness became particularly pronounced due to the upheaval and displacement caused by the conflict (Siegelberg, 2020). Various populations were rendered stateless due to changes in national borders, the collapse of empires, and the persecution of specific ethnic and social groups (Harff & Gurr, 1988).

Moreover, European governments' conduct may seem like a natural consequence of the war today, but at the time, mass denationalisation was an entirely new and unforeseen phenomenon (Gibney, 2020). They presupposed a state structure that, if not entirely totalitarian, was already demonstrating an inability to tolerate any opposition, preferring to lose its citizens rather than harbour them with opinions that differed from the prevailing one. Arendt (1958) revealed, in addition to the above, that war was not necessary for the sovereignties of neighbouring countries to come into conflict and that this could develop in ideological terms not only in the extreme case of war but also during peace.

Arendt (1958) also recalls that there was hardly a country left on the European continent that had not, between the two wars, "[...] passed legislation formulated in such a way as to allow a large number of its inhabitants to be rejected at any opportune moment, even if this right was never used" (Arendt, 1958, p. 310). After all, Arendt goes on to say that no paradox of contemporary politics is as painfully ironic

Table 1 Stateless populations by country after World War II

Country	Stateless population	Description	Estimated numbers
Germany	Jewish Holocaust Survivors, Displaced Persons (DPs)	Holocaust survivors and DPs were left stateless due to border changes and a lack of welcome in their home countries.	500,000+
Poland	Ethnic Germans	Expelled ethnic Germans were rendered stateless due to border changes.	1,000,000+
Soviet Union	Ethnic Minorities (Crimean Tatars, Chechens, Volga Germans)	Deportations and loss of citizenship for various ethnic groups.	1,500,000+
Czechoslovakia	Sudeten Germans	Expulsion of Sudeten Germans post-war, leading to statelessness until absorption by Germany.	3,000,000+
Hungary	Ethnic Germans, Jews	Post-war expulsion and Holocaust survivors left stateless.	200,000+
Yugoslavia	Ethnic Italians, Germans	Expulsion of ethnic groups following border changes, leading to statelessness.	300,000+
Austria	Jewish Refugees	Holocaust survivors were unable to return to Eastern Europe and were left stateless.	150,000+
Italy	Ethnic Slovenians, Croatians	Border changes and population transfers result in statelessness.	100,000+
China	White Russians	Russian refugees from the Bolshevik Revolution were left stateless after the Chinese Civil War.	50,000+
Palestine	Jewish Refugees, Palestinian Arabs	Jewish refugees pre-Israel establishment and Palestinian Arabs post-1948 were left stateless.	700,000+
Japan	Koreans in Japan	Koreans brought during colonial rule were left stateless until naturalisation reforms.	600,000+
Romania	Ethnic Germans, Jews	Post-war expulsion and Holocaust survivors rendered stateless.	200,000+

Source: Authors' elaboration based on data from Siegelberg (2020)

as the discrepancy between the efforts of well-meaning idealists, who stubbornly persisted in considering the rights enjoyed by citizens of civilised countries to be "inalienable", and the situation of human beings with no rights at all (Arendt, 1958).

The expression "stateless peoples" at least recognised that some people had lost the protection of their government and needed international agreements to safeguard their legal status (Berger, 1947). Since then, the term "displaced persons" has been forged during WWII to liquidate the problem of statelessness once and for all through the simple expedient of ignoring its very existence. In this vein, Arendt

(1958) points out that the failure to recognise that a person could be "stateless" led the authorities, whatever they were, to attempt to repatriate them, i.e., deport them to their country of origin, even if the latter refused to recognise the prospective returnee as a citizen or, on the contrary, wanted them back just to punish them. However, it is at least known that while there were one million "recognised" stateless people, there were more than ten million de facto stateless people, albeit ignored (Kulshrestha, 2022). Worse, the number of potentially stateless people continues to rise. According to the United Nations High Commissioner for Refugees (UNHCR), by 2023, at least 4.4 million people in 95 countries are considered stateless or of undetermined nationality. The global figure is widely recognised as significantly higher, given the relative invisibility of stateless people in national statistics.

In this sense, Arendt also considers that the stateless person, without the right to live and work, naturally had to live in constant transgression of the law (Kulshrestha, 2022). They were liable to go to jail without ever committing a crime. Moreover, the whole hierarchy of values in civilised countries was inverted in his case. Since this situation was an anomaly not foreseen in the general law, it was better that he became the anomaly it did foresee: the criminal (Arendt, 1958). Similarly, Arendt (1958) points out that you can identify whether someone has been excluded from the protection of the law by asking whether committing a crime could benefit them legally, even temporarily. If small thieves can improve their legal position, they have been deprived of their human rights. Therefore, crime becomes the only viable way to seek some form of human equality, even if it is recognised as an exception to the norm. That said, stateless people have lost not only their rights as citizens but also their human rights, a double loss that, for Arendt, is developed from the term "without rights", the meaning and significance of which is still far from obvious. For Arendt (1958), there is only one genuinely human right that is not reducible to those that are ours as human beings: the "right to have rights" or the "right to belong to some kind of organised community" (Arendt, 1958, p. 376).

In light of Lafer's (1997) reflections, it can be considered that the terms above are related to the historical experience of *displaced people, which* led Arendt to conclude that citizenship is the right to have rights because equality in dignity and rights for human beings is not a given. It is a collective coexistence construct that requires access to a common public space. In short, this access to public space—the right to belong to a political community—allows for constructing a common world through asserting human rights. Stateless people became "disenfranchised" because they were deprived of this specific human right, which Arendt sees as a condition for the possibility of exercising other rights meaningfully. In her more advanced analyses, Arendt goes on to clarify that her perspective of "without rights" has to do with a fundamental condition that cannot be reduced to the denial or violation of specific rights. She uses the example of soldiers who take part in war and says that even if their right to life is deprived, they still belong to the political community that can recognise them, so they are not "without rights".

Even if stateless people are guaranteed some rights, such as the right to life and freedom of opinion, they are "without rights" because they are dependent on the charity or goodwill of others (Arendt, 1958). Therefore, the notion of "without rights" denotes an indispensable condition that can make even rights that are already guaranteed extinguishable, a condition with legal, political, and ontological dimensions. From the legal perspective, there is a loss of personality, making the person unable to insert themselves into the legal community of the nation-states that receive them and deprived of the rights established by international law (Storey, 2024). Stateless people have thus been subjected to illegality (Jain, 2022). However, Arendt's "without rights" terminology indicates a legal challenge and draws attention to their loss's political and ontological dimensions. There was not just a loss of personality but of the entire social fabric in which they had been born and established a distinct place in the world for themselves.

In *We Refugees*, Arendt (1943) considers that stateless people were deprived of their homes, families, occupations, and language; in other words, they had lost the familiarity of everyday life and experienced the disruption of their private lives. In addition, they were deprived of a political community that considered their ideas, opinions, and actions relevant and significant. This makes Arendt (1958) claim that losing home and political *status* could become synonymous with losing humanity. Arendt's statement concerns the aforementioned ontological implications about the "disenfranchised" as a condition that eliminates the human *status of* someone capable of action and speech. To the extent that the "disenfranchised" cannot find a place in the world and are confined to camps, they are also deprived of the possibility of intersubjective participation in a world forged through proper human activities, especially action and discourse.

For Arendt (1943), it is with words and deeds that we insert ourselves into the human world so humans can perfectly well live without working, forcing others to work for them. They can simply use and enjoy the world of things without adding a helpful object. On the other hand, "[...] life without speech and action, on the other hand—and this is the only way of life that in earnest has renounced all appearance and all vanity in the biblical sense of the word—is dead to the world; it has ceased to be a human life because it is no longer lived among men" (Arendt, 1958, p. 189). Statelessness, therefore, denies humanity itself, since when it comes to politics, as Arendt (2003) reminds us, there is no such thing as "man" in the singular, but only "men" in their plurality; it is only human that there are plural men who, to stand out from nature, need to appear in each other's presence.

In addition, Arendt (1958) analyses the failure of international efforts to deal with statelessness. By considering nationality as the organising norm of the international system, these efforts have come to consider stateless people as an unfortunate exception within an otherwise healthy and normal regime. Statelessness is defined as an anomaly, a problem that could be solved through exceptional and provisional measures, an approach that has also been manifested in the minority treaties and the practices of repatriation and naturalisation. The first solution, in this case, was based on the idea that every person has a national home (homeland) to which they can be sent, without considering, however, the new challenge of statelessness in the twentieth century, which created a group that was simply unportable, given that there was no country in the world where they could enjoy the right of residence. As for naturalisation, which was reserved only for individual cases, it came to a standstill with the arrival of millions of stateless people (Arendt, 1958). Finally, even the option of asylum, which Arendt goes on to describe as the only right that effectively symbolised human rights in the sphere of international relations (Arendt, 1958), was unable to deal with stateless people during the twentieth century, as the emergence of massive contingents of stateless people practically put an end to this aid, which had always been guaranteed in exceptional cases.

In this sense, the above measures aggravated the problem by considering statelessness as an anomaly. Treating it as an exceptional and transitory phenomenon was relatively "desirable" precisely because it left the system intact (Arendt, 1958), avoiding radical questioning of its organisational principles. Nevertheless, it is these very principles that should have been subjected to critical scrutiny because, as Arendt (1958, p. 236) writes, the "[…] *explosion of 1914 and its serious consequences of instability had destroyed the façade of the political system —enough to leave its skeleton exposed, and with this*, [...]" the suffering of an ever-increasing number of groups of people became visible, to whom the rules of the world around them suddenly no longer applied. Therefore, writing against the grain, against the simple and reduced tendency to treat stateless people as an "unfortunate exception", Arendt considered them a symptom manifested in the problems of these hidden structures and offered a critical examination that brought to light the constitutive tensions of nation-states and human rights. Her critical approach suggested that statelessness is a systemic problem produced by the main organisms of the international order.

3 The Right to Have Rights: A Human Perspective or Another Frontier for CSR Practices

This subsection builds on the expression "the right to have rights" found in *Origins of Totalitarianism*, namely that eighteenth-century man has emancipated himself from history. History and nature have both become alien to us in the sense that the essence of man can no longer be understood in terms of one or the other. On the other hand, *humanity*, which for the eighteenth century, in Kantian terminology, was nothing more than a regulating idea, has become de facto inescapable today (Kant, 1795). This new situation, in which 'humanity' has de facto taken over a role previously attributed to nature or history, would mean in this context that the *right to have rights*, or the right of each individual to belong to humanity, should be guaranteed by humanity itself.

Nothing assures us that this is possible (Arendt, 1958). This context brings to light two points: the idea of humanity and the meaning that Arendt (1958) intends to give to the expression "right to have rights", which, according to Lafer (1988, p. 166), is based on "[…] *full access to the legal order that only citizenship offers*". In Arendt's (1958) view, humanity has to do with human plurality, with the fact that "[...] *men, and not Man, live on Earth and inhabit the world*" (Arendt, 1951, p. 8).

Given the above, it is necessary to consider the Arendtian notion of plurality from the point of view of a society that is capable of supporting and protecting subjects, safeguarding them as subjects of rights and duties, in other words, within a legal order that will allow them to live in harmony and security and share the human world. The loss of human rights that affects stateless people results in the loss of their ability to appear in the common space, thus making them invisible and disposable due to their exclusion from plurality. Arendt (1958) starts from this premise and states that "the right to have rights" means living in a structure in which the actions and opinions of an individual have the right to belong to some form of organised community.

Belonging to a community is therefore essential if individuals are no longer to be seen as abstract, generic beings but as men and women who are recognised and judged based on their actions and words. In this way, participation in a shared world becomes necessary for having the freedom to act and think manifested in the world and made real. At this point, Arendt's concern to restore the dignity of public space becomes clearer. If she exalts this space, it is formed by dialogue and debate about the human world, allowing individuals to reveal who they are and what the world represents.

Alves Neto (2009) explains that for Arendt, the radical loss of the protection of an organised political community has resulted in the exclusion of millions of people from any legitimacy, a phenomenon as unprecedented as the loss of one's own space in the world. For this reason, the stability of this political community guarantees men the "right to have rights", providing them with a place defined and recognised by others, where their opinions are meaningful and their actions effective. In addition, it is essential to note that, as Benhabib (2005) highlights, there is a divide between the expression "right to have rights" into two areas, the first linked to the meaning of the term "right" and the second to the meaning of "rights". In this way, the expression "right" has to do with humanity, claiming to belong to a human group and evoking a moral imperative of belonging, namely that all human beings should be treated as people belonging to a human group to which protection corresponds.

Following this analysis, Benhabib (2005) explains that the term "rights" has to do with prior belonging to a political community, which means, in other words, that it is only possible to have rights when you are a member of a community, part of the legal apparatus that guarantees belonging and the expected coexistence in a political community that protects the acquisition of civil and legal rights. In other words, the first aspect of the "right to have rights" refers to the fundamental guarantee that every human being must have, protecting them against abuses such as those experienced in totalitarian regimes, especially the Nazi (National Socialist German Workers' Party) regime, which denied citizenship status to individuals, making them foreigners anywhere in the world and depriving them of a recognised legal existence (Lafer, 1988). Furthermore, the ability to have rights is fundamental to enabling individuals to participate fully in the public arena. In both cases, citizenship must be protected by each Constitution, which Hannah Arendt describes as a stable, objective, and enduring document, emerging from the foundation of the body politic and not subject to alteration or amendment "according to circumstances" (Arendt, 2011, p. 207). According to Arendt (1958), the lack of belonging to a community, i.e., the absence of political status in the individual, results in the deprivation of access to public space and leads to the consequent confinement in private, meaningless life. In extreme situations, such as totalitarian regimes, this culminates in a lack of protection and abandonment, leading to a condition of superfluity for those who are relegated to the heaps of minorities, refugees, and stateless people (Buoli & Giannuli, 2017).

The approaches presented here in the light of Arendt also allow for dialogue with other classical authors, including Kant, since the proximity between Arendt's thesis and Kant's right to visit begins with the understanding that it is humanity's responsibility to guarantee a common right to coexist on Earth (Kant, 1795). In Arendt, the foundation of politics and action is ontologically rooted in humanity, not as a single species, but as a plurality of singular beginnings, including all individuals. It is possible to bring Kant's right to visit closer to the concept of citizenship based on belonging, seen as a duty of humanity. It should be added that in both Kant and Arendt, the right to visit and belong is not seen as a privilege or a benevolent concession based on individual characteristics, such as the genius of Jews. On the contrary, belonging is guaranteed as a right inherent to each person. According to Kant (1795), this is not about philanthropy but about the right, and hospitality here means the right of a foreigner because he arrives in the land of another, not to be treated hostilely by the latter.

Another aspect that demonstrates the affinity between the two authors refers to the principles of republican government and constitution, which internally ensure an organised community and externally the recognition of "belonging" or the right to visit everyone simply because they are human (Kant, 1795). It is also important to mention that both authors reject the idea of a world state with unified legislation. Kant proposes a world federation of free states, an idea also discussed by Benhabib (2005). Kant imagines a world condition in which all members of the human race participate in a civil order and enter into a condition of legal association with each other. However, this civil condition of legal coexistence is not equivalent to being a member of a republican entity. Kant's cosmopolitan citizens still need their republics to be citizens. Kant carefully distinguishes between a world government and a world federation. A world government, which he argues will only result in a universal monarchy, would be a soulless despotism. At the same time, a federative union would still allow citizenship to be exercised within circumscribed communities.

Arendt (1958) also opposes the global/world state idea. It is, therefore, essential to note that she rejects the idea of universal equality, which underpins her resistance to any totalising form of government. Although Arendt did not express her opinion on a world federation, an example of her perspective can be interpreted in her proposal for cooperation between Jews and Arabs in Palestine, with a government based on a federation or a Jewish-Arab binational government structure (Bernstein, 1996). In this sense, Arendt's proposal favours a relationship of solidarity between peoples, mediated by a supranational entity capable of offering legal and social security in relations between borders. However, such a solution presupposes the recognition of a nation-state on the one hand and the strengthening of public international law on the other. The bond of nationality is essential for affirming the dignity of the human person. In contrast, the bonds of solidarity between states would be necessary for ensuring peace in transnational relations. This perspective is not content with formality in legal relations and imposes a commitment to social and political transformation.

For Arendt, the "right to have rights" and the right to belong to the common human world are essential to realise the possibility of being a subject of rights, and these are the most basic rights of a human being. In reality, as Arendt proposes, the life of someone without a homeland, of a refugee, and of any type of person whose belonging to a state of origin is denied represents not only their vulnerability to possible rights violations, which are no longer part of the subject's life, but also something much more severe and perverse: the expulsion of humanity itself. Human dignity needs a new guarantee since, as we have seen, the Kantian guarantee of a natural right respected within a peaceful republic has been destroyed (Kant, 1795).

Therefore, the right to have a right is not just a philosophical proposition to base a new conception of human rights but also a political indication that, according to Kant and Arendt, humanity cannot exclude *displaced persons* and any form of marginalisation of individuals. This right represents the negation of totalitarianism, including its classic forms, such as Nazism and Stalinism, as well as its persistent vestiges that remain in contemporary liberal democracies, which resist effective international supervision of human rights. Nonetheless, Arendt's proposition of the "right to have rights" highlights the significance of political action in cultivating new forms of political responsibility and solidarity as responses to the problems that challenge the "disenfranchised" and as a way of extending the fragile institutional guarantees of equality and freedom.

4 Concluding Remarks: Rethinking Statelessness in the Age of Migration and Globalised Businesses

Contemporary processes and practices related to globalisation, especially international immigration, have deprived millions of people of the protection that comes with citizenship rights. It is estimated that more than 100 million people live outside their country of origin, and approximately another 53 million are internally displaced people (UNHCR, 2023). Arendt's endeavour to understand statelessness as a "mass phenomenon" (Arendt, 1958) is more topical than ever, and the involvement of MNEs in addressing the issue of stateless individuals through CSR represents a significant development in the broader discourse on human rights and corporate responsibility. By leveraging their resources, influence, and global reach, MNEs can play a crucial role in providing stateless individuals with opportunities for social and economic inclusion, thereby contributing to a more equitable and just world (Scherer & Palazzo, 2007).

MNEs have historically played a significant role in advancing CSR initiatives, focusing on sustainable development, ethical labour practices, and community engagement. Recently, there has been a growing recognition of MNEs' role in addressing the plight of stateless individuals, similar to how they have tackled other social issues through CSR (Van Tulder & Van Mil, 2022). CSR provides a framework for MNEs to address the challenges stateless individuals face. By adopting policies that promote the inclusion of stateless people in their operations and supply chains, MNEs can help mitigate some of the adverse effects of statelessness, such as lack of employment and social exclusion. For instance, companies can offer stateless individuals job training and employment opportunities, ensuring they can access livelihoods and integrate into society. Several MNEs have already begun integrating stateless individuals' needs into their CSR strategies (Barnard et al., 2019; Reade et al., 2019). For example, some technology companies have developed educational programs targeting refugees and stateless individuals to improve their digital literacy and employment prospects (Harima et al., 2021). Similarly, MNEs in the manufacturing sector have created inclusive hiring practices that give preference to stateless individuals and refugees, providing them with stable employment and a pathway to social inclusion (Reade et al., 2019) (Table 2).

Engaging in initiatives supporting stateless individuals fulfils ethical and social responsibilities and offers strategic benefits for MNEs (Kingston, 2019). By building a reputation as socially responsible entities, MNEs can enhance their brand image, attract talent, and foster customer loyalty. Furthermore, such initiatives can lead to a more stable and productive workforce, as employees who feel valued and included are more likely to be motivated and engaged (Maggioni et al., 2019).

It is considerable that Hannah Arendt's analysis reaches a new importance in understanding today's scourge of those in asylums, refugees, undocumented immigrants, and even citizens who have become stateless thanks to social exclusion, misery, wars, poverty, and helplessness. If we think in the light of Arendt's writings, the existence of a world makes every refugee, outcast, and stateless person, among others, ask ourselves: How can I not belong to it? Given that there is something so relevant to human life, to confer and welcome one's identity, that allows one to be and appear, surely it must be desired. How can there not be someone in it? How can there be someone who doesn't belong to it? The anguish of stateless people is the unsettling feeling of knowing that there is a world and that it is impossible to inhabit it, in the sense of belonging to a political community and fully enjoying the rights of action, existence, and, in short, becoming someone.

Currently, in the context of migration, despite the particularities of the crises faced in different regions of the world, most of the policies adopted do not challenge the structure of the modern state. Nevertheless, since the sixteenth century, this state has been based on territorial domination, administrative control, the consolidation of a shared cultural identity, and political legitimacy through mechanisms of representation. Since the end of WWII, as analysed by Arendt (1958), and especially in recent decades, we have witnessed the global dissemination of norms and mechanisms dedicated to human rights. These rights are no longer seen merely as abstract commitments but are recognised as fundamental to a new conception of state

Table 2 MNEs' strategies and actions to mitigate the adverse effects of statelessness (Source: Authors' elaboration)

Strategy	Action	Companies	Arendt's perspective
Employment Opportunities	Implement inclusive hiring policies that do not discriminate against stateless individuals. Offer training and development programs for skill enhancement.	*Starbucks*: Partners with NGOs to provide employment opportunities for refugees, including stateless individuals. *Hilton Hotels*: Offers hospitality training to refugees.	Employment opportunities alone cannot fully integrate stateless individuals into public life, as their lack of legal status often excludes them from political participation.
Advocacy and Policy Influence	Advocate for legal and policy reforms to address statelessness—support public awareness campaigns about the impacts of statelessness.	*IKEA Foundation*: Advocates for refugee rights and supports legal reforms. *UNHCR's Private Sector Partnerships*: Companies like Zain promote awareness of statelessness.	Advocacy by MNEs is crucial, yet the increasing focus on market-driven solutions often sidelines deeper political engagement needed to fundamentally address statelessness.
CSR Initiatives	Fund or support educational programs for stateless individuals. Provide or support access to healthcare and essential services.	Google.org: Funds educational programs for refugees. *Johnson & Johnson*: Provides healthcare services to displaced populations, including the stateless.	While beneficial, CSR initiatives often substitute for robust public policies and civic actions that address the root causes of statelessness.
Partnerships and Collaborations	Collaborate with NGOs specialising in helping stateless individuals. Work with governments to develop and implement strategies to reduce statelessness.	*Airbnb*: Partners with IRC to provide housing for refugees. *Tent Partnership for Refugees*: MNEs like Pfizer and PepsiCo collaborate to support refugees.	Partnerships can lead to positive change, but strong public institutions and civic engagement are necessary for these efforts to be sustainable.

(continued)

Table 2 (continued)

Strategy	Action	Companies	Arendt's perspective
Support for Documentation and Legal Aid	Offer legal aid to help stateless individuals obtain the necessary documentation. Sponsor initiatives for acquiring identity documents.	*Microsoft*: Supports legal aid clinics for refugees. *Western Union*: Partners with NGOs to assist refugees in obtaining identification documents.	Legal aid is essential for individual empowerment but must be coupled with political action to ensure systemic change and the inclusion of stateless people in the public sphere.
Ethical Business Practices	Ensure non-discrimination in all business practices. Maintain non-discriminatory practices in supply chain management.	*Unilever*: Commits to non-discriminatory practices and diverse hiring. *Patagonia*: Ensures ethical practices in its supply chain, supporting the rights of all workers.	Ethical practices help mitigate immediate issues but do not replace the need for collective action and policy changes to integrate stateless individuals into society fully.

legitimacy. This new perspective argues that a state does not have exclusive authority over the fundamental rights of its citizens.

In this sense, these developments are considered crucial by various categories of immigrants. Although we do not fully agree with the strong assertion that the rise of the universal human rights framework has led to the decline of national citizenship and the emergence of post-national affiliation, there is a consensus that there has been a significant shift in the importance attributed to national identity in favour of a universal basis for rights. According to Benhabib (2005), this has allowed immigrants to claim many of those rights previously associated exclusively with citizenship status.

In addition, the developments can also lead to misperceptions that the condition of the "disenfranchised", as described by Arendt, has completely disappeared from the contemporary scene. However, given the complexities of human rights, as analysed by Arendt, these issues manifest differently in international human rights regulations. Furthermore, contemporary events, such as the increasingly restrictive global immigration control apparatus, have made navigating the paradoxes of human rights increasingly complex for millions of immigrants. Similarly, Hannah Arendt, examining the fundamental tension between the concepts of "human being" and "citizen" in the light of the Declaration of 1789, argues that this tension was not eliminated with the transition to "human" rights. For Arendt (1958), the complexities that arise from the intersection between the universality of the human condition and territorial sovereignty in international human rights contexts are relevant. Article 14 of the Universal Declaration of Human Rights guarantees victims of persecution the right to "seek and enjoy asylum in other countries". This formulation does not automatically guarantee the right to asylum. Still, it balances it with the sovereign right to control borders, emphasising that the human right to leave does not automatically imply the right to enter another country.

In this way, the precarious concert that the human rights framework achieves between universal personality and territorial sovereignty has been jeopardised by worrying developments that have successively led to difficulties in accessing rights for various categories of immigrants. In the contemporary context of restrictive immigration controls, measures ranging from detention to deportation, previously reserved for states of emergency, have become the norm. The camps have re-emerged as the only practical substitute for a homeland that does not exist (Arendt, 1958). Yet, they have appeared in new forms, such as detention centres and waiting zones, but they continue to be designed by arbitrary rules or by the very absence of laws and norms.

If that weren't enough, to stop immigrants before they reach the country's borders and escape human rights laws, countries and states have resorted to extreme measures, such as building border walls, intercepting them on the high seas, stripping parts of their territories of their jurisdiction, as well as signing readmission treaties that risk returning those who request shelter and asylum. In addition, this scenario of extreme measures has been exacerbated by the growing incidence of racism and xenophobia in various countries receiving immigrants, which makes it difficult not only for permanent residents to exercise their rights but also for citizens belonging to racial minorities. Furthermore, systemic discrimination and restrictive policies not only create legal and social barriers to the full integration of these groups into the host society but also widen socio-economic disparities and promote a climate of marginalisation that undermines the principles of diversity and equality that should be fundamental to harmonious and fair coexistence.

Given the developments mentioned so far, Arendt's assertion that stateless people are in a fundamental condition of "without rights" has a new meaning. In the current context, this term does not necessarily denote a complete absence or lack of legal rights but rather a growing vulnerability. Individuals classified as "without rights" highlight the precarious legal, political, and ontological situation many immigrants face in a period characterised by the spread of human rights and the globalisation of migration control.

Finally, these challenging problems highlight that the universal framework of human rights fails to guarantee the "right to have rights", or we could also say the right to citizenship, to humanity, and its respective humanisation. Faced with the inability of the existing institutional order to fully guarantee this prerogative, various political struggles led by immigrants have emerged as urgent movements. Hannah Arendt's proposition about the "right to have rights" offers a fruitful starting point for understanding the limiting challenges of the current human rights framework and exploring new political possibilities for reinventing it in response to the pressing problems faced by the "disenfranchised".

References

Alves Neto, R. (2009). *Alienações do mundo: Uma interpretação da obra de Hannah Arendt*. Rio de Janeiro: PUC-Rio.

Arendt, H. (1943). We refugees. *The Menorah Journal, 31*, 69–77. Reprinted in Anderson, M. M. (Ed.) (1999). *Hitler's exiles: Personal stories of the flight from Nazi Germany to America* (pp. 253–262). The New Press.

Arendt, H. (1951). *The origins of totalitarianism*. Harcourt Brace Jovanovich.

Arendt, H. (1958). *The human condition*. University of Chicago Press.

Arendt, H. (1963). *On revolution*. Viking Press.

Arendt, H. (1968). *Men in Dark Times*. Harcourt, Brace & World.

Arendt, H. (2003). *Was ist Politik? Fragmente aus dem Nachlass*. Piper Verlag GmbH.

Barnard, H., Deeds, D., Mudambi, R., & Vaaler, P. M. (2019). Migrants, migration policies, and international business research: Current trends and new directions. *Journal of International Business Policy, 2*, 275–288.

Benhabib, S. (2005). *Los derechos de los otros: Extranjeros, residentes y ciudadanos*. Gedisa.

Berger, J. A. (1947). Displaced persons: A human tragedy of world war II. *Social Research, 14*(1), 45–58.

Bernstein, R. J. (1996). *Hannah Arendt and the Jewish question*. MIT Press.

Blitz, B. K., & Lynch, M. (2009). *Statelessness and the benefits of citizenship: A comparative study*. Geneva Academy of International Humanitarian Law and Human Rights and International Observatory on Statelessness.

Buoli, M., & Giannuli, A. S. (2017). The political use of psychiatry: A comparison between totalitarian regimes. *International Journal of Social Psychiatry, 63*(2), 169–174.

Butler, J. (1990). *Gender trouble: Feminism and the subversion of identity*. Routledge.

Butler, J. (2004). *Precarious life: The powers of mourning and violence*. Verso.

Canovan, M. (1992). *Hannah Arendt: A reinterpretation of her political thought*. Cambridge University Press.

Chauí, M. (2019). *Neoliberalismo: a nova forma do totalitarismo*.

Foucault, M. (1977). *Discipline and punish: The birth of the prison*. Pantheon Books.

Foucault, M. (1975). *Surveiller et Punir: Naissance de la prison*. Gallimard Editions.

Foucault, M. (1991). Governmentality. In G. Burchell, C. Gordon, & P. Miller (Eds.), *The Foucault effect: Studies in governmentality*. University of Chicago Press.

Gibney, M. J. (2020). Denationalisation and discrimination. *Journal of Ethnic and Migration Studies, 46*(12), 2551–2568.

Habermas, J. (1984). *The theory of communicative action: Reason and the rationalization of society*. Beacon Press.

Habermas, J. (1989). *The structural transformation of the public sphere: An inquiry into a category of bourgeois society*. MIT Press.

Harff, B., & Gurr, T. R. (1988). Toward an empirical theory of genocides and politicides: Identification and measurement of cases since 1945. *International Studies Quarterly, 32*(3), 359–371.

Harima, A., Periac, F., Murphy, T., & Picard, S. (2021). Entrepreneurial opportunities of refugees in Germany, France, and Ireland: Multiple embeddedness framework. *International Entrepreneurship and Management Journal, 17*(2), 625–663.

Horst, C., & Lysaker, O. (2021). Miracles in dark times: Hannah Arendt and Refugees as 'Vanguard'. *Journal of Refugee Studies, 34*(1), 67–84.

Jain, N. (2022). Manufacturing statelessness. *American Journal of International Law, 116*(2), 237–288.

Kant, I. (1795). *Perpetual Peace (Zum ewigen Frieden; Ein philosophischer Entwurf)*. Friedrich Nicolovius.

Kerber, L. K. (2007). The stateless as the citizen's other: A view from the United States. *The American Historical Review, 112*(1), 1–34.

Kingston, L. N. (2019). Conceptualizing statelessness as a human rights challenge: Framing, visual representation, and (partial) issue emergence. *Journal of Human Rights Practice, 11*(1), 52–72.
Kulshrestha, S. (2022). Delving into struggles: Ensuring access to justice for stateless persons. *Jus Corpus Law Journal, 3*, 301.
Lafer, C. (1988). *A reconstrução dos direitos humanos: um diálogo com o pensamento de Hannah Arendt*. Companhia das Letras.
Lafer, C. (1997). A reconstrução dos direitos humanos: a contribuição de Hannah Arendt. *Estudos Avançados, 11*(30), 55–65.
Maggioni, D., Santangelo, G. D., & Koymen-Ozer, S. (2019). MNEs' location strategies and labor standards: The role of operating and reputational considerations across industries. *Journal of International Business Studies, 50*, 948–972.
Namondo, S. (2020). *Who is a migrant, a refugee, a stateless person and an asylum seeker?* The Organization for World Peace.
Reade, C., McKenna, M., & Oetzel, J. (2019). Unmanaged migration and the role of MNEs in reducing push factors and promoting peace: A strategic HRM perspective. *Journal of International Business Policy, 2*(4), 377–396.
Scherer, A. G., & Palazzo, G. (2007). Toward a political conception of corporate responsibility: Business and society seen from a Habermasian perspective. *Academy of Management Review, 32*(4), 1096–1120.
Siegelberg, M. L. (2020). *Statelessness: A modern history*. Harvard University Press.
Storey, H. (2024). *The refugee definition in international law*. Oxford University Press.
Van Tulder, R., & Van Mil, E. (2022). *Principles of sustainable business: Frameworks for corporate action on the SDGs*. Taylor & Francis.

Jenerton Arlan Schütz is a doctor and master in science education (Unijuí). Degree in Sociology, History, and Pedagogy. He is a lecturer and permanent researcher in the Stricto Sensu Masters and Doctorate in Education Programme at the Catholic University of Brasilia (UCB/DF). Post-doctoral student in Professional and Technological Education (Instituto Federal Goiano). He is also a lecturer and visiting researcher in the Professional Master's Programme in Professional and Technological Education at the Federal Institute of Goiás, Ceres Campus.

Marcos de Moraes Sousa is a full professor at the Goiano Federal Institute and a permanent professor of the Administration course (Master and PhD) at the Federal University of Goiás—with a research line in Public Administration and Public Policies. Marcos is also a permanent professor at the Goiano Federal Institute in the area of Professional and Technological Education—with a research line in organisation, evasion, and public policies—and a permanent professor in the administration course—at the Goiano Federal Institute Rio Verde Campus—with a research line in strategy and innovation. Marcos took a post-doctorate at the University of Bern, Kompetenzzentrum für Public Management. He is a member of the Brazilian Society of Public Administration. He is the leader of the Justice Administration research group at the Federal University of Goiás and the Innovation and Strategy research group at the Federal Institute of Goiás. Marcos has published widely in journals of international relevance and is currently the academic editor of *PLOS One* and *Societal Impacts* of the Elsevier Editorial Group.

Miguel M. Torres, PhD in Economics, is a Lecturer in International Business at Kent Business School (KBS). His career includes consulting services to the European Commission and World Bank, applied research and teaching roles at Leeds, Milano, Aveiro, KU Leuven, and Uppsala universities, with appointments at the University Federal of Goiás, IF Goiano, and the University Catolica of Portugal. He has held multiple visiting professorships and is an INSOL International and Marie Skłodowska-Curie Fellow. He completed executive programs at Harvard, Carnegie Mellon University (USA), and University College London (UK). His research focuses on the intersection of international business, public policy, and strategic management, focusing on multinational management models, corporate political risk, and negotiation strategies, particularly emphasizing the complexities of cross-border restructuring and foreign direct investment drivers using cross-sectional, time-series, and panel data approaches. His work has been published in the *International Business Review, Journal of Business Ethics, Multinational Business Review, Journal of Economic Methodology, Arbitration International, International Insolvency Review, PLOS One, and AIB Insights*. He co-edited the books *Global Perspectives on Climate Change, Inequality, and Multinational Corporations* (Springer, 2025) and *Dead Firms: Causes and Effects of Cross-Border Corporate Insolvency* (Emerald, 2017).

Transforming Africa Byte by Byte: AI's Role in Achieving SDG7

Hicham Nachit, Rachid Hachad, Saad Laraqui, and Miguel M. Torres

1 Introduction

The transformative potential of artificial intelligence (AI) in advancing sustainable development goals (SDG) is a critical area of exploration. The motivation behind examining the role of AI in achieving SDG7 is driven by the urgent need to address climate change, decarbonize critical sectors, and reduce environmental pollution through sustainable energy solutions (Jayachandran et al., 2022). Concerned by the continuous decline in the quality of life, poverty, environmental degradation, and escalated war and conflicts, the United Nations in 2015 instituted the 17 SDGs and 169 targets. Access to clean, modern, and affordable energy, also known as SDG7, is one of these goals. Universal access to clean, modern, and affordable electricity is one of the metrics for measuring a good life, as it fundamentally affects education, healthcare, food security, job creation, and other socioeconomic indices. The Sustainable Development Agenda of 2030, adopted by all United Nations member states, provides a strategic blueprint for peace and prosperity for all people, encompassing these SDGs that call all countries to action regardless of economic status. Current approaches to achieving SDG7 often lack a comprehensive integration of

H. Nachit (✉) · S. Laraqui
ESCA Ecole de Management, Casablanca, Morocco
e-mail: hnachit@esca.ma; slaraqui@esca.ma

R. Hachad
Hassan II University of Casablanca, Casablanca, Morocco
e-mail: rachid.hachad-etu@etu.univh2c.ma

M. M. Torres
Kent Business School, University of Kent, Canterbury, UK
e-mail: m.torres@kent.ac.uk

J. Bento, M. M. Torres (eds.), *Global Perspectives on Climate Change, Inequality, and Multinational Corporations*, CSR, Sustainability, Ethics & Governance,
https://doi.org/10.1007/978-3-031-80797-8_10

all its components, such as access, affordability, reliability, sustainability, and modernity of energy sources (Matenga, 2022).

Africa currently faces significant challenges in the energy sector, remaining largely underserved despite its vast energy resources, which leaves many of its populations in energy poverty. Projections indicate that most African nations will need help to meet the targets for universal electricity access by 2030, with only Mauritius, Morocco, and South Africa expected to achieve these goals. This situation underscores the broader issues of limited consumption, pervasive reliability challenges, prohibitive prices, and financial stress on utilities (Blimpo & Cosgrove-Davies, 2019). While North African countries have made strides in achieving universal electricity access, their adoption of renewable energy has yet to catch up. Egypt, although investing in renewable energy, still prioritises fossil fuels. Morocco is leading the way with ambitious renewable energy targets and is expected to meet the 42% renewable energy share by 2030 (Sanusi, 2023).

Morocco presents a particularly compelling case for study due to its lack of conventional energy resources, which makes it heavily dependent on the international primary energy market. Most of its energy supply is imported. To address this, Morocco has implemented a comprehensive energy strategy to support the transition to renewable energy and energy efficiency across all consumer sectors of the economy, including housing, transport, and industry. The Moroccan government has liberalised the renewable energy market and adopted financial mechanisms to stimulate private sector involvement and facilitate public–private partnerships. In this context, the Moroccan Agency for Sustainable Energy (MASEN) was created to accompany and drive the clean energy transition. Morocco's renewable capacity increased by 1856 MW between 2011 and 2020, significantly contributing to Africa's wind power capacity alongside Egypt. In 2011, Morocco and Egypt contributed 809 MW of Africa's wind capacity of 995 MW. By 2020, they contributed 2785 MW of the continent's total of 6496 MW (IRENA, 2021). Despite these efforts, Morocco still faces significant challenges, particularly technical and financial barriers, in developing renewable energy and achieving energy efficiency. AI technologies can play a critical role in overcoming some of these technical barriers.

This chapter aims to provide a comprehensive review of the evolving capabilities of AI and its pivotal role in the contemporary sustainability-driven business context. It investigates the confluence of AI and sustainability in Africa, emphasising the renewable energy sector, and examines how AI can address the continent's distinctive challenges and opportunities. By analysing theoretical frameworks and existing literature, we establish a foundation for understanding the global progression of AI technologies and their specific applications in renewable energy. The case study of MASEN displays AI applications in solar, hydrogen, and wind energy optimisation, predictive maintenance, and grid management, demonstrating quantifiable benefits such as enhanced efficiency, cost reductions, and improved energy access. Integrating AI into the renewable energy sector in Africa represents a significant opportunity to advance SDG7, ensuring access to affordable, reliable, sustainable, and modern energy for all. The case study of MASEN suggests that AI has been instrumental in optimising the design and operation of solar and wind energy systems, enhancing grid management, and supporting decentralised energy solutions. While AI-driven innovations offer tremendous potential for advancing sustainable energy access in Africa, it is essential to address ethical and environmental considerations and foster a supportive regulatory environment to ensure the equitable and sustainable realisation of these benefits (Awogbemi et al., 2024).

2 AI and Electrification in Africa

Africa presents unique challenges and opportunities in advancing electrification (Li et al., 2023b). The continent has abundant renewable energy resources, including solar, wind, and hydropower. However, many African countries grapple with significant barriers to energy access, such as inadequate infrastructure, limited financial resources, and regulatory challenges (Mungai et al., 2022). By 2022, the number of people worldwide without access to electricity reached 760 million. Recent crises stemming from the impacts of the COVID-19 pandemic and Russia's invasion of Ukraine have pushed this number back to 2019 levels, with Sub-Saharan Africa accounting for 80% of those lacking electricity access. Specifically, the number of Sub-Saharan Africans without electricity rose from 580 million in 2019 to 600 million in 2022, approaching historic highs (IEA, 2023).

Less than a fifth of African countries have set targets to achieve universal electricity access by 2030. An additional 45% of the continent has set targets for electricity access, but these are less ambitious than the goals under SDG7 (IEA, 2023). AI can play a pivotal role in addressing these obstacles by optimising the use of existing resources, reducing costs, and improving the reliability of energy systems (Ahmad et al., 2021; Quest et al., 2022; Qin et al., 2024). AI algorithms can enhance predictive maintenance of energy infrastructure, reducing downtime and prolonging the lifespan of equipment. Smart grids powered by AI can balance supply and demand more effectively, integrating renewable energy sources seamlessly into the grid. Additionally, AI-driven data analytics can identify the most efficient locations for new renewable energy projects, maximising their impact.

AI's ability to analyse and interpret large datasets can help identify the most effective strategies for expanding energy access in remote and underserved areas. For example, AI can help design and manage microgrids, providing reliable electricity to communities not connected to the national grid (Ibegbulam et al., 2023). Moreover, AI-driven predictive maintenance can enhance the sustainability and cost-effectiveness of renewable energy installations, reducing operational costs and attracting investments (Moghayedi et al., 2023). These advancements demonstrate how AI can support sustainable development and improve African living standards. Table 1 shows that AI also has the potential to transform the energy industry and other economic sectors. The potential for renewable energy capacity additions in Africa is significant, as depicted in Fig. 1. The chart illustrates the projected growth

Table 1 The transformative potential of AI in Africa

Sector	Aim	Application	Benefit	Challenge
Energy	Enhancing efficiency and accessibility of renewable energy systems	Predictive maintenance for solar and wind farms	Reduced operational costs Improved energy access Enhanced grid stability	Technological infrastructure Investment gaps Regulatory frameworks
Agriculture	Increasing productivity and sustainability of farming practices	Precision agriculture using AI-driven sensors and drones	Higher crop yields Efficient use of resources Improved food security	Access to AI technology Data privacy concerns Need for skilled workforce
Healthcare	Improving diagnostics, treatment, and healthcare management	AI-based diagnostic tools and predictive analytics for disease outbreaks	Early disease detection Personalised treatment plans Efficient healthcare delivery	Data security and privacy Integration with existing systems Ethical considerations
Education	Enhancing learning experiences and accessibility to quality education	AI-powered personalised learning platforms and intelligent tutoring systems	Tailored education plans Access to education in remote areas Improved learning outcomes	Digital divide Acceptance of AI in education Cost of implementation
Infrastructure	Developing smart cities and improving public services	AI for traffic management and predictive maintenance of infrastructure	Reduced congestion Lower maintenance costs Improved urban planning	Infrastructure investment Technical expertise Coordination among stakeholders
Environmental Management	Monitoring and managing natural resources and environmental conservation efforts	AI for climate modelling and monitoring deforestation and wildlife	Better resource management Enhanced conservation efforts Real-time environmental monitoring	Data accuracy and reliability Ethical implications Funding for AI initiatives
Economic Development	Driving economic growth through innovative AI applications across various sectors	AI-driven financial services and AI for market analysis and forecasting	Increased financial inclusion Better market predictions Enhanced economic planning	Regulatory challenges Ensuring fairness and transparency Bridging the skills gap in AI

(continued)

Sector	Aim	Application	Benefit	Challenge
Social Development	Addressing social challenges and improving quality of life	AI for social welfare programs and Analysis of social trends and needs	Improved social services Targeted welfare programs Enhanced decision-making	Privacy concerns Ethical use of AI Ensuring inclusivity and equity in AI applications

Source: Authors' elaboration

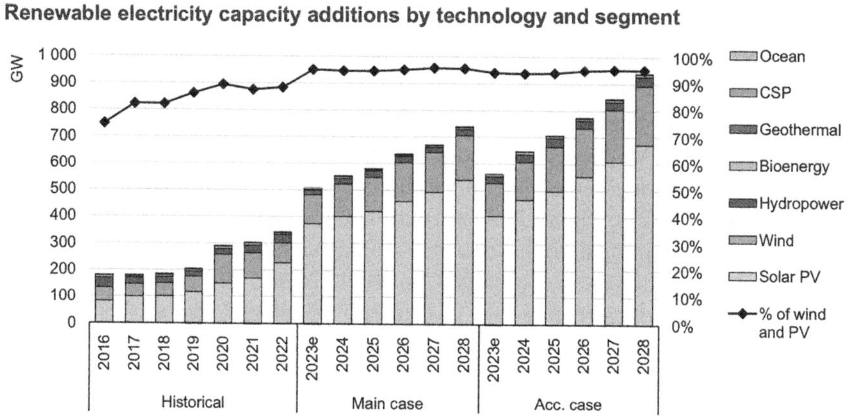

Fig. 1 Renewable electricity capacity additions by technology and African segment. (Source: International Energy Agency, IEA, 2024). Notes: CSP = concentrated solar power. Capacity additions refer to net additions. Historical and forecast solar PV capacity may differ from previous editions of the renewable energy market report. This year, PV data for all countries have been converted to DC (direct current), increasing capacity for countries reporting in AC (alternating current). Conversions are based on an IEA survey of more than 80 countries and interviews with PV industry associations. Solar PV systems work by capturing sunlight using photovoltaic cells and converting it into DC electricity. The DC electricity is then usually converted using an inverter, as most electrical devices and power systems use AC. Until about 2010, AD and DC capacity in most PV systems were similar, but with developments in PV system sizing, these two values may now differ by up to 40%, especially in utility-scale installations. Solar PV and wind additions include capacity dedicated to hydrogen production

in renewable electricity capacity by technology and segment from 2016 to 2028, highlighting the dominant roles of solar PV and wind. This dominance is well-aligned with the African context, where plentiful solar and wind resources make these technologies particularly advantageous for the continent's renewable energy development.

AI's role extends to addressing financial barriers associated with electrification projects. By facilitating innovative financing models, reducing operational costs, and attracting investments, AI contributes to creating sustainable and economically viable electrification solutions. The focus extends to decentralised energy systems and microgrids, empowering remote and underserved communities with reliable access to electricity (Ibegbulam et al., 2023).

While Africa is making strides in adopting AI in the renewable energy sector, it still needs to catch up to developed regions such as Europe and North America. Despite some progress, Africa needs to catch up in AI development and integration. According to the 2022 global Government Artificial Intelligence Readiness Index compiled by Oxford Insights, the top five African countries in AI development are Mauritius (57th in the world), Egypt (65th), South Africa (68th), Tunisia (70th), and Morocco (87th). Mauritius is the most advanced sub-Saharan country in AI development, while South Africa, which has the only commercially available 5G infrastructure, follows closely (Burke, 2023). These rankings are attributed to their scores in the 'Government' pillar, which includes criteria such as the existence of a national vision for AI, the development of online services, the existence of data protection and privacy laws, and the establishment of cybersecurity strategies (Rogerson et al., 2022). Figure 2 shows that while Sub-Saharan countries show a medium level in adopting digital skills, they still lag significantly behind in AI compared to developed European countries (Choi et al., 2020). This disparity is primarily due to gaps in infrastructure and investment. Developed regions benefit from advanced technological infrastructure, substantial financial resources, and supportive regulatory frameworks that facilitate the integration of AI into renewable energy systems (Ahmad et al., 2021). The challenge for Africa is to scale up AI integration to bridge this gap and fully leverage AI's potential to transform the renewable energy sector and achieve SDG7.

3 AI and SDG7 in Africa

Artificial intelligence has emerged as a powerful tool for advancing SDGs. According to Vinuesa et al. (2020), AI can accomplish 134 out of the 169 targets across all the SDGs. This is particularly evident in the Society group of goals (SDG1, SDG2, SDG3, SDG4, SDG5, SDG6, SDG7, SDG11, and SDG16), where AI-based technologies have the potential to benefit 67 targets (82%) by enhancing the provision of essential services such as food, health, water, and energy. In the context of SDG7, which focuses on ensuring access to affordable, reliable, sustainable, and modern energy for all, AI is critical in optimising energy systems and infrastructure planning.

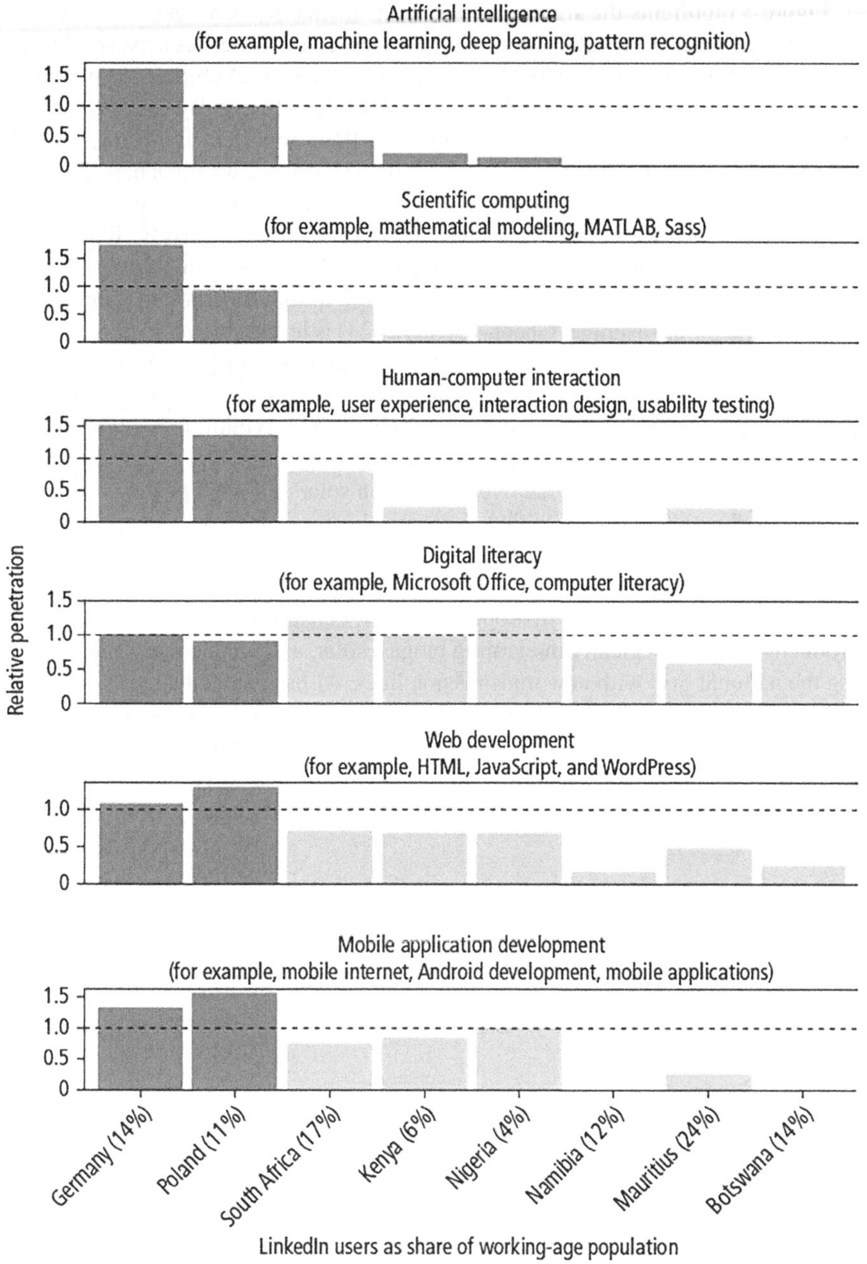

Fig. 2 Adoption of various skills in selected Sub-Saharan African countries and benchmark countries in Europe. (Source: Adapted from Choi et al., 2020). Note: On the relative penetration axis, 1 represents the global average adoption of a skill. Benchmark countries were chosen based on similar shares of LinkedIn members in their working-age population to the shares in selected Sub-Saharan African countries. Parentheses show the share of LinkedIn users in each country's total working-age population

Figure 3 highlights the significant variations in electricity access across African countries from 2000 to 2022, revealing a stark contrast between North African countries, with an impressive 99.8% electrification rate, and Sub-Saharan countries, where only 49.4% of the population has access to electricity. By 2022, the global average, as the International Energy Agency (IEA) reported, is 90.2%. Notable advancements have been made in countries like Mauritius, which achieved universal access decades ago, and Morocco and South Africa, which have significantly progressed from 71.6% to 99.8% and 77.3% to 95%, respectively. Remarkable improvements are also seen in Gabon, which soared from 31% in 2000 to 93.3% in 2022, and Ghana, which rose from 45% to 88.6% in the same period. Now almost matching the global average, Ghana (Byrne, 2024) is leveraging AI to play a crucial role in advancing SDG7 by optimising the planning and integration of renewable energy sources. AI-driven models simulate the impacts of various infrastructure investments on energy, water, and food systems, using evolutionary algorithms to find optimal solutions that maximise benefits while minimising negative impacts. These models have revealed that overreliance on solar and wind power could cause dangerous fluctuations in river flows, negatively impacting ecosystems and communities. With renewable energy accounting for about 30% of global energy generation in 2023, up from 18% in 2000, Ghana aims to enhance its renewable energy mix, currently dominated by hydropower, which constitutes one-third of its energy production. By strategically integrating biogas, solar, and wind energy and improving the national grid with new transmission lines, AI has enabled a more stable and sustainable energy transition, ensuring that by the 2030s, Ghana can achieve a significant portion of its energy from renewable sources, aligning with SDG7 while mitigating potential adverse effects on the environment and public health. While Ghana is on track to achieve a significant portion of its energy from renewable

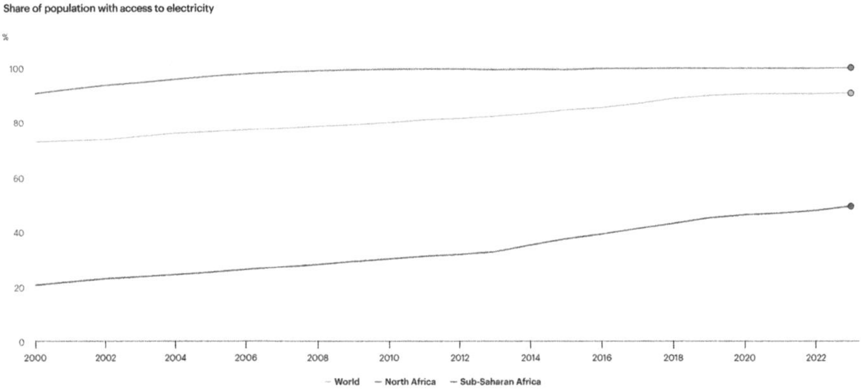

Fig. 3 SDG7 Data and Projection: Share of population with access to electricity in African countries (2000–2020). (Source: International Energy Agency, IEA, 2023). Note: IEA produces its own estimates for access to electricity based on administrative data collected from ministries of energy. This data may differ from what the World Bank reports in the Tracking SDG7 Report, which uses survey data

sources, other Sub-Saharan countries remain significantly below the average access to electricity, with Benin at 38%, Niger at 19.7%, and Chad at just 7.8%.

AI has evolved significantly from its initial applications in automating routine tasks and managing data to advanced systems capable of complex data analysis and decision-making (Duan et al., 2019). Early AI technologies were limited in scope, focusing primarily on automating basic processes. However, advancements in machine learning (ML), deep learning (DL), and natural language processing (NLP) have enabled AI to process vast datasets, identify intricate patterns, and make accurate predictions. These capabilities are valuable in sustainability, where AI can optimise energy usage, reduce waste, and enhance the efficiency of renewable energy systems (Awogbemi et al., 2024). Globally, AI applications in sustainability have rapidly expanded, optimising energy grids, forecasting demand, and predicting maintenance needs for energy infrastructure. Google's AI subsidiary, DeepMind, developed a neural network in 2019 to increase the accuracy of wind power forecasts for its 700 MW renewable fleet. By predicting output up to 36 h in advance, Google's model improved forecast accuracy, allowing the company to sell power in advance and increasing the financial value of its wind power by 20% (Rozite et al., 2023). AI-driven predictive analytics maximise the output of solar and wind farms by analysing weather patterns, ensuring a stable and efficient energy supply (Zhou, 2022). Additionally, AI enhances grid management by predicting fluctuations and ensuring stability, which is essential for integrating intermittent renewable energy sources (Shi et al., 2020). These advancements showcase AI's potential to support the transition to a more sustainable and resilient energy system.

Several studies highlight AI's transformative potential in achieving SDG7, which aims to ensure access to affordable, reliable, sustainable, and modern energy for all. AI technologies can significantly enhance the efficiency and effectiveness of renewable energy sources, making clean energy more affordable and accessible. AI's role in energy demand forecasting is becoming increasingly indispensable. By analysing vast datasets, AI can accurately predict energy demand, allowing for the optimisation of renewable energy production. For example, machine learning algorithms can analyse historical weather data to forecast solar and wind energy generation, thereby improving the reliability of these energy sources (Ahmad and Chen, 2020). AI also plays a vital role in grid management, enhancing the stability and efficiency of energy distribution networks by predicting and mitigating fluctuations (Wang et al., 2020).

AI-driven predictive maintenance reduces downtime and extends the lifespan of renewable energy installations, making them more cost-effective (Sun & Medaglia, 2019). The potential of AI in achieving SDG7 is further demonstrated through its applications in optimising renewable energy sources. AI systems can efficiently manage energy storage solutions, ensuring a balanced supply and demand. AI can enhance battery storage performance by predicting energy consumption patterns and optimising charge cycles, thus extending battery life and reducing costs (Muqeet et al., 2022).

AI's transformative potential in achieving SDG7 is particularly relevant for Africa, where access to reliable and affordable energy remains a significant challenge. AI can play a central role in optimising energy production and consumption, significantly enhancing the efficiency of renewable energy systems (Ahmad et al., 2021). By leveraging advanced algorithms and data analytics, AI can address various challenges in the renewable energy sector, from improving energy generation to managing distribution and consumption. This is particularly relevant for achieving SDG7, which aims to ensure access to affordable, reliable, sustainable, and modern energy for all, especially in Africa. Solar energy optimisation is one of AI's primary applications in renewable energy. Solar energy systems depend heavily on weather conditions, which can be unpredictable and variable. AI algorithms, particularly machine learning models, can analyse historical weather data and real-time information to forecast solar radiation levels. This predictive capability allows solar farms to adjust the positioning of solar panels to maximise energy capture, even under suboptimal weather conditions (Quest et al., 2022; Awogbemi et al., 2024). AI can optimise the angle and orientation of solar panels based on the predicted intensity and duration of sunlight, ensuring that the maximum possible energy is harvested throughout the day. AI's predictive analytics enable solar farms to forecast potential energy production, allowing grid operators to manage energy supply and demand better. This increases the efficiency of solar energy systems, enhances grid stability, and reduces the need for backup power sources (Hannan et al., 2021). Furthermore, advanced AI models can manage solar energy storage systems, optimising the charging and discharging cycles to align with consumption patterns, thus extending battery life and reducing operational costs (Frank, 2021).

AI is also instrumental in maintaining and managing wind turbines (Chatterjee & Dethlefs, 2021). Wind turbines are complex machines that require regular maintenance to operate efficiently. Traditional maintenance approaches are often reactive, addressing issues only after they occur, leading to costly downtime and repairs. Powered by AI, predictive maintenance transforms this paradigm by anticipating maintenance needs before failures occur. Machine learning algorithms contribute to analysing data from sensors embedded in wind turbines, monitoring various parameters such as vibration, temperature, and rotational speed. By identifying patterns and anomalies in this data, AI can predict potential failures and schedule maintenance activities proactively (Ibegbulam et al., 2023; Qin et al., 2024). This approach minimises unplanned downtime, extends the lifespan of wind turbines, and reduces maintenance costs. Predictive maintenance also ensures that wind turbines operate optimally, maximising energy production and contributing to a more reliable renewable energy supply (Awogbemi et al., 2024). Moreover, by reducing the frequency and severity of maintenance-related disruptions, AI-driven predictive maintenance can enhance wind energy projects' overall sustainability and economic viability (Wu et al., 2022).

Beyond solar and wind energy, AI has broader applications in renewable energy systems. AI can optimise the integration of different renewable energy sources into the grid, balancing supply and demand in real time. This is particularly important in regions with high renewable energy production variability (Miskiewicz, 2022). AI also enhances energy storage solutions, such as batteries, by predicting energy usage patterns and optimising charging and discharging cycles. This ensures that stored energy is used efficiently, reducing waste and improving the reliability of renewable energy systems. Moreover, AI-driven microgrid management systems can provide reliable electricity to remote and underserved areas, which is important for expanding energy access in Africa (Ibegbulam et al., 2023). These AI applications improve operational efficiency and facilitate the integration of decentralised energy resources, promoting energy resilience and reducing the environmental footprint of energy systems (Di Vaio et al., 2020).

Decentralised energy systems, such as microgrids, benefit significantly from AI integration. AI can manage these systems more effectively by predicting energy consumption and production patterns, optimising the use of local renewable resources, and reducing dependency on central power grids. This is particularly advantageous in remote areas of Africa where grid infrastructure is limited (Goralski & Tan, 2020). AI-driven microgrids can leverage localised renewable energy sources such as solar and wind to provide stable and reliable electricity to communities otherwise off the main grid. This decentralised approach improves energy access and enhances community resilience to power outages and grid failures (Ibegbulam et al., 2023). Furthermore, AI facilitates innovative financing models for electrification projects by improving operational efficiency and attracting investments. This makes renewable energy projects more economically viable and sustainable, addressing financial barriers to energy access in Africa (Quest et al., 2022; Ibegbulam et al., 2023). By enabling more accurate demand forecasting and dynamic pricing models, AI can also help optimise the economic performance of decentralised energy systems, ensuring that they remain financially sustainable over the long term (Hannan et al., 2021).

AI-driven electrification initiatives in Africa also have significant socio-economic impacts. They can create jobs, drive economic growth, and improve living standards. For example, AI can help develop local expertise and build capacity in managing and maintaining renewable energy systems. This local empowerment is essential for sustainable and inclusive development (Ibegbulam et al., 2023). Moreover, AI technologies enable more effective data management and analysis, facilitating better decision-making processes. This can lead to more efficient policy implementation and regulatory frameworks, fostering a conducive environment for renewable energy investments and advancements (Hannan et al., 2021). Deploying AI in renewable energy projects can spur innovation and entrepreneurship, creating new business opportunities and enhancing the overall economic landscape (Di Vaio et al., 2020). The social benefits of improved energy access, such as enhanced healthcare, education, and overall quality of life, are significant. Reliable electricity can power medical equipment, improve vaccine refrigeration, and enable online education platforms, contributing to broader socio-economic development goals (Miskiewicz, 2022).

Several African countries are making significant strides in adopting AI in the renewable energy sector. Morocco has implemented ambitious renewable energy projects through MASEN, showcasing the continent's potential to leverage AI for sustainable development (Moudene et al., 2023). In Nigeria, AI is enhancing the efficiency and reliability of small hydropower systems (Ibegbulam et al., 2023).

Additionally, Omdena's AI community has developed an interactive map identifying the top Nigerian regions for solar power installations.[1] This map provides valuable insights for government and policymakers on resource allocation, crucial for sustainable energy planning in a country where over 100 million people lack stable access to electricity (Rutenberg et al., 2021). These examples showcase the potential of AI to drive substantial improvements in renewable energy generation and management across Africa, which is crucial for achieving SDG7. Sub-Saharan Africa stands to benefit significantly from AI to overcome challenges and accelerate the transition, thus compensating for the existing access disparities. To reach SDG7, around 110 million new connections are needed each year from 2022 (IEA, 2023). With the pace seen in recent years, this target remains far off track. Meanwhile, North African countries, already advanced in electrification and renewable energy production compared to their Sub-Saharan counterparts, can use AI to further strengthen their positions as leaders in renewable energy adoption and efficiency.

4 Economic Impacts of AI-Driven Renewable Energy

4.1 AI and Job Creation in Africa

AI is playing a transformative role in the renewable energy sector, driving significant economic growth through job creation and the evolution of existing roles. Implementing AI in renewable energy systems requires a skilled workforce to develop, manage, and maintain these technologies, leading to the emergence of new technical roles such as data scientists, AI specialists, and software engineers (Di Vaio et al., 2020). In AI-driven solar farm management, AI systems analyse large datasets from solar panels to optimise performance and predict maintenance needs, creating new jobs in data monitoring and analysis while also enhancing the roles of engineers and technicians who now need to interpret AI-generated data to improve efficiency and reliability (Awogbemi et al., 2024). AI applications in small hydropower systems in Nigeria have optimised energy production and generated employment opportunities in local communities (Ibegbulam et al., 2023).

AI-driven job creation is significantly influenced by the diverse economic contexts of African countries, as reflected in their varying unemployment rates. Countries with relatively lower unemployment rates, such as Kenya (5.44%) and Nigeria (5.79%), may be better positioned to integrate AI technologies into their economies, potentially leveraging these advancements to create new job opportunities and enhance existing roles. Conversely, countries with higher unemployment

[1] Omdena is a collaborative AI platform that operates globally, including significant projects in Africa. It focuses on solving real-world challenges by bringing together AI engineers, data scientists, and domain experts. In Africa, Omdena has been involved in various projects aimed at enhancing clean energy access, predicting infrastructure needs, and optimising renewable energy systems.

rates, such as South Africa (29.83%), Tunisia (16.09%), and Morocco (10.33%), may face more significant challenges in harnessing AI for job creation. In these nations, the existing high levels of unemployment could make it more difficult to quickly adapt to and benefit from AI-driven advancements, as the workforce may require substantial upskilling and retraining to meet the demands of new AI-related roles.

To maximise the benefits of AI-driven job creation across the continent, it is crucial to consider these economic disparities and tailor strategies accordingly. Countries with higher unemployment rates may need to focus on enhancing digital literacy and providing targeted vocational training to equip their populations with the skills required for AI-driven jobs in renewable energy. Meanwhile, nations with lower unemployment rates should continue investing in AI technologies and related infrastructure to sustain and expand the job opportunities emerging from this sector. Moreover, as AI technologies proliferate, there is a multiplier effect on related industries, including manufacturing, IT services, and education, all of which support the development and deployment of AI solutions in renewable energy (Hannan et al., 2021). By addressing these diverse needs, African countries can better position themselves to harness the potential of AI in creating jobs, driving economic development, and advancing sustainable energy initiatives, directly contributing to the achievement of SDG7.

4.2 Economic Benefits

Integrating AI into the power supply chain promises substantial economic benefits, optimising cost savings and investment returns. AI automates complex processes, reducing the need for extensive human intervention and lowering operational costs by minimising errors and enhancing efficiency (Ahmad et al., 2021). This optimisation leads to increased profitability, as AI-driven systems manage and predict energy loads with precision, reducing waste and optimising resource allocation. AI catalyses the development of innovative business models within the energy sector. These models include dynamic pricing, which utilises AI algorithms to adjust prices in real time based on supply and demand dynamics (IEA, 2017), and sophisticated energy trading platforms that exploit AI's analytical strength to predict market trends and optimise trading strategies (O'Leary, 2013). These advancements significantly create new revenue streams and strengthen the financial frameworks of energy firms, enabling them to operate with greater agility and respond more effectively to market dynamics (Sodhro et al., 2019).

Furthermore, deploying AI in predictive maintenance and grid management models reduces long-term capital expenditures and operational costs, thus improving the return on investment for energy stakeholders. The economic impact of AI extends further to facilitate the integration of renewable energy sources into the grid more effectively. By accurately forecasting energy production from renewable sources, AI supports these renewables' stable and efficient integration, mitigating the typical unpredictability of their outputs (Rodríguez et al., 2018). AI's capability to manage complex data structures in renewable energy enhances energy transition and resource scheduling, improving grid integration and performance (Zhang et al., 2022; Li et al., 2023a; Olabi et al., 2023).

AI's influence also extends to enhancing the financial viability of renewable projects. For example, AI applications in energy storage management optimise the use and discharge cycles based on predictive consumption patterns and grid demands, significantly extending battery life and reducing replacement costs, which are fundamental for the economics of energy storage systems (Wei et al., 2019). In Morocco, the Noor Ouarzazate Solar Complex employs AI-driven optimisation techniques to enhance the efficiency of its solar power generation.[2] The AI systems at Noor Ouarzazate help predict energy output and manage energy storage more effectively, ensuring grid stability and maximising financial returns from renewable investments. Furthermore, AI-driven optimisation in solar and wind farms enhances their output efficiency, translating into better financial returns for renewable energy investors. Maintaining grid stability by optimising energy consumption across various sectors to handle fluctuations ensures optimal grid performance, thus improving the economic landscape for renewable investments (Debnath & Mourshed, 2018; Qin et al., 2024).

5 Social Impacts of AI-Driven Renewable Energy

5.1 Enhancing Energy Access

AI is expected to address social challenges related to energy access and poverty significantly. By optimising energy production, distribution, and consumption, AI ensures more reliable and equitable access to electricity, particularly in remote and underserved areas. This has profound implications for improving the quality of life and fostering inclusive development. AI improves grid management by predicting energy demand and supply fluctuations, ensuring stable and reliable electricity delivery.

This is particularly important in regions with intermittent energy sources or limited grid infrastructure (Miskiewicz, 2022). In many African countries, rural and remote areas often face significant challenges in accessing stable electricity. AI can analyse patterns in energy consumption and production, enabling grid operators to

[2] The Noor Ouarzazate Solar Complex is the largest concentrated solar power (CSP) plant in the world, located near Ouarzazate, Morocco. With a total capacity of 580 MW, the complex is a key component of Morocco's ambitious solar energy program. It consists of several phases: Noor I, Noor II, and Noor III, which utilise different CSP technologies, and Noor IV, a photovoltaic (PV) plant. The complex contributes significantly to Morocco's renewable energy goals, helping to offset hundreds of thousands of tonnes of CO_2 emissions annually and providing substantial employment opportunities during both the construction and operational phases.

anticipate and respond to changes in energy demand. For example, AI can help balance the load on the grid by predicting peak usage times and adjusting the energy distribution accordingly. This ensures that even remote areas receive a consistent electricity supply, reducing the frequency of blackouts and improving overall grid reliability (Ibegbulam et al., 2023). By enhancing grid stability, AI helps bridge the energy access gap, ensuring that more communities can benefit from reliable electricity. This, in turn, supports economic activities, education, healthcare, and other essential services that depend on a stable power supply (Goralski & Tan, 2020). In Morocco, AI-driven solutions are being utilised to enhance the efficiency and reliability of solar energy projects, thereby improving energy access for underserved communities (Moudene et al., 2023).

5.2 Reducing Energy Poverty

AI-driven renewable energy systems can significantly reduce energy poverty by lowering energy production costs and making clean energy more affordable. AI enables more efficient operation and maintenance of renewable energy installations, reducing operational costs and ultimately lowering the price of electricity for consumers (Ibegbulam et al., 2023). In regions where traditional grid infrastructure is not feasible, AI can facilitate deploying and managing microgrids and off-grid renewable energy systems. These AI-optimised systems can provide reliable electricity to isolated communities, enhancing their economic opportunities and living standards. AI can help design and manage solar microgrids that provide affordable and sustainable energy to remote villages, reducing their reliance on expensive and polluting diesel generators (Awogbemi et al., 2024).

The social impacts of AI-driven renewable energy extend beyond immediate energy access. By providing stable and affordable electricity, AI helps unlock broader socio-economic benefits, including improved educational outcomes, better healthcare services, and enhanced economic opportunities for marginalised communities (Di Vaio et al., 2020). In Kenya, Apollo Agriculture leverages AI to analyse satellite imagery, soil data, farmer behaviour, and crop yield models. This allows the company to offer farmers tailored financing options and customised seed and fertiliser packages, significantly improving agricultural productivity and economic stability (Rutenberg et al., 2021). Moreover, AI's role in enhancing energy efficiency and sustainability also contributes to environmental preservation, which has long-term benefits for the health and well-being of communities. By reducing the reliance on fossil fuels and promoting clean energy sources, AI helps mitigate the adverse effects of climate change, thus supporting the overall goal of sustainable development (Goralski & Tan, 2020).

6 Challenges and Risks in AI Implementation for SDG7

6.1 Fairness and Biases in AI

While robust AI systems are not immune to bias, they can perpetuate existing inequalities, particularly in vulnerable communities, if not carefully designed and managed. The research by Ding et al. (2024) explores the potential of artificial intelligence to alleviate energy poverty, demonstrating that AI can significantly reduce energy poverty through economic effects, technological progress, and human capital development. However, the study also reveals the uneven impact of AI across different income groups. While AI effectively mitigates energy poverty in high-income and lower-middle-income countries, this positive influence is absent in upper-middle-income countries. The impacts in low-income countries, where energy access is most limited, also remain unreported, highlighting the unequal distribution of AI's benefits globally.

This disparity underscores the importance of addressing biases inherent in AI systems. These biases often arise from the data used to train AI models, which can reflect historical and societal inequalities. If AI algorithms are trained on data biased against certain demographic groups, the resulting AI systems may produce discriminatory outcomes, exacerbating social inequalities. In the context of renewable energy, AI systems with bias could lead to the unequal distribution of resources, where certain regions or communities are unfairly prioritised or neglected. This could hinder efforts to achieve equitable access to affordable and clean energy, as outlined in SDG7.

These concerns align with those raised by Gehl Sampath (2021), who argues that AI can reinforce structural inequalities in the global South. She identifies four critical challenges: the increasing vulnerability of users, a disconnect in the data-led industrialisation narrative, the opacity of governments, and the rise of the "political" private sector. These issues suggest that while AI has the potential to address energy poverty, its application in developing countries must be handled with caution. Without careful consideration, AI could exacerbate existing inequalities, making it crucial to establish policy frameworks that prioritise transparency, fairness, and inclusivity. To mitigate these biases, it is essential to implement strategies that ensure AI systems are transparent, fair, and inclusive. Regular audits, the use of diverse training datasets, and active stakeholder involvement in designing and deploying AI technologies are crucial steps toward reducing bias. Continuous efforts must be made to refine AI models, including thorough examinations of the data and algorithms used to process it, to eliminate bias and ensure that AI contributes positively to global development goals (Ibegbulam et al., 2023; Awogbemi et al., 2024).

6.2 AI and Structural Challenges

Implementing AI technologies to achieve SDG7 in Africa presents unique structural challenges, including infrastructure deficits, regulatory hurdles, and skill gaps. Many African countries need more infrastructure, such as limited internet connectivity and insufficient technological facilities, which can significantly impede the deployment and effective use of AI in renewable energy systems (Ibegbulam et al., 2023). For example, only about 36% of Africa's population has access to broadband internet, and the cost of mobile internet is prohibitively high, averaging 10.5% of monthly income.[3] This makes digital connectivity a major hurdle for implementing AI, which relies heavily on robust internet infrastructure for real-time data processing and smart grid management.

Another significant challenge is the scarcity of comprehensive and reliable data, which is crucial for operating AI systems. Many African countries are "endemically data poor" (Mutiso, 2024), particularly in the energy sector, where the availability of detailed and timely data is limited. This shortage of data hampers the development of AI models that require large datasets for accurate forecasting, optimisation, and energy management. Additionally, the cost of accessing advanced AI and cloud computing resources is often too high for many African researchers and startups, making it difficult for them to develop and implement AI solutions effectively.

The regulatory environment in many African countries further complicates the adoption of AI. The existing regulations often need to be updated or more developed to address the complexities of AI technologies, which can create uncertainties and discourage investment in AI-driven renewable energy projects (Miskiewicz, 2022; Mutiso, 2024). Moreover, the skills gap in the region, where there is a shortage of professionals with expertise in AI and renewable energy, presents another barrier to successfully implementing AI technologies. This lack of skilled labour can slow down the deployment and maintenance of AI systems, limiting the potential of AI to contribute meaningfully to achieving SDG7 (Awogbemi et al., 2024). Addressing these structural challenges requires a coordinated approach from governments, private sector stakeholders, and international organisations. Investing in improving digital infrastructure, updating regulatory frameworks to accommodate AI technologies better, and developing capacity-building programs to enhance local expertise in AI and renewable energy (Hannan et al., 2021). By tackling these issues, African countries can better integrate AI into their energy systems, ensuring that the benefits of AI are distributed equitably and that progress toward SDG7 is accelerated across the continent.

[3] See World Bank. (2023). From Connectivity to Services: Digital Transformation in Africa. World Bank. (2024). Digital Transformation Drives Development in Africa.

6.3 Ethical Implications

Exploring AI is deeply connected to ethical implications. Implementing AI in renewable energy systems also raises important ethical considerations, including privacy concerns and the autonomy of decision-making processes. AI systems often rely on large amounts of data, including sensitive personal information. Ensuring the privacy and security of this data is paramount to protect individuals' rights and maintain public trust in AI technologies. Robust data protection policies, anonymisation techniques, and transparent data handling practices are necessary to mitigate privacy risks (Goralski & Tan, 2020).

The increasing reliance on AI for decision-making in energy management can lead to concerns about the loss of human autonomy and accountability. Decisions made by AI systems can sometimes be opaque, making it difficult to understand the rationale behind them. This "black box" nature of AI can lead to ethical dilemmas, especially when critical decisions about energy distribution and access are made by algorithms (Falk & van Wynsberghe, 2023). It is important to ensure that AI systems are designed and implemented with ethical principles to address such ethical implications. This includes maintaining human oversight, ensuring transparency in AI decision-making processes, and establishing clear accountability mechanisms for AI-driven decisions (Arogyaswamy, 2020). By embedding ethical considerations into the development and deployment of AI technologies, stakeholders can help build trust and ensure that AI contributes positively to the goal of sustainable and equitable energy access.

Integrating AI into the energy value chain presents a series of multifaceted challenges and barriers, illustrating the dual nature of its impact. While AI offers profound operational efficiency and energy management advantages, its deployment demands substantial regulatory adjustments focused on safety, privacy, and security (Ahmad et al., 2021). This requirement highlights the delicate balance that must be maintained between leveraging cutting-edge technology and safeguarding stakeholders' interests. As AI becomes necessary for competitive energy operations, the sector faces the burden of continuously evolving business strategies to incorporate these advanced technologies without compromising regulatory compliance or operational security (Quest et al., 2022; Heymann et al., 2024).

The operational demands of AI technologies can significantly increase energy consumption, adding to the carbon footprint of information technology. Training sophisticated AI models, such as those used in predictive analytics and system management, requires considerable energy, which can exacerbate carbon emissions if not managed with renewable energy sources (Lau et al., 2023; Gaur et al., 2023). Strubell et al. (2020) highlight that training an AI model could emit as much carbon dioxide as a trans-American flight, potentially counteracting the carbon reduction achieved through AI-driven optimisation of energy systems (Nishant et al., 2020). This paradoxically undermines the sustainability goals that AI aims to support. One effective approach is to power AI operations with renewable energy sources. Transitioning data centres and computational facilities to run on solar, wind, or hydroelectric power can significantly reduce their carbon footprint (Wu et al., 2022). Moreover, AI algorithm advancements focus on improving computational efficiency, reducing the energy required for training and running AI models (Goralski & Tan, 2020). These efforts aim to make AI technologies more sustainable, ensuring that their environmental costs do not offset the benefits of AI in renewable energy optimisation.

Thus, adopting AI in the power value chain also presents the potential for a rebound effect, where initial decreases in energy consumption and costs may paradoxically result in lower energy prices and increased consumption. This unexpected outcome can undermine long-term sustainability objectives by negating the environmental benefits initially sought through AI integration. Complicating this issue further are the potential systemic risks associated with excessive reliance on AI, known as green swan events (Campiglio et al., 2018), which are highly unpredictable and potentially severe. These complexities require vigilant oversight from policymakers and industry leaders to balance the proliferation of AI technologies with comprehensive regulatory measures and risk management strategies. Such efforts are essential to safeguard the sector's resilience and ensure that AI's integration aligns with broader environmental goals.

7 Africa's Strategic Roadmap for AI-Driven Sustainability in SDG7

Africa is increasingly central to the global AI supply chain, particularly in early production, with countries like Egypt, Rwanda, and Mauritius leading by publishing comprehensive AI strategies (Adams, 2022). Despite this progress, the region remains dominated by foreign AI technologies that often fail to align with local developmental priorities. Policy responses must build on national digital agendas to bolster local AI capacity and support local economies, emphasising inclusive digital, data, and computing infrastructure and skills development. For example, Egypt's national AI strategy highlights the importance of avoiding unfavourable foreign agreements that do not align with inclusive developmental goals (Adams, 2022). In this context, the Continental AI Strategy adopted in 2024 aims to guide African countries in effectively utilising AI to achieve development goals and enhance the well-being of their populations.[4] This strategy advocates for a people-centred, inclusive approach to building AI capabilities in infrastructure, talent, datasets, innovation, and partnerships while emphasising ethical usage and risk management. By tailoring AI systems to reflect Africa's unique diversity, languages, culture, and contexts, the strategy seeks to establish a competitive AI market that aligns with the

[4] See African Union. (2024). African ministers adopt the landmark Continental Artificial Intelligence Strategy, the African Digital Compact, to drive Africa's development and inclusive growth.

continent's aspirations. Alongside this, the African Digital Compact envisions a future where digital technologies drive sustainable development, economic growth, and societal well-being, positioning Africa as a global digital leader focusing on talent development and public-private collaborations. Both initiatives are key components of the African Union's Digital Transformation Strategy (2020–2030) and Agenda 2063, supporting Africa's long-term development objectives.[5]

Decoupling economic growth from environmental degradation through sustainable consumption and production policies has become crucial in global post-pandemic recovery strategies. The circular economy model, as an alternative to the linear economy, has gained traction for its potential to support sustainable development. In Africa, AI is increasingly seen as a tool to enhance circularity, promoting inclusive and sustainable industrial development. AI is a crucial enabler for CE, aiding in designing robust and sustainable products, facilitating new circular business models, and supporting the broader infrastructures needed to scale circularity (Roberts et al., 2024). Evidence from 27 European Union member states links AI to the circular economy and the achievement of SDG12: Responsible Consumption and Production (Platon et al., 2024). However, leveraging AI for circularity in Africa faces challenges such as infrastructure gaps, limited technical capacity, and the dominance of foreign AI technologies that may not align with local priorities. Additionally, AI poses risks like algorithmic bias and increased inequities; thus, its positive impact depends on the framework in which it is used and the existing ecosystems (Ahmed, 2022). To overcome these challenges, an AI-driven policy framework aligned with circular economy principles can support renewable energy development, promote energy efficiency, and incentivise resource reduction, reuse, and recycling, ultimately enhancing sustainability and creating closed-loop systems (Danish & Senjyu, 2023).

Africa must adopt a multifaceted strategy to fully tap into AI's potential in advancing renewable energy and achieving SDG7. Initially, conducting a thorough assessment of the current capabilities and needs across different regions and industries is essential. This evaluation helps identify gaps and areas where AI can significantly impact, such as optimising resource allocation in regions with abundant solar resources but limited grid infrastructure (Ibegbulam et al., 2023; Awogbemi et al., 2024). Setting clear, measurable objectives for AI integration in renewable energy is also crucial. These targets should focus on increasing energy access, improving efficiency, and reducing costs, thus providing a clear roadmap for stakeholders and ensuring alignment towards common goals (Hannan et al., 2021).

Developing a robust infrastructure to support AI applications is another critical aspect. Investing in data collection and processing capabilities, high-performance computing facilities, and reliable internet connectivity forms the backbone of successful AI implementation. Platforms for data sharing and collaboration among stakeholders can enhance the precision and effectiveness of AI applications in

[5] See African Union. The Digital Transformation for Africa (2020–2030). African Union. *Agenda 2063: The Africa We Want.*

renewable energy projects (Miskiewicz, 2022). Capacity building and education are equally important. Creating partnerships with educational institutions to offer specialised courses and training programs can develop a skilled workforce capable of driving AI and renewable energy innovations. Initiatives like workshops, certification courses, and degree programs focused on AI and renewable energy help build local expertise and ensure that training aligns with the practical needs of the sector (Goralski & Tan, 2020).

Pilot projects are vital for testing and refining AI-driven renewable energy solutions. These projects serve as testbeds for innovative solutions, allowing stakeholders to assess their feasibility and impact before scaling them up. Successful pilot projects can be replicated across different regions, maximising their benefits and contributing to broader energy goals (Ibegbulam et al., 2023). Effective policies also play a vital role. Ethical AI guidelines ensure that AI applications are transparent, fair, and accountable, which helps build trust among users and stakeholders (Jobin et al., 2019). Additionally, incentivising companies to adopt sustainable AI practices through tax breaks, grants, or subsidies can foster a culture of sustainability within the energy sector (Miskiewicz, 2022).

Developing regulatory frameworks that address the unique challenges of AI implementation in renewable energy is necessary for safe and effective technology use. These regulations should ensure data security, standardise AI technologies, and facilitate innovation while protecting public interests (Falk & van Wynsberghe, 2023). Promoting public awareness of AI's benefits and risks in renewable energy is also essential. Engaging communities and stakeholders in the development and implementation process ensures their needs and concerns are addressed, fostering trust and cooperation (Goralski & Tan, 2020). International cooperation and local innovation are crucial for developing and deploying AI-driven renewable energy solutions in Africa. Partnerships with international organisations, technological companies, and governments can provide access to advanced AI technologies and best practices worldwide, enhancing local projects' capacity (Hannan et al., 2021). Establishing innovation hubs and incubators to support local startups and entrepreneurs working on AI and renewable energy solutions is vital. These hubs can provide mentorship, funding, and access to technology to help bring innovative ideas to market, driving homegrown solutions tailored to the specific needs of African communities (Ibegbulam et al., 2023).

Encouraging collaborative research projects between African institutions and international partners can facilitate the exchange of ideas and accelerate the development of AI applications tailored to the unique needs of the African context. Joint research initiatives can focus on addressing local energy challenges using AI, thereby contributing to sustainable development (Awogbemi et al., 2024). Collaboration between academia, government, and industry is essential for driving AI innovation in the renewable energy sector. Joint research initiatives and funding programs can support collaborative projects that bring together academic researchers, government agencies, and industry experts (Miskiewicz, 2022).

Creating public–private partnerships to leverage the strengths of both sectors is important for advancing AI in renewable energy. Governments can provide regulatory support and funding, while private companies can offer technological expertise and market insights. These partnerships facilitate the deployment of AI solutions and ensure their scalability, bridging the gap between policy and practice (Goralski & Tan, 2020). Developing knowledge transfer programs to facilitate the exchange of skills and expertise between different sectors is essential. Internships, fellowships, and collaborative training programs can help build a skilled workforce capable of sustaining and expanding AI innovations (Hannan et al., 2021).

Forming alliances between academic institutions and industry is indispensable to align research and education with industry needs. This can help ensure that graduates have the skills required by the renewable energy sector and that research projects address practical challenges faced by the industry. Such alliances drive innovation and support sustainable development, ensuring academic research is grounded in real-world applications and enhancing its relevance and impact (Ibegbulam et al., 2023).

Overall, Africa's strategic approach to AI-driven sustainability in renewable energy involves a comprehensive framework that addresses infrastructure, education, policy, and collaboration. By leveraging AI to enhance circularity and promoting inclusive, sustainable industrial development, Africa can make significant strides towards achieving SDG7. This requires a concerted effort to build local capacities, develop robust infrastructures, and foster international partnerships, ensuring that AI technologies are deployed responsibly and effectively to benefit the entire continent.

8 MASEN's AI Integration in Renewable Energy Systems

Established in 2010, MASEN plays a crucial role in Morocco's transition to sustainable energy, significantly contributing to the national goal of achieving 52% of its energy mix from renewable sources by 2030. As the entity responsible for managing and developing Morocco's renewable energy resources, MASEN had, by the end of 2020, successfully brought 1200 MW of wind power, 1770 MW of hydro, and over 700 MW of solar power into operation. In addition to these accomplishments, MASEN is actively exploring the potential of AI to optimise its renewable energy operations, enhance efficiency, and drive innovation. By integrating AI technologies, MASEN aims to overcome existing challenges, improve energy production and management, and ensure a sustainable and resilient future for Morocco and Africa, setting a strong example for other nations in the region.

8.1 Solar Energy

MASEN plans to leverage AI technologies to enhance various aspects of its solar energy projects. AI offers solutions to some of the most challenging aspects of managing renewable energy sources in different areas.

System Design Optimisation AI algorithms analyse vast datasets on the performance of existing solar installations to optimise the design of new systems. This includes the optimal placement of solar panels to maximise sunlight capture, considering local conditions such as topography and weather patterns. According to the interview with MASEN's CEO, Tarik Hamane,[6] "AI enables MASEN to design solar systems that are not only efficient but also tailored to specific environmental conditions, ensuring maximum energy output". MASEN plans to use AI to develop intelligent solar tracking systems that adjust the angle of solar panels throughout the day to follow the sun's path. AI will also help select the best locations for new solar farms by analysing factors such as solar irradiance, land availability, and proximity to the grid. These AI-driven optimisations are expected to significantly enhance the efficiency and output of MASEN's solar energy projects.

Predictive Maintenance AI-driven predictive maintenance systems monitor the performance of solar panels in real time. These systems can significantly reduce downtime and maintenance costs by detecting anomalies and predicting potential failures before they occur. This ensures that solar farms operate at peak efficiency with minimal disruptions. "With AI, we can foresee and prevent issues before they arise, saving time and reducing costs."

Energy Demand Forecasting Machine learning techniques will analyse historical and real-time energy consumption patterns. This enables accurate predictions of future energy demand, allowing operators to adjust production accordingly. Such forecasting is important for balancing supply and demand, particularly in regions with fluctuating energy needs. AI's ability to predict energy demand helps align production schedules with actual consumption patterns, reducing waste and improving efficiency. MASEN is also exploring the use of AI to develop virtual power plants that aggregate the output of multiple small-scale solar installations. By predicting energy production and demand, these virtual power plants can optimise the use of distributed energy resources, providing a more reliable and efficient energy supply. This innovative approach not only enhances energy management but also supports the integration of renewable energy into the national grid.

---

[6] We interviewed Tarik Hamane, the former acting CEO of MASEN, who has been instrumental in advancing Morocco's renewable energy projects, particularly the Noor Ouarzazate Solar Complex. With extensive experience in managing large-scale energy initiatives, Tarik Hamane's leadership has been key in positioning Morocco as a leader in renewable energy in Africa. In June 2024, he was nominated by King Mohammed VI as the head of the National Office of Electricity and Drinking Water (ONEE), further solidifying his influence in the country's energy transition.

8.2 Wind Energy

Integrating AI in wind energy systems is a cornerstone of MASEN's comprehensive strategy to enhance renewable energy production, improve operational efficiency, and drive Morocco's ambitious renewable energy goals. By harnessing the power of AI, MASEN aims to optimise every aspect of wind energy generation, from precise forecasting and real-time monitoring to dynamic adjustments and strategic planning, ensuring that wind farms operate at peak performance. In 2019, a study conducted by MASEN demonstrated that using advanced AI techniques in wind farm management could increase wind production capacity by up to 10%, thereby reducing operational costs and strengthening the reliability of the national electric grid.[7]

Enhanced Energy Production Forecasting AI can analyse large volumes of historical and real-time weather data to forecast wind speeds and atmospheric conditions accurately. These predictions allow for better energy production planning, ensuring wind turbines operate optimally to maximise electricity generation. "Accurate wind forecasting is essential for optimising turbine operation and ensuring consistent energy production." AI will be used to develop advanced meteorological models that provide precise wind forecasts in advance. These models will consider local topography, seasonal weather patterns, and real-time data from weather stations and satellites. Improved forecasting accuracy will enable better scheduling of maintenance activities and energy dispatch, thereby enhancing the overall efficiency of wind farms.

Real-Time Turbine Monitoring AI systems will monitor turbine conditions in real time, analysing data on vibrations, temperatures, and electrical performance. Early detection of anomalies through AI analysis can prevent significant failures and reduce maintenance costs by scheduling timely interventions. "Real-time monitoring of turbines allows us to maintain high performance and prevent costly breakdowns." MASEN plans to implement AI-powered condition monitoring systems that continuously analyse data from sensors embedded in wind turbines. These systems use machine learning algorithms to detect patterns indicative of potential faults, such as unusual vibration frequencies or temperature fluctuations. By addressing issues before they escalate, MASEN can extend the lifespan of its wind turbines and reduce maintenance costs, ensuring uninterrupted energy production.

Dynamic Optimisation AI can dynamically adjust the orientation of wind turbines based on real-time wind direction and speed forecasts. This maximises energy capture and improves the overall efficiency of wind farms. Such optimisation is essential for achieving higher energy yields from existing infrastructure. "Dynamic optimisation of turbine orientation using AI can significantly enhance energy

---

[7] Afrique Media. (2024). *L'Intelligence Artificielle au Service* de l'Energie Renouvelable au Maroc : Avancées, Applications et Perspectives.

output." Additionally, AI will optimise the spacing and placement of wind turbines within a wind farm to minimise turbulence and maximise energy capture. This involves creating digital twins of wind farms to simulate different configurations and identify the most efficient layout. By leveraging AI, MASEN aims to increase the productivity and profitability of its wind energy projects, contributing to Morocco's renewable energy goals.

8.3 Hydrogen Energy

Hydrogen energy, though still in its developmental stages, represents a transformative opportunity for MASEN to diversify its renewable energy portfolio and contribute to a more sustainable energy future. By integrating AI into the hydrogen production process, MASEN is positioning itself at the forefront of innovation, leveraging advanced technologies to enhance efficiency, reduce costs, and ensure the long-term viability of hydrogen as a key energy source. Through real-time visualisation, flexible system design, and process optimisation, AI will play a crucial role in shaping the future of hydrogen energy at MASEN, making it a cornerstone of Morocco's energy strategy.

Real-Time Visualisation and Simulation AI enables real-time visualisation and simulation of hydrogen production processes. This accelerates the design phase by allowing rapid iterations and identifying potential issues before construction begins. "AI simulations provide us with a detailed understanding of the hydrogen production process, allowing for efficient and effective design." MASEN will use AI to model different hydrogen production methods, such as electrolysis and thermochemical processes. By simulating various scenarios, AI can help identify the most cost-effective and energy-efficient methods for large-scale hydrogen production. These simulations aid in optimising plant design and operations, ensuring that hydrogen production is economically viable and environmentally sustainable.

Flexible System Design AI can guide the design of modular and flexible hydrogen production plants that can adapt to future technological advancements and market changes. This flexibility is vital for scaling production and integrating new processes without significant overhauls. "AI-designed modular systems ensure that our hydrogen plants can evolve with technological advancements."

Optimisation of Electrolysis Processes AI can optimise the electrolysis process, which is critical for hydrogen production. AI can reduce costs and improve energy efficiency by predicting the best times to utilise renewable energy sources like solar and wind for electrolysis. "Optimising electrolysis with AI not only saves costs but also enhances energy efficiency." MASEN plans to deploy AI-powered electrolysis systems that use machine learning algorithms to continuously monitor and adjust operating parameters, such as temperature, pressure, and current, to maximise

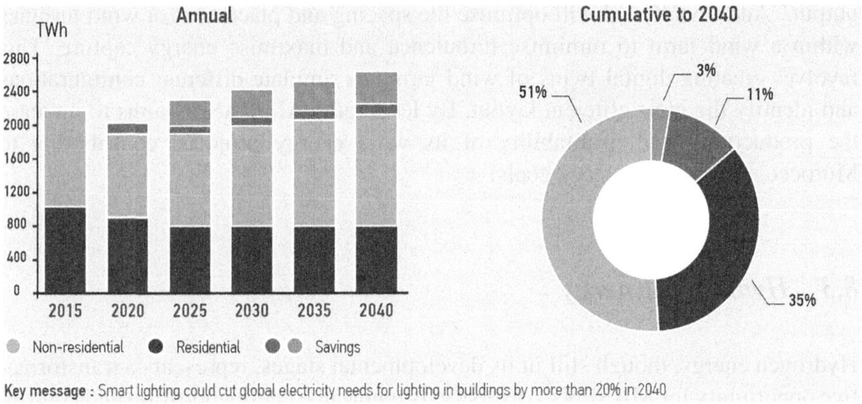

Fig. 4 Potential electricity savings from smart lighting in buildings to 2040. (Source: Moroccan Agency for Sustainable Energy - MASEN). Note: The electricity savings potential from smart lighting, including sensors and digital controls, is calculated as the energy use reductions beyond the estimated savings from expected deployment of LED lighting already estimated in the IEA Central Scenario

hydrogen yield and minimise energy consumption. This optimisation is essential for making hydrogen a competitive and sustainable energy carrier in the future. Similarly, smart lighting technology could cut global electricity needs for lighting in buildings by more than 20% by 2040. Figure 4 projects the broad potential of AI-driven efficiencies in energy savings from smart lighting in buildings.

AI will play a pivotal role in optimising the integration of hydrogen production with renewable energy sources, further positioning Africa as a leader in green hydrogen development. By utilising AI algorithms to predict periods of excess wind or solar power, hydrogen production can be scheduled to take advantage of these surplus energy periods, thereby reducing costs and improving overall efficiency. This integrated approach ensures the effective use of renewable energy and supports the sustainability and economic viability of hydrogen production. Significant strides have already been made to establish Africa at the forefront of green hydrogen developments, highlighted by the formation of the Africa Green Hydrogen Alliance in 2022, which includes key countries such as Morocco, Egypt, and South Africa. These efforts align with the European Union's REPowerEU Plan, which aims to import up to ten million tonnes of green hydrogen annually, much of it from Africa.[8] The European Investment Bank projects that Africa could produce over 50 million tonnes of green hydrogen annually by 2035 at a highly competitive cost, positioning countries like Morocco to become major exporters in the growing European hydrogen market.[9] This AI-driven integration of hydrogen production with renewable

[8]European Commission. (2022). REPowerEU: Affordable, *secure,* and sustainable energy for Europe.

[9]Radford, C., Field, A. (2023). *Green hydrogen in Africa: A continent of possibilities?.* White & Case.

energy contributes to Africa's energy independence and advances the global goals of SDG7, ensuring access to affordable, reliable, sustainable, and modern energy for all.

8.4 Electrical Grid Management

MASEN's commitment to enhancing the stability and efficiency of electrical grids marks a key step in its renewable energy strategy, with AI playing a central role in this transformation. By integrating advanced AI technologies into grid management, MASEN aims to create a more resilient, reliable, and intelligent energy infrastructure. This forward-thinking approach supports the stability and security of Morocco's electrical grid and sets the stage for innovative energy management solutions that can be scaled across the continent.

Grid Stability Assessment AI contributes to the assessment of grid stability by addressing various stability issues, including transient, frequency, and voltage stability. AI algorithms can predict potential instabilities and suggest corrective actions to prevent power outages. "Ensuring grid stability with AI is essential for reliable energy supply." AI will be used to develop advanced grid management systems that can detect and respond to grid imbalances in real time. These systems will analyse data from smart metres, weather stations, and energy production sites to predict and mitigate potential disruptions. By ensuring grid stability, MASEN aims to provide consumers with a reliable and uninterrupted energy supply.

Fault Detection AI algorithms will monitor the smart grid to detect anomalies in the data flow from various metres. Differentiating between network problems and fluctuations in renewable energy production allows for timely and accurate responses to potential issues. "AI-based fault detection ensures quick and precise identification of issues, maintaining grid reliability." MASEN will implement AI-powered fault detection systems that use machine learning to analyse data from sensors and smart metres. These systems will identify patterns indicative of faults, such as sudden drops in voltage or abnormal power flows, and trigger alerts for immediate intervention. This proactive approach enhances the reliability and security of the electrical grid.

Real-Time Management AI's real-time data processing capabilities ensure optimal energy management, including rectifying anomalies and dynamic pricing strategies. This enhances the security of the grid and promotes transparent energy markets. "Real-time management of the grid with AI leads to efficient energy distribution and improved market transparency." AI will also enable dynamic pricing models that adjust electricity prices based on real-time supply and demand. This will encourage consumers to use electricity during off-peak periods, reducing strain on the grid and promoting more efficient energy use. Moreover, AI will facilitate the integration of distributed energy resources, such as rooftop solar panels and home batteries, into the grid, further enhancing grid resilience and sustainability.

MASEN's forward-thinking approach to integrating AI into its renewable energy projects exemplifies AI's transformative potential in the energy sector. By optimising system design, enhancing predictive maintenance, improving energy forecasting, and managing grid stability, AI technologies can significantly boost the efficiency and reliability of renewable energy sources. These AI-driven innovations contribute to Morocco's renewable energy goals and serve as a model for other African nations seeking to harness AI for sustainable development. The integration of AI into renewable energy projects aligns directly with the objectives of SDG7, which aims to ensure access to affordable, reliable, sustainable, and modern energy for all. The specific AI applications outlined in MASEN's strategy contribute to this goal in several ways. By optimising the design and operation of solar and wind energy systems, AI helps to increase the efficiency and reliability of renewable energy sources, ensuring a more stable and consistent energy supply, which is essential for meeting the growing energy demands in Morocco and beyond. AI-driven predictive maintenance and real-time monitoring reduce operational costs by preventing equipment failures and minimising downtime, making renewable energy more affordable for consumers. Besides, AI's ability to accurately forecast energy demand and optimise energy distribution ensures that renewable energy is used effectively, improving the overall management of energy resources and supporting the integration of renewable energy into the national grid.

MASEN contributes to developing innovative and sustainable energy solutions by promoting them through technological advancements such as hydrogen energy and smart grid management. These advancements support Morocco's energy transition and provide scalable models for other countries aiming to achieve SDG7. Although these plans are currently in the future stages of development, MASEN's initiatives underscore the critical role of AI in achieving sustainable energy goals and driving economic growth in Africa. Through strategic investments in AI and continuous innovation, MASEN is well-positioned to create a sustainable and resilient energy future for Africa.

9 Conclusion

Integrating AI into Africa's renewable energy sector holds immense potential for advancing SDG7, ensuring access to affordable, reliable, sustainable, and modern energy for all. Throughout this chapter, we've explored AI's transformative role in optimising energy systems, improving efficiency, and driving innovation across various energy sources, focusing on MASEN. The challenges Africa faces in the energy sector, from limited access to unreliable infrastructure, underscore the importance of innovative solutions like AI.

MASEN's initiatives demonstrate how AI can enhance energy production, optimise grid management, and support decentralised energy systems, contributing significantly to sustainable energy goals. AI has proven instrumental in optimising the design and operation of renewable energy systems, particularly in managing solar and wind energy. By analysing vast datasets, AI algorithms can predict energy production, enhancing efficiency and reliability. For example, AI-driven predictive maintenance systems reduce downtime and maintenance costs, ensuring that renewable energy sources operate at peak efficiency. Moreover, AI's role in managing decentralised energy systems, such as microgrids, is particularly advantageous in remote areas, where it can leverage local renewable energy sources to provide stable and reliable electricity to underserved communities, improving their quality of life and fostering economic development.

However, implementing AI technologies in Africa is fraught with structural challenges, including infrastructure deficits, regulatory hurdles, and skill gaps. These challenges are compounded by limited internet connectivity and insufficient technological facilities, which impede the deployment and effective use of AI in renewable energy systems. Additionally, the scarcity of comprehensive and reliable data, particularly in the energy sector, hampers the development of AI models essential for accurate forecasting and optimisation. Financial barriers also persist, where AI can help optimise costs and improve the economic viability of renewable energy projects by attracting investments and ensuring these projects are more financially sustainable.

Addressing these structural challenges requires coordinated efforts from governments, private sector stakeholders, and international organisations. It is crucial to invest in improving digital infrastructure, updating regulatory frameworks to accommodate AI technologies, and developing capacity-building programs to enhance local expertise in AI and renewable energy. By tackling these issues, African countries can better integrate AI into their energy systems, ensuring equitable distribution of AI benefits and accelerating progress toward SDG7. Private and foreign investments are crucial in overcoming these barriers, driving innovation, and facilitating the deployment of AI solutions at scale.

Ethical considerations are also paramount in the implementation of AI. If not carefully designed and managed, AI systems can perpetuate inequalities, particularly in vulnerable communities. Biases in AI can lead to unequal distribution of resources, hindering efforts to achieve equitable access to clean energy. It is essential to implement strategies that ensure AI systems are transparent, fair, and inclusive, with regular audits, diverse training datasets, and active stakeholder involvement in designing and deploying AI technologies. Furthermore, AI-driven renewable energy systems raise environmental concerns, particularly regarding energy consumption. Managing these impacts through renewable energy sources and improving AI's computational efficiency is essential to ensuring that the environmental benefits of AI are realised.

The economic benefits of AI in the renewable energy sector, particularly in job creation and the evolution of existing roles, are significant. As AI technologies proliferate, there is a multiplier effect on related industries, including manufacturing,

IT services, and education, all of which support the development and deployment of AI solutions in renewable energy. These economic benefits are particularly relevant in Africa, where the potential for AI-driven job creation can help address some of the continent's socio-economic challenges. Additionally, by driving sustainable energy solutions, AI helps mitigate the impacts of climate change and supports global environmental goals. The involvement of multinationals in AI-driven projects further enhances technological innovation and investment, which can reduce inequality by providing access to modern energy services in underserved regions.

Africa is becoming increasingly central to the global AI supply chain. However, the region remains dominated by foreign AI technologies that may not align with local developmental priorities. To build local AI capacity and support local economies, policy responses must emphasise inclusive digital, data, and computing infrastructure and skills development. The Continental AI Strategy adopted in 2024 aims to guide African countries in effectively utilising AI to achieve development goals and enhance the well-being of their populations. By tailoring AI systems to reflect Africa's unique diversity, languages, culture, and contexts, this strategy seeks to establish a competitive AI market that aligns with the continent's aspirations.

Furthermore, integrating AI into the renewable energy sector aligns with global efforts to decouple economic growth from environmental degradation. AI is a crucial enabler for the circular economy, aiding in designing sustainable products, facilitating new circular business models, and supporting the broader infrastructures needed to scale circularity. However, leveraging AI for circularity in Africa faces challenges, including infrastructure gaps, limited technical capacity, and the dominance of foreign AI technologies. An AI-driven policy framework aligned with circular economy principles can support renewable energy development, promote energy efficiency, and incentivise resource reduction, reuse, and recycling, ultimately enhancing sustainability and creating closed-loop systems.

Africa's strategic approach to AI-driven sustainability in renewable energy involves a comprehensive framework that addresses infrastructure, education, policy, and collaboration. By leveraging AI to enhance circularity and promoting inclusive, sustainable industrial development, Africa can make significant strides towards achieving SDG7. This requires concerted efforts to build local capacities, develop robust infrastructures, and foster international partnerships, ensuring that AI technologies are deployed responsibly and effectively to benefit the entire continent. By aligning AI strategies with broader goals such as climate change mitigation, reducing inequality, and advancing sustainable development, Africa can play a pivotal role in shaping the future of global energy systems and contribute meaningfully to achieving the Sustainable Development Goals.

References

Adams, R. (2022). AI in Africa: Key concerns and policy considerations for the future of the continent. *Policy Brief*, (8). https://afripoli.org/ai-in-africa-key-concerns-and-policy-considerations-for-the-future-of-the-continent

Ahmad, T., & Chen, H. (2020). A review on machine learning forecasting growth trends and their real-time applications in different energy systems. *Sustainable Cities and Society, 54*, 102010.

Ahmad, T., Zhang, D., Huang, C., Zhang, H., Dai, N., Song, Y., & Chen, H. (2021). Artificial intelligence in the sustainable energy industry: Status Quo, challenges and opportunities. *Journal of Cleaner Production, 289*, 125834.

Ahmed, S. (2022). *Artificial intelligence and the circular economy in Africa: Key considerations for a just transition*. Research ICT Africa. https://researchictafrica.net/2022/03/15/artificial-intelligence-ai-and-the-circular-economy-in-africa-key-considerations-for-a-just-transition

Arogyaswamy, B. (2020). Big tech and societal sustainability: An ethical framework. *AI & SOCIETY, 35*(4), 829–840.

Awogbemi, O., Von Kallon, D. V., & Kumar, K. S. (2024). Contributions of artificial intelligence and digitization in achieving clean and affordable energy. *Intelligent Systems with Applications, 22*, 200389.

Blimpo, M. P., & Cosgrove-Davies, M. (2019). *Electricity access in Sub-Saharan Africa: Uptake, reliability, and complementary factors for economic impact*. World Bank Publications.

Burke, J. (2023). 'A goldmine at our fingertips': The promise and perils of AI in Africa. *The Guardian*. Retrieved from https://www.theguardian.com/technology/2023/oct/30/a-goldmine-at-our-fingertips-the-promise-and-perils-of-ai-in-africa

Byrne, D. (2024). How artificial intelligence is helping Ghana plan for a renewable energy future. *Nature*.

Campiglio, E., Dafermos, Y., Monnin, P., Ryan-Collins, J., Schotten, G., & Tanaka, M. (2018). Climate change challenges for central banks and financial regulators. *Nature Climate Change, 8*(6), 462–468.

Chatterjee, J., & Dethlefs, N. (2021). Scientometric review of artificial intelligence for operations & maintenance of wind turbines: The past, present and future. *Renewable and Sustainable Energy Reviews, 144*, 111051.

Choi, J., Dutz, M. A., & Usman, Z. (Eds.). (2020). *The future of work in Africa: Harnessing the potential of digital technologies for all*. World Bank Publications.

Danish, M., & Senjyu, T. (2023). Shaping the future of sustainable energy through AI-enabled circular economy policies. *Circular Economy, 2*(2), 100040.

Debnath, K. B., & Mourshed, M. (2018). Forecasting methods in energy planning models. *Renewable and Sustainable Energy Reviews, 88*, 297–325.

Di Vaio, A., Palladino, R., Hassan, R., & Escobar, O. (2020). Artificial intelligence and business models in the sustainable development goals perspective: A systematic literature review. *Journal of Business Research, 121*, 283–314.

Ding, T., Li, H., Liu, L., & Feng, K. (2024). An inquiry into the nexus between artificial intelligence and energy poverty in the light of global evidence. *Energy Economics, 136*, 107748.

Duan, Y., Edwards, J. S., & Dwivedi, Y. K. (2019). Artificial intelligence for decision making in the era of Big Data–evolution, challenges and research agenda. *International Journal of Information Management, 48*, 63–71.

Falk, S., & van Wynsberghe, A. (2023). Challenging AI for sustainability: What ought it mean? *AI and Ethics*, 1–11.

Frank, B. (2021). Artificial intelligence-enabled environmental sustainability of products: Marketing benefits and their variation by consumer, location, and product types. *Journal of Cleaner Production, 285*, 125242.

Gaur, L., Afaq, A., Arora, G. K., & Khan, N. (2023). Artificial intelligence for carbon emissions using system of systems theory. *Ecological Informatics, 76*, 102165.

Gehl Sampath, P. (2021). Governing artificial intelligence in an age of inequality. *Global Policy, 12*, 21–31.

Goralski, M. A., & Tan, T. K. (2020). Artificial intelligence and sustainable development. *The International Journal of Management Education, 18*(1), 100330.

Hannan, M. A., Al-Shetwi, A. Q., Ker, P. J., Begum, R. A., Mansor, M., Rahman, S. A., Dong, Z. Y., Tiong, S. K., Indra Mahlia, T. M., & Muttaqi, K. M. (2021). Impact of renewable energy utilization and artificial intelligence in achieving sustainable development goals. *Energy Reports, 7*, 5359–5373.

Heymann, F., Quest, H., Garcia, T. L., Ballif, C., & Galus, M. (2024). Reviewing 40 years of artificial intelligence applied to power systems–a taxonomic perspective. *Energy and AI, 15*, 100322.

Ibegbulam, C. M., Aigbovbiosa, O. J., Olowonubi, J. A., & Fatounde, S. A. (2023). Role of artificial intelligence in electrification of Africa. *Engineering Science & Technology Journal, 4*(6), 456–472.

IEA. (2017). *Digitalisation and energy*. IEA. https://www.iea.org/reports/digitalisation-and-energy. Licence: CC BY 4.0

IEA. (2023). *SDG7: Data and projections*. IEA. https://www.iea.org/reports/sdg7-data-and-projections

International Renewable Energy Agency (IRENA). (2021). *Renewable energy statistics 2020*. IRENA. https://irena.org/publications/2021/Aug/Renewable-energystatistics-2021

Jayachandran, M., Gatla, R. K., Rao, K. P., Rao, G. S., Mohammed, S., Milyani, A. H., Azhari, A. A., Kalaiarasy, C., & Geetha, S. (2022). Challenges in achieving sustainable development goal 7: Affordable and clean energy in light of nascent technologies. *Sustainable Energy Technologies and Assessments, 53*, 102692.

Jobin, A., Ienca, M., & Vayena, E. (2019). The global landscape of AI ethics guidelines. *Nature Machine Intelligence, 1*(9), 389–399.

Lau, C. K., Gozgor, G., Mahalik, M. K., Patel, G., & Li, J. (2023). Introducing a new measure of energy transition: Green quality of energy mix and its impact on CO_2 emissions. *Energy Economics, 122*, 106702.

Li, J., Herdem, M. S., Nathwani, J., & Wen, J. Z. (2023a). Methods and applications for Artificial Intelligence, Big Data, Internet of Things, and Blockchain in smart energy management. *Energy and AI, 11*, 100208.

Li, D., Bae, J. H., & Rishi, M. (2023b). Sustainable development and SDG-7 in Sub-Saharan Africa: Balancing energy access, economic growth, and carbon emissions. *The European Journal of Development Research, 35*(1), 112.

Matenga, Z. (2022). Assessment of energy market's progress towards achieving sustainable development goal 7: A clustering approach. *Sustainable Energy Technologies and Assessments, 52*, 102224.

Miskiewicz, R. (2022). Clean and affordable energy within sustainable development goals: The role of governance digitalization. *Energies, 15*(24), 9571.

Moghayedi, A., Hübner, D., & Michell, K. (2023). Achieving sustainability in south African commercial properties: The impact of innovative technologies on energy consumption. *Facilities, 41*(5/6), 321–336.

Moudene, K., El-Oud, R., Ejbari, R., & Amedjar, A. (2023). Renewable energy and economic growth in Morocco. *Journal of Human Resource and Sustainability Studies, 11*(2), 401–413.

Mungai, E. M., Ndiritu, S. W., & Da Silva, I. (2022). Unlocking climate finance potential and policy barriers—A case of renewable energy and energy efficiency in Sub-Saharan Africa. *Resources, Environment and Sustainability, 7*, 100043.

Muqeet, H. A., Javed, H., Akhter, M. N., Shahzad, M., Munir, H. M., Nadeem, M. U., Bukhari, S., & Huba, M. (2022). Sustainable solutions for advanced energy management systems of campus microgrids: Model opportunities and future challenges. *Sensors, 22*(6), 2345.

Mutiso, R. M. (2024). AI in Africa: Basics over buzz. *Science, 383*(6690), eado8276.

Nishant, R., Kennedy, M., & Corbett, J. (2020). Artificial intelligence for sustainability: Challenges, opportunities, and a research agenda. *International Journal of Information Management, 53*, 102104.

O'Leary, D. E. (2013). Artificial intelligence and big data. *IEEE Intelligent Systems, 28*(2), 96–99.

Olabi, A. G., Abdelghafar, A. A., Maghrabie, H. M., Sayed, E. T., Rezk, H., Al Radi, M., Obaidan, K., & Abdelkareem, M. A. (2023). Application of artificial intelligence for prediction, optimization, and control of thermal energy storage systems. *Thermal Science and Engineering Progress, 39*, 101730.

Platon, V., Pavelescu, F. M., Antonescu, D., Constantinescu, A., Frone, S., Surugiu, M., Mazilescu, R., & Popa, F. (2024). New evidence about artificial intelligence and eco-investment as boosters of the circular economy. *Environmental Technology & Innovation, 35*, 103685.

Qin, M., Hu, W., Qi, X., & Chang, T. (2024). Do the benefits outweigh the disadvantages? Exploring the role of artificial intelligence in renewable energy. *Energy Economics, 131*, 107403.

Quest, H., Cauz, M., Heymann, F., Rod, C., Perret, L., Ballif, C., Virtuani, A., & Wyrsch, N. (2022). A 3D indicator for guiding AI applications in the energy sector. *Energy and AI, 9*, 100167.

Roberts, H., Zhang, J., Bariach, B., Cowls, J., Gilburt, B., Juneja, P., Tsamados, A., Ziosi, M., Taddeo, M., & Floridi, L. (2024). Artificial intelligence in support of the circular economy: Ethical considerations and a path forward. *AI & SOCIETY, 39*(3), 1451–1464.

Rodríguez, F., Fleetwood, A., Galarza, A., & Fontán, L. (2018). Predicting solar energy generation through artificial neural networks using weather forecasts for microgrid control. *Renewable Energy, 126*, 855–864.

Rogerson, A., Hankins, E., Fuentes Nettel, P., & Rahim, S. (2022). *Government AI readiness index*. Oxford Insights. Retrieved from https://www.unido.org/sites/default/files/files/2023-01/Government_AI_Readiness_2022_FV.pdf

Rozite, V., Miller, J., & Oh, S. (2023). *Why AI and energy are the new power couple*. IEA. https://www.iea.org/commentaries/why-ai-and-energy-are-the-new-power-couple

Rutenberg, I., Gwagwa, A., & Omino, M. (2021). Use and impact of artificial intelligence on climate change adaptation in Africa. In *African handbook of climate change adaptation* (pp. 1107–1126). Springer International Publishing.

Sanusi, Y. A. (2023). Achieving the SDG 7, universal access to energy in Africa: Progress, obstacles, and prospects. In *SDGs in Africa and the Middle East region* (pp. 1–33). Springer International Publishing.

Shi, Z., Yao, W., Li, Z., Zeng, L., Zhao, Y., Zhang, R., Tang, Y., & Wen, J. (2020). Artificial intelligence techniques for stability analysis and control in smart grids: Methodologies, applications, challenges and future directions. *Applied Energy, 278*, 115733.

Sodhro, A. H., Pirbhulal, S., & De Albuquerque, V. H. C. (2019). Artificial intelligence-driven mechanism for edge computing-based industrial applications. *IEEE Transactions on Industrial Informatics, 15*(7), 4235–4243.

Strubell, E., Ganesh, A., & McCallum, A. (2020, April). Energy and policy considerations for modern deep learning research. In *Proceedings of the AAAI conference on artificial intelligence* (Vol. 34, No. 09, pp. 13693–13696).

Sun, T. Q., & Medaglia, R. (2019). Mapping the challenges of artificial intelligence in the public sector: Evidence from public healthcare. *Government Information Quarterly, 36*(2), 368–383.

Vinuesa, R., Azizpour, H., Leite, I., Balaam, M., Dignum, V., Domisch, S., Felländer, A., Langhans, S. D., Tegmark, M., & Fuso Nerini, F. (2020). The role of artificial intelligence in achieving the sustainable development goals. *Nature Communications, 11*(1), 1–10.

Wang, R., Hsu, S. C., Zheng, S., Chen, J. H., & Li, X. I. (2020). Renewable energy microgrids: Economic evaluation and decision-making for government policies to contribute to affordable and clean energy. *Applied Energy, 274*, 115287.

Wei, N., Li, C., Peng, X., Zeng, F., & Lu, X. (2019). Conventional models and artificial intelligence-based models for energy consumption forecasting: A review. *Journal of Petroleum Science and Engineering, 181*, 106187.

Wu, C.-J., Raghavendra, R., Gupta, U., Acun, B., Ardalani, N., Maeng, K., Chang, G., Aga, F., Huang, J., Bai, C., Gschwind, M., Gupta, A., Ott, M., Melnikov, A., Candido, S., Brooks, D., Chauhan, G., Lee, B., Lee, H.-H., Akyildiz, B., Balandat, M., Spisak, J., Jain, R., Rabbat, M., & Hazelwood, K. (2022). Sustainable AI: Environmental implications, challenges and opportunities. *Proceedings of Machine Learning and Systems, 4*, 795–813.

Zhang, L., Ling, J., & Lin, M. (2022). Artificial intelligence in renewable energy: A comprehensive bibliometric analysis. *Energy Reports, 8*, 14072–14088.

Zhou, Y. (2022). Artificial intelligence in renewable systems for transformation towards intelligent buildings. *Energy and AI, 10*, 100182.

Hicham Nachit, PhD, is a lecturer and researcher at ESCA Ecole de Management, specialising in business administration with a focus on strategy & innovation. He teaches courses in the International Business program, including International Trade and Digital Business Models, while coordinating research projects. Actively involved in the Casablanca Climate Leadership Forum, part of the Business Schools for Climate Leadership Africa initiative, his research centres on sustainability in international business, focusing on the adoption of the UN Sustainable Development Goals (SDGs).

Rachid Hachad is a PhD candidate at the School of Law and Economics, Hassan II University of Casablanca, and a part-time lecturer in business and management disciplines at the same university. His research focuses on the intersection of management, innovation, and family business, and examines the role of family governance structures, innovation strategies, and socio-emotional wealth in driving business performance and sustainability.

Saad Laraqui, PhD, is the Associate Dean of Research at ESCA Ecole de Management. He has two decades of experience as a professor and director at the University of Maryland Global Campus. Previously, he directed the Youth Development Program at the OCP Foundation, enhancing youth integration into Morocco's economy, and played a key role in establishing Mohammed VI Polytechnic University. Saad also served as the founding dean of the business school at the International University of Rabat. His research focuses on foreign direct investment, economic integration, and sustainability.

Miguel M. Torres, PhD in Economics, is a Lecturer in International Business at Kent Business School (KBS). His career includes consulting services to the European Commission and World Bank, applied research and teaching roles at Leeds, Milano, Aveiro, KU Leuven, and Uppsala universities, with appointments at the University Federal of Goiás, IF Goiano, and the University Catolica of Portugal. He has held multiple visiting professorships and is an INSOL International and Marie Skłodowska-Curie Fellow. He completed executive programs at Harvard, Carnegie Mellon University (USA), and University College London (UK). His research focuses on the intersection of international business, public policy, and strategic management, focusing on multinational management models, corporate political risk, and negotiation strategies, particularly emphasizing the complexities of cross-border restructuring and foreign direct investment drivers using cross-sectional, time-series, and panel data approaches. His work has been published in the *International Business Review, Journal of Business Ethics, Multinational Business Review, Journal of Economic Methodology, Arbitration International, International Insolvency Review, PLOS One, and AIB Insights*. He co-edited the books *Global Perspectives on Climate Change, Inequality, and Multinational Corporations* (Springer, 2025) and *Dead Firms: Causes and Effects of Cross-Border Corporate Insolvency* (Emerald, 2017).

The Role Played by Blockchain Technologies in Supply Chains in Fighting Inequality

Ovidiu Niculae Bordean, Ioana Natalia Beleiu, Irina-Iulia Salanță, and Emil Lucian Crișan

1 Introduction

The United Nations' Sustainable Development Goals (SDGs) were formulated with the deliberate intention of making the world more sustainable. The United Nations (UN) defined 17 SDGs and 169 associated targets representing an urgent call for action for all countries. Besides the annual reports issued by the UN, many studies have been conducted on the impact of the actions taken by companies to ensure sustainability. Out of these very important studies are those that focus on the adoption of technology for supporting the SDGs. Recently, it has become evident that blockchain technology plays an important role in ensuring sustainable development. Not only does blockchain technology positively impact the SDGs, but it also influences supply chains by providing transparency, trust, and security (Li et al., 2019).

Blockchain, a distributed database organised as a list of ordered blocks, where the committed blocks are immutable, has disrupted traditional business processes given that the applications and transactions are decentralised, with all having the same level of certainty (Casino et al., 2019). Previous research has already pointed out that within the Resource-Based View theory, blockchain technology is an enabler of various capabilities that can disrupt supply chain operations (Madhani, 2021). According to this theory, resources and capabilities must be aligned and coordinated in such a way as to lead to a competitive advantage. As a resource that is once deployed in the supply chain, blockchain technology will determine

O. N. Bordean · I. N. Beleiu (✉) · I.-I. Salanță · E. L. Crișan
Babeș-Bolyai University, Faculty of Economics and Business Administration, Cluj-Napoca, Romania
e-mail: ovidiu.bordean@econ.ubbcluj.ro; ioana.beleiu@econ.ubbcluj.ro; irina.salanta@econ.ubbcluj.ro; emil.crisan@econ.ubbcluj.ro

J. Bento, M. M. Torres (eds.), *Global Perspectives on Climate Change, Inequality, and Multinational Corporations*, CSR, Sustainability, Ethics & Governance,
https://doi.org/10.1007/978-3-031-80797-8_11

systematic reconfigurations that will develop unique capabilities, enabling superior advantages over the competition. Supply chain applications that use blockchain technology are said to improve productivity, lead times, customer service, and relationships with supply chain members (Jum'a, 2023).

Initially used in cryptocurrencies (e.g., Bitcoin, Ethereum, etc.), nowadays blockchain technology has revolutionised many industries such as banking, operations, real estate, insurance, healthcare, music, and renewable energy (Dutta et al., 2020). Revising 178 papers on blockchain technology use in various industrial sectors, Dutta et al. demonstrated the positive effects that blockchain technology might have on organisations (Dutta et al., 2020). Moreover, Li et al. (2019) identified several characteristics of blockchain technology that make it extremely useful for any application, regardless of the sector of activity: decentralisation, transparency, immutability, irreversibility, autonomy, anonymity, open source, ownership, provenance, and contract automation (Li et al., 2019).

Even though there have been attempts to map blockchain technology-related interventions to specific SDGs (Adams et al., 2018; Griffiths & Baudier, 2022), there is still little knowledge about how blockchain technology can address and aid the sustainability of supply chains. Moreover, pressure from governments, communities, and consumers to meet sustainability goals has prompted further investigation of this relationship. In the past, researchers were interested in understanding how blockchain technology adopted in supply chains contributes to various SDGs (Chandan et al., 2023). However, we know very little about the potential impact of using blockchain technology within supply chains on the SDG's "Reduced inequalities", as the existing literature is fragmented and often focused on a specific aspect (Parmentola et al., 2022). SDG10 aims to establish equality among nations and within countries themselves. At the same time, technology and information systems have become part of individuals' lives and influence not only people and communities, but also businesses and economies on a global scale.

To reduce this theoretical gap, our paper aims to investigate in detail the complex relationship between blockchain technology, supply chains, and SDG10. This approach was done with the help of a systematic literature review using the CIMO framework. This framework is used primarily in the field of management and organisational studies to systematically analyse and evaluate different phenomena and their impact. The acronym CIMO stands for Context, Intervention, Mechanism, and Outcome. The CIMO framework helps researchers and practitioners understand how certain interventions produce specific outcomes in particular contexts through the operation of underlying mechanisms. CIMO analysis offers a comprehensive and structured approach to researching the application of blockchain technology in supply chains. It also helps gain deep insights into how blockchain technology can transform supply chains and contribute to SDG10 by providing the iterative testing and refinement of blockchain technology interventions. Finally, CIMO analysis enables the comparison of different blockchain technologies across different contexts, facilitating the identification of best practices. The following research questions guided this research:

RQ1: Which are the mechanisms through which blockchain technology applied in supply chains influences SDG10 outcomes?
RQ2: What are the triggers that lead to SDG10-related outcomes when supply chains use various blockchain technology interventions?
RQ3: What kind of interventions that explain blockchain technology's various uses in supply chains lead to SDG10-related outcomes?
RQ4: What are the SDG10-related outcomes associated with blockchain technology use within supply chains?

The contributions of the current paper are threefold. First, it clarifies the role of blockchain technology within supply chains in terms of SDG10 by identifying the mechanism through the CIMO framework. This is a very useful approach given the fact that there is a paucity of studies that address the relationship between blockchain technology, supply chains, and SDG10. Second, it provides a clear research agenda that could serve as a starting point for researchers interested in sustainability issues. Third, the paper maps the SDG10-related goals, distinguishes between attained and expected goals, and identifies the negative outcomes that blockchain technology adoption within supply chains may have on SDG10 outcomes.

The remainder of the paper is organised as follows. After the introduction, the literature review section explains the role of blockchain technology in the supply chain and emphasises the impact that blockchain technology might have for the SDGs in general and SDG10 in particular. Section 3 details the methodology used, while Sect. 4 presents the results of the CIMO analysis. Section 5 deals with the implications that the findings might have for both theoreticians and practitioners, including the limitations of the research. Section 6 concludes.

2 Literature Review

2.1 *Blockchain Technology and Supply Chains*

Supply chains refer to the network of organisations, activities, resources, and technologies involved in the production, distribution, and delivery of goods and services to end consumers. They encompass all the steps and processes required to transform raw materials into finished products and deliver them to customers. Multinational enterprises (MNEs) often have complex supply chains that span across various countries and involve multiple suppliers, manufacturers, distributors, and retailers. These supply chains are difficult to manage as it is almost impossible to evaluate information and cope with the risks within such complex and intricate networks (Ivanov et al., 2018; Sarpong, 2014). Moreover, added difficulties associated with managing these supply chains come from cultural differences, regulatory compliance, logistics coordination, and lack of traceability (Fawcett et al., 2009). Recent scandals prove the vulnerability of supply chains and the increased risks that MNEs are taking when using supply chains. One example is that of the salmonella outbreak associated with Maradol-brand papayas, which affected hundreds of people in the United States. Despite the Centres for Disease Control and Prevention identifying the source of the contaminated papayas, not all shipments could be traced or recalled, leading to ongoing safety concerns and detrimental consequences. Another example is the 2015 *E. coli* outbreak at Chipotle Mexican Grill outlets that resulted in numerous customers falling ill, triggering significant reputational damage for the company, reflected in a stock price decline of up to 42%. Chipotle faced challenges related to transparency and accountability in its supply chains, as well as limitations in monitoring multiple suppliers in real-time, hindering effective contamination prevention even post-discovery (Saberi et al., 2019). Unfortunately, these are not isolated examples; MNEs need to find ways through which they can overcome situations when their supply chains are affected and find solutions that could lead to improving supply chain transparency, security, and process integrity.

More and more it becomes evident that a potential answer to these problems might be the application of new technological advancements, such as blockchain technology, in terms of supply chain operations. It has been acknowledged that the application of new technologies can improve the visibility, flexibility, and integration of global supply chains and logistics processes by effectively managing demand volatility and handling cost fluctuations (Wang et al., 2016). Blockchain technology is a distributed digital ledger technology that allows for the secure and transparent recording of transactions across multiple computers or nodes (Crosby et al., 2016). It operates on a decentralised network, where each participant has a copy of the entire blockchain, ensuring transparency, traceability, and security. Hence, blockchain technology enables a distributed consensus framework that informs the participating entities about every event and transaction by creating an irrefutable record. This framework is responsible for the trustworthiness of the blockchain technology, as the records added in blocks are linked through hash values, and the decision to add a new block must be approved through a consensus approach on the part of all the actors involved. The blockchain technology architecture comprises five distinct modules that contribute to the creation of the protocols for blockchain applications (Choi et al., 2020; Min, 2019): data source module, transaction module, block creation module, consensus module, and connection and interface module.

Within the blockchain-based supply chains, four different entities can be identified (Saberi et al., 2019). These identities are not necessarily part of traditional supply chains: (i) registrars—they will allocate the identities within the network; (ii) standards—for example, Fairtrade standards for sustainable supply chains; (iii) certifiers—they provide certifications to actors; and (iv) actors—such as manufacturers, retailers, and customers.

Blockchain technology can provide several benefits to supply chains. First, blockchain technology enables the recording of transactions in a transparent and immutable manner. This transparency allows supply chain participants to track and verify the movement of goods, ensuring authenticity and reducing the risk of fraud or counterfeiting. Authorised actors of the blockchain-based supply chains will retain the same copy of the ledger, and the list of transactions that will be visible to all actors of the network (Kouhizadeh & Sarkis, 2018). Second, with blockchain technology, every transaction or event related to a product can be recorded and traced back to its origin. This feature is particularly valuable in industries where provenance and traceability are critical, such as food and pharmaceuticals. It helps identify and address issues such as product recall, quality control, and compliance with regulations. Third, blockchain technology uses cryptographic algorithms to ensure the security of transactions and data. The decentralised nature of the blockchain technology makes it difficult for malicious actors to tamper with or alter information. This enhances the security and integrity of supply chain data, reducing the risk of data breaches and unauthorised access. Fourth, by eliminating the need for intermediaries and manual record-keeping processes, blockchain technology can streamline supply chain operations and reduce administrative costs. Smart contracts, which are self-executing contracts stored on the blockchain, can automate and enforce contractual agreements, improving efficiency and reducing the need for intermediaries. Fifth, blockchain technology enables trust and collaboration among supply chain participants. The shared and transparent nature of the blockchain allows for increased trust and accountability between parties. It also facilitates secure and efficient sharing of information, reducing the need for intermediaries and improving collaboration across the supply chain network. Sixth, blockchain technology can support sustainability initiatives in supply chains. It can help to track and verify sustainable practices, such as responsible sourcing, fair trade, and environmental certifications. This transparency and accountability can incentivise and reward sustainable behaviour throughout the supply chain. Overall, blockchain technology has the potential to revolutionise supply chains by enhancing data access, transparency, traceability, security, efficiency, and collaboration. It can address many of the challenges faced by supply chains, such as information asymmetry, lack of trust, and inefficiencies in record-keeping and data sharing.

2.2 Blockchain Technology and Sustainable Development Goals

Since the initial elaboration of the UN 2030 Agenda for Sustainable Development (United Nations, 2015), many studies have been conducted on the 17 SDGs and their 169 associated targets to identify ways in which countries can fight poverty and inequality, but also protect the planet and ensure peace and prosperity for all the people by 2030. There is a general belief that information and communications technologies (i.e., artificial intelligence, Internet of Things, big data, and blockchain technology) have beneficial effects and can be a source of sustainable development.

Recent studies have focused on the potential that blockchain technology has with regard to sustainability and consequently with regard to the various SDGs put in place by the UN. For example, Griffiths and Baudier (2022) have conducted interviews with 10 experts from the banking and blockchain technology domains and concluded that blockchain technology can enable transparency and trust among the stakeholders of the banking industry, thus facilitating the implementation of the UN-promoted Principles of Responsible Banking (Griffiths & Baudier, 2022). Saberi et al. (2019) have analysed the extent to which blockchain technology may serve as a solution to addressing problems within supply chains (Saberi et al., 2019). Their findings are valuable for understanding how blockchain technology's traceability can help unmask unethical suppliers and better ensure human rights and fair working practices. A conceptual framework that addresses various ways through which businesses can play a role in facilitating the measurement of the SDGs with the help of blockchain technology was proposed by De Villiers et al. (2021). The framework is useful for both mature and start-up companies and focuses on the ability of blockchain technology to accurately record data and then report it transparently by distributing it to a network, thus ensuring trust in the accuracy of the data. A similar viewpoint is also shared by Adams et al. (2018) who refer to the role of blockchain technology as they draw attention to the socially and environmentally beneficial outcomes framed in terms of the UN SDGs.

There have been several attempts within the literature to connect blockchain technology solutions to the SDG spectrum. For instance, Aysan et al. (2021) used the Blockchain Ledger Impact database (Goldstein & Tillemann, 2020) to map the contribution of blockchain technology projects to the SDGs. According to their findings, blockchain technology projects contribute to the largest number of SDGs within East Asia and the Pacific. The findings also indicate that SDG16 topped the list of the SDGs addressed by blockchain technology projects. Through a systematic literature review and a thematic analysis that uses 45 articles that were published between 2016 and 2022, Chandan et al. (2023) have mapped the relationship between the food supply chains, blockchain technology, and SDGs.

3 Methodology

This review has been made with regard to the practices theorised by Tranfield et al. (2003). It includes three steps: (i) planning, which refers to the clarification of the research question, (ii) searching for relevant sources and the analysis, and (iii) presenting the results. A summary of the search and selection of the sources is presented in Fig. 1.

Considering the model proposed by Crişan and Marincean (2023), we have initially established pairs of relevant terms related to our research questions (I1): blockchain adoption in supply chains, which generates impact concerning SDG10. For this purpose, in October 2023, we performed searches (A1) on Web of Science (WoS) databases—one source recognised as comprehensive for academic sources in the management field (Hausberg & Korreck, 2018), considering terms for all three constructs of our analysis: (1) for blockchain technology we have used "4.0", "blockchain" and "distributed ledger" as alternative search terms; (2) for supply chain we have used only the term "supply chain"; while for (3) the SDG10 has been transposed in our searches by considering the following terms: "assist*" (for

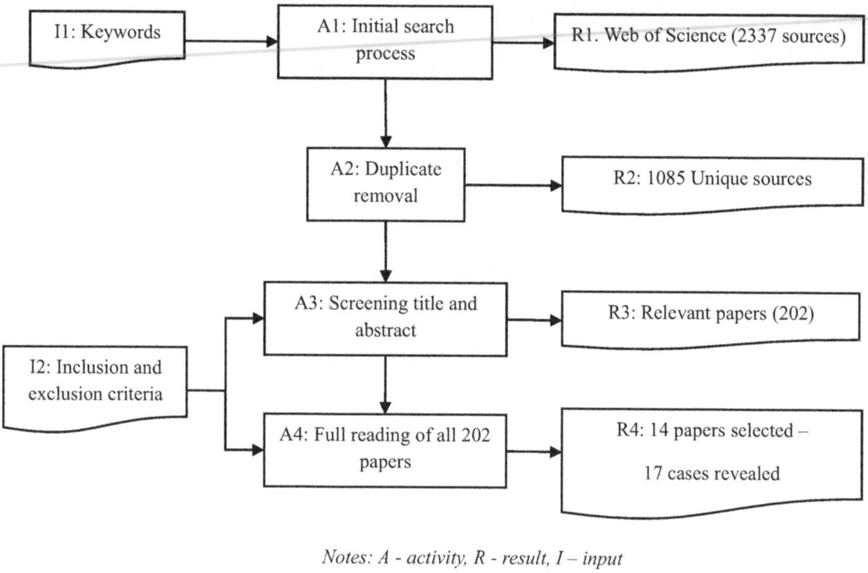

Fig. 1 The search and selection process. (Source: Authors' elaboration; Notes: A Activity, R Result, I Input)

assistance and additional alternatives), "discrimin*" (for discrimination alternatives), "equalit*" (for equality alternatives), "inclusion", "income", "inequalit*" (for inequality alternatives), "invest*" (for investment alternatives), "migra*" (for migration alternatives), "regulat*" (for regulation alternatives), "remittance*" (for remittance alternatives), "representation", and "treatment*" (for treatment alternatives). These terms correspond to the SDG10-related targets as established by the UN. Table 1 records the number of results for each search of term combinations in WoS (2337 total—R1).

Duplicate sources were eliminated (A2) (1252), and the retained 1085 (R2) papers were screened in terms of their titles and abstracts (A3). Inclusion (sources presenting cases of blockchain technology use in supply chains that have SDG10-related outcomes) and exclusion criteria (sources with a different topic, reviews) were considered (I2). All authors participated in this process; in cases of blockchain technology, the article was revised and discussed until a consensus was achieved. After screening, 202 papers were kept as being relevant to our research question (R3). The 202 papers were further read (A4). Since we discovered that a single paper might cover more than one case, we shifted the focus of the CIMO analysis from the paper as a whole to the individual cases presented within each paper. We retained only sources with cases that: (a) have detailed real blockchain adoption in supply chains that relate to SDG10-related goals or (b) have tested/proposed solutions that involved a theoretical blockchain adoption in supply chains, and for these proposals/tests, SDG10-related goals are considered. Finally, we retained 14 papers

Table 1 Web of Science (WoS) search results

Pairs of terms	WoS	Pairs of terms	WoS
"4.0" "Supply chain" "assist*"	110	"Blockchain" "Supply chain" "invest*"	678
"4.0" "Supply chain" "discrimin*"	14	"Blockchain" "Supply chain" "migra*"	8
"4.0" "Supply chain" "equalit*"	1	"Blockchain" "Supply chain" "regulat*"	311
"4.0" "Supply chain" "inclusion"	22	"Blockchain" "Supply chain" "remittance*"	2
"4.0" "Supply chain" "income"	14	"Blockchain" "Supply chain" "representation"	14
"4.0" "Supply chain" "Inequalit*"	12	"Blockchain" "Supply chain" "treatment*"	31
"4.0" "Supply chain" "invest*"	623	"Distributed ledger" "Supply chain" "assist*"	9
"4.0" "Supply chain" "migra*"	16	"Distributed ledger" "Supply chain" "discrimin*"	0
"4.0" "Supply chain" "regulat*"	118	"Distributed ledger" "Supply chain" "equalit*"	1
"4.0" "Supply chain" "remittance*"	1	"Distributed ledger" "Supply chain" "inclusion"	3
"4.0" "Supply chain" "representation"	23	"Distributed ledger" "Supply chain" "income"	2
"4.0" "Supply chain" "treatment*"	50	"Distributed ledger" "Supply chain" "Inequalit*"	0
"Blockchain" "Supply chain" "assist*"	107	"Distributed ledger" "Supply chain" "invest*"	65
"Blockchain" "Supply chain" "discrimin*"	4	"Distributed ledger" "Supply chain" "migra*"	1
"Blockchain" "Supply chain" "equalit*"	2	"Distributed ledger" "Supply chain" "regulat*"	39
"Blockchain" "Supply chain" "inclusion"	20	"Distributed ledger" "Supply chain" "remittance*"	0
"Blockchain" "Supply chain" "income"	27	"Distributed ledger" "Supply chain" "representation"	0
"Blockchain" "Supply chain" "Inequalit*"	9	"Distributed ledger" "Supply chain" "treatment*"	2

Source: Authors' elaboration

that describe 17 cases concerning blockchain technology adoption in supply chains for further analysis (R4). To analyse the validity, rigour, and reliability of the selected sources, we have used the Critical Appraisal Skills Program (CASP) scale, considering the work of Dybå and Dingsøyr (2008). CASP refers to the use of eight quality criteria (research objective, research design, sampling, data collection, data analysis, reflexivity, findings, and research value) to establish the acceptability of the sources. Two authors independently evaluated each paper in terms of the CASP criteria, with a second round of analysis being conducted if needed until a common agreement was established by both authors.

The 17 cases were analysed by considering the CIMO logic employed in similar papers (Crişan & Marincean, 2023). In the rather fragmented management field (Turnbull James & Denyer, 2009), this framework is capable of generating logical explanations (mechanisms) on how particular interventions (I) create specific outcomes (O) in different contexts (C). The mechanisms (M), or the interpretations of how contexts, interventions, and outcomes are related, are the main results of the CIMO analysis, results that can be further used by different authors to analyse the same field (Halminen et al., 2021), the use of blockchain technology in supply chains, and their impact on SDG10-related goals. Regarding the analysis, the 17 cases were revised by considering each C-I-O dimension, as detailed: (i) for context, we analysed the supply chain domain, the country in which blockchain technology was used, the triggers for blockchain technology adoption in supply chains (if there were any), and the stakeholders involved in blockchain technology adoption at the level of the supply chain; (ii) for intervention, we analyse the nature of blockchain technology adoption in supply chains (if this involved practical measures or theoretical/proposals), the extent of blockchain technology use by supply chain members, and the challenges of blockchain technology adoption in supply chains; and (iii) for outcomes, we identified the SDG10-related goals, whether these goals were primary or secondary goals, and if other goals were identified.

4 Results

4.1 Sources

To conduct the analysis, we initially included all relevant sources (14 articles). Further analysis revealed that two of these sources (Benstead et al., 2022; Gong et al., 2022) provided more cases describing the use of blockchain technology in supply chains, and each of these was analysed separately. Thus, 17 cases were finally analysed.

4.2 Context

Context refers to the external and internal environmental factors that influence behavioural change (Denyer et al., 2008). In our case, the context includes (i) the domain in which the blockchain technology was used, (ii) the country in which blockchain technology use was revealed, (iii) the trigger that influenced the blockchain adoption, and (iv) the stakeholders involved in the supply chain that used the modern technology. The 17 analysed cases present the context in different levels of detail, a few of them including all four aspects of the context.

Blockchain technology is applicable in various domains. The most frequently encountered fields where blockchain technology is used in supply chains are the fashion industry (4 cases), followed by the pharmaceutical industry (3), agriculture (2), finance (2), recycling of marine plastic debris (2), production of green composite wind turbine blades (1), MNEs across several industries (1), healthcare (1), and transportation/logistics (1). The country in which blockchain technology was implemented is mentioned in 8 cases, from the United States and Canada to Western Europe (Italy, Germany, Sweden) and to India and Thailand. The other 9 analysed cases didn't mention a specific geographic area since the proposed blockchain models represent prototypes or theoretical propositions that were not yet implemented in the form of a real situation.

The adoption of blockchain in supply chains has been influenced by various triggers, depending on industry specifics and the current economic and social realities. In the pharmaceutical industry, for instance, the digitalisation and online sale of pharmaceutical products have led to the production of counterfeit drugs and a decline in product quality. Consumers have little influence over the supply chain due to a lack of transparency, which erodes trust in pharmaceutical products' production, sales, and movement (Abdallah & Nizamuddin, 2023). As the number of entities in the pharmaceutical supply chain has increased, the complexity of drug delivery from manufacturer to consumer has also grown, raising concerns about counterfeit medication, incorrect or incomplete information reaching consumers, and significant distribution delays that can negatively impact both consumer safety and business growth (Bapatla et al., 2023). The absence of a centralised monitoring system to provide market control and real-time data on prices, availability, and authentication has also been mentioned (Humayun et al., 2022).

In the fashion industry, there is a recognised need for better control over production and traceability (Caldarelli et al., 2021), as well as for verifying the authenticity of items and engaging customers (Benstead et al., 2022). Additionally, a vision to enhance supply chain transparency to ensure improved working conditions and more sustainable manufacturing practices is mentioned (Benstead et al., 2022). Similarly, the transportation industry also highlights the need for transparency and traceability (Tan & Sundarakani, 2021).

Agriculture in developing countries faces numerous challenges, such as price fluctuations and a lack of uniformity, the presence of untrusted intermediaries in the supply chain, crop wastage due to inadequate storage facilities or long transportation links, lack of proper monitoring (Bhadra et al., 2022), and dependence on the local distributors (Kittipanya-Ngam & Tan, 2020). In the case of multinational enterprises (MNEs), multifaceted solutions are required to incorporate both top-down approaches led by focal companies and bottom-up approaches that empower potential victims of modern slavery and reduce vulnerability, as there are numerous fraudulent behaviours in hiring migrants (Christ & Helliar, 2021). Modern slavery and human trafficking are critical issues mentioned in one case that focuses on cryptocurrencies (Naheem, 2021). Given these challenges, there is an urgent need for a robust supply chain setup that can eliminate blind spots and ensure transparency.

The stakeholders involved in blockchain technology adoption by supply chains also vary, depending on the industry. Blockchain technology adoption is associated with a reduction in the number of involved stakeholders. For example, in the pharmaceutical industry, the classic model involves stakeholders such as pharmaceutical manufacturers, distributors, hospitals or pharmacies, and consumers or patients.

The blockchain-based model has reduced the layers of communication between sellers and consumers by using IoT containers (Abdallah & Nizamuddin, 2023). In agriculture, farmers, government officials, cold storage service providers, distributors, retailers, and consumers are usually involved in the supply chain (Bhadra et al., 2022).

4.3 Intervention

Due to the emergent nature of blockchain technology in the supply chain, in most cases, the model that was analysed is theoretical or describes a prototype (10 cases). Only seven of the analysed cases describe tested models when blockchain technology was adopted in the supply chain. Nevertheless, the theoretical models of implementation and the prototypes are equally important, as the real-case situations explore and highlight the potential for using blockchain technology (Christ & Helliar, 2021). By analysing how the implementation of the blockchain technology was adopted in the supply chain, we classified the intervention as being narrow, extended, or complex, taking into account the categories of stakeholders involved and the scope of the intervention.

Narrow intervention is mentioned in five of the analysed cases. It implies an app that relies on blockchain technology that facilitates access to supply chain information for the customers or interested parties through scanning a smart label with an NFC, RFID tag, or a QR code (Benstead et al., 2022). Most cases that fit this type of intervention come from the fashion industry (4 cases). The case of the retraced business model claims to be based on two pillars: internal transparency (data collection) and external transparency (communication with customers) (Benstead et al., 2022), while another analysed company states that the objective is to bring more visibility to the brand (Caldarelli et al., 2021). These apps do not add data to the blockchain. Rather, they only access information on the blockchain. In the case of recycling marine plastic debris (1 case), such apps are used as a trading platform to exchange digital tokens (Gong et al., 2022). The scope of this type of intervention is to provide access to information.

Extended intervention refers to the situation in which many supply chain entities are interacting with each other in many different possible ways using blockchain technology, aiming to share information in such a way as to ensure transparency, traceability, and safety for all those involved. Most of the analysed cases (8) describe this type of intervention. Three of these cases describe the application of the model in the pharmaceutical industry, using technological solutions to provide a transparent supply chain (e.g., distributed ledger technology) and smart contracts (Abdallah & Nizamuddin, 2023; Bapatla et al., 2023; Humayun et al., 2022). The use of smart contracts ensures data provenance, reduces the number of intermediaries, and provides all interested parties with a secure transaction history (Humayun et al., 2022). Similarly, the model applies to healthcare, involving stakeholders such as doctors and medical care providers (1 case) (Garcia et al., 2022). This type of intervention is also present in agriculture (2 cases), facilitating the creation of an open market that uses smart contracts for the validation of parties that participate in the supply chain, ensures transparency of government schemes, reduces corruption that can influence the price setting of agricultural produce, guarantees payments between stakeholders, ensures traceability of the products and the utilisation of cold storage spaces (Bhadra et al., 2022). Another case mentions the use of blockchain technology to track the recycling of marine plastic debris and to verify the working conditions of individuals collecting plastic. As a consequence, the use of forced labour or the provision of substandard income can be reduced (Gong et al., 2022). A similar model applies to MNEs across industries (1 case) that aim to protect vulnerable migrant employees from modern slavery and illicit labour practices (Christ & Helliar, 2021).

Complex intervention describes the situation in which the model is generally accepted and applicable by the interested parties, aiming to provide access to information, ensure transparency and traceability, and mostly increase safety and facilitate regulation (4 cases). One case refers to a prototype model that uses blockchain technology based on distributed ledgers and smart contracts for a shipping company and its supply chain partners in order to promote transparency, real-time traceability, and integrity of the information, based on a decentralised system accepted by the entire network (Tan & Sundarakani, 2021). Providing access to information and connecting all the stakeholders in one place (e.g., through a distributed shared ledger) is also applicable in finance (2 cases), but not only. In the case of the green composite wind turbine blade supply chain, the theoretical model presents a pioneering quality control system based on blockchain technology, ensuring that the quality of the product can not only be examined and verified by a regulator but also be monitored by other related nodes (Yu et al., 2021).

Although blockchain technology provides numerous advantages, it faces several challenges and barriers. In the fashion industry, the lack of awareness is mentioned as an issue that can influence the adoption of the technology, as is the scepticism and reluctance to accept the increased level of transparency and the lack of understanding of the technology (Benstead et al., 2022). The complexity of the solutions and the need for resources, together with the lack of a solid digital foundation and unclear regulations in the field, are identified as challenges (Benstead et al., 2022). The need for change in the mindset concerning data sharing is also mentioned in the transportation field (Tan & Sundarakani, 2021). The need to recover the investment is also considered an issue, as the technology is still at the adoption stage, while key stakeholders are waiting for it to reach its full potential. In addition, the information stored on the blockchain may cause high transaction fees when the data volume is too large (Yu et al., 2021). The ultimate benefits can only be realised if the technology is adopted by all relevant stakeholders, but there is a lack of motivation at the industry level to adopt the new technology (Tan & Sundarakani, 2021).

4.4 Outcomes

Based on the information available in the 17 cases analysed, we extracted relevant outcomes derived from the context and interventions of the proposed or existing blockchain platforms. Even if not all papers mention outcomes, we differentiated between the ones we identified according to the following taxonomy: attained versus expected SDG10-related outcomes, primary versus secondary outcomes, and non-SDG10-related outcomes. In addition, the least expected category was the negative outcomes we managed to identify in our research.

4.4.1 Attained Versus Expected SDG10-Related Outcomes

In the attained outcomes, the most frequently mentioned one refers to *transparency*. This is mentioned solely in relation to the sustainable fashion industry, for regular products in one case, but also for luxury brands and for shoe production (Benstead et al., 2022) relating to the regulations target of the SDG10. Another case mentions transparency along with traceability outcomes in the agriculture supply chain by covering product *traceability* in the supply chain, elimination of counterfeit products, easily accessible cold storage facilities, fair payment procedures between buyers and sellers, a transparent process of distributing government financial benefits to farmers, and overall government monitoring (Bhadra et al., 2022), relating also to the regulation target of the SDG10. Other outcomes that include transparency and relate to one of the SDG10 targets also add *trust* among actors and *safety* for workers, as SDG10 equality-related outcomes in the sustainable Italian fashion industry. It does this by excluding unsustainable choices such as unethical waste management, employee overexploitation, and irrational production (Caldarelli et al., 2021). Other less frequently encountered, yet equally relevant, attained outcomes related to the SDG10 on *representation* in the case of the Plastic Bank, which is committed to solving ocean plastic waste and global poverty (Gong et al., 2022). Finally, we identified *equality* as an SDG10 outcome in the case of Dell's blockchain platform, which tracks the recycling of marine plastics. It also strengthens the protection of plastic waste collectors by using blockchain solutions to verify whether individuals collecting plastic suffer forced labour or receive a substandard income (Gong et al., 2022).

The expected SDG10-related outcomes were more frequently described in our sample of cases, starting with the one on *regulations* discussed most in the pharmaceutical industry with a focus on the authenticity of the medicine received by the end consumer (Abdallah & Nizamuddin, 2023; Bapatla et al., 2023), a pioneering quality control system for the green composite wind turbine blade supply chain based on blockchain technology (Yu et al., 2021), and air cargo financing (Yadav et al., 2022). Another expected SDG10-related outcome is targeting *equality* alone (Garcia et al., 2022) in the medical prescription system and the medication supply chain (Humayun et al., 2022), or in combination with the *migrant remittance* SDG10 target in combating the problem of unethical migrant recruitment (Christ & Helliar, 2021) and in curbing modern slavery and human trafficking (Naheem, 2021). One other expected outcome relates to the SDG10 policies and mentions the challenges facing Thai rice exporters (Kittipanya-Ngam & Tan, 2020).

4.4.2 Primary Versus Secondary Outcomes

In terms of primary versus secondary outcomes related to the SDG10 targets, most of the cases present secondary outcomes, except for the cases discussed by Christ and Helliar (2021) and Garcia et al. (2022).

4.4.3 Non-SDG10 Related Outcomes

Another category of outcomes, including non-SDG10-related ones, are worth mentioning as they are anchored in the specifics of the blockchain technology. The cases we analyse present some non-SDG10-related outcomes in the pharmaceutical industry, such as avoiding communication delays, eliminating counterfeiting, recording transactions, reducing costs of drug recalls, ensuring high accuracy and reducing human errors, tracing mechanisms, and a real-time decision support tool (Bapatla et al., 2023) involving a traceable supply chain from the source of raw materials to the end user, monitoring medicine supply and transparency for sales reps with access to pertinent information, and quality insurance (Humayun et al., 2022). Transparency in freight booking (Tan & Sundarakani, 2021), along with traceability, privacy, and scalability, are frequently mentioned in the case of green composite wind turbines (Yu et al., 2021), mutual trust in air cargo financing (Yadav et al., 2022), facilitated social audits, fraud prevention, and whistleblowing encouragement in the recruitment process (Christ & Helliar, 2021).

4.4.4 Negative Outcomes

A least expected category of outcomes includes the negative ones as mentioned by Parmentola et al. (2022), which are rarely researched. The negative outcomes we identified are centred on corruption (Christ & Helliar, 2021) in the international recruitment process, as the NGO staff or union members involved may be subject to corruption in select jurisdictions. Switching the field to medical prescription, we noted another negative outcome on the performance of the blockchain, as this may be degraded as the network complexity increases (Garcia et al., 2022). Finally, researchers believe that there is a risk of data tampering (Benstead et al., 2022) in the fashion industry.

4.5 Mechanisms

Mechanisms are defined as key results of synthesising research, representing the theory of why specific outcomes emerge (Denyer et al., 2008). In the context of the blockchain technology of supply chains in fighting inequality, mechanisms describe the processes that explain their specific functionality, use, and reason for creation. The identification of these mechanisms is not always straightforward. Consequently, we compiled information based on context, intervention, and outcomes to generate five mechanisms that explain the role played by blockchain technology associated with supply chains in fighting inequality. For most of the analysed cases, we do not have just one mechanism. This is because implementing blockchain technology in supply chains when it comes to fighting inequality implies undergoing complex operations with a multitude of stakeholders and solutions. Consequently, for all cases, we identified a combination of mechanisms. In Fig. 2 we explain the 5 mechanisms under a pyramidal form, as most cases depict a more basic mechanism at the bases of the pyramid, such as access, moving gradually towards the top with regulation one being the most rare and difficult to attain.

Access, with an incidence of 15, is by far the most frequently revealed mechanism, explained by the fact that most existing blockchain platforms or models are mentioned as triggers of mistrust in collaboration and involving unwanted consequences unfolding along the supply chain. The access mechanism is the most basic one, aligned with all three types of intervention and present in terms of a wide variety of outcomes, as this is the most common purpose of blockchain platforms associated with supply chains in fighting inequality.

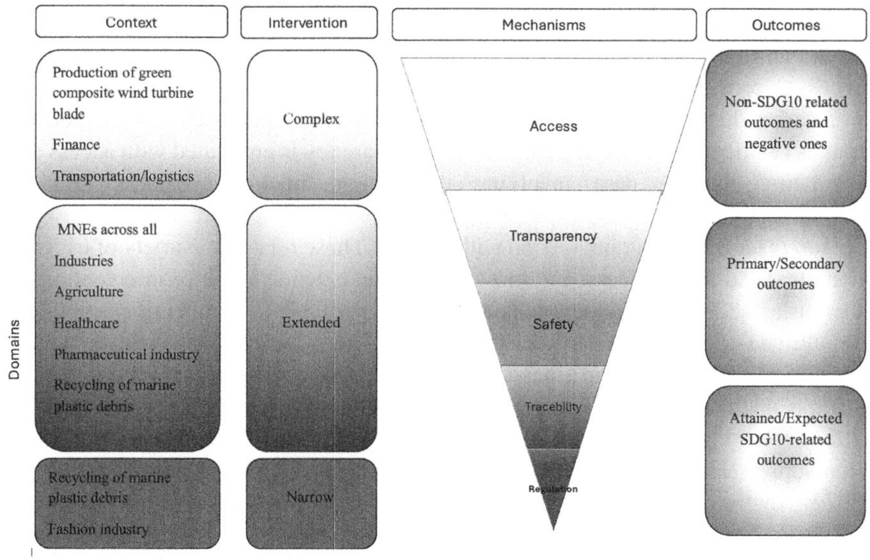

Fig. 2 The identified mechanisms (Source: Authors' elaboration)

Access, as a mechanism, is explained alone in the case of redefining the value of plastic waste (Gong et al., 2022) and the verification of supply chain information in the case of fashion products (Benstead et al., 2022) with three cases. Furthermore, access is combined with safety to reduce employee overexploitation (Caldarelli et al., 2021), adding traceability in the case of the pharmaceutical industry by improving integrity, reliability, and security among the stakeholders in the supply chain (Abdallah & Nizamuddin, 2023) and also safety in the pharma industry by enhancing control in the medication supply chain (Humayun et al., 2022) and medical prescriptions (Garcia et al., 2022). In the case of recruiting migrant workers (Christ & Helliar, 2021), this mechanism adds a new layer of functionality by facilitating social audits and protecting better stakeholders' interests, replacing transparency with traceability. Access is combined with either transparency and regulation focusing on increased inclusiveness for all members of the supply chain in a freight booking application (Tan & Sundarakani, 2021) or traceability and regulation giving the possibility for regulators to easily monitor all actions in the case of a green composite wind turbine (Yu et al., 2021). Access combined with both transparency and traceability is mentioned in the agrifood industry, creating an open marketplace to meet the demand and supply of agricultural products by including the validation of parties who participate in the agriculture supply chain (Bhadra et al., 2022) and protecting exporters of Thai rice products (Kittipanya-Ngam & Tan, 2020) or adding safety in the case of counterfeit medication (Bapatla et al., 2023).

Transparency as a mechanism is present in 8 cases. This feature often facilitates information flow among the supply chain stakeholders and is most often combined with access and safety in the pharma industry related to e-prescriptions (Garcia et al., 2022), the medication supply chain (Humayun et al., 2022), migrant issues (Christ & Helliar, 2021), unfair treatment of workers (Gong et al., 2022), or counterfeit medication by also adding the traceability mechanism (Bapatla et al., 2023). In the agrifood industry, this mechanism combines with that of traceability and access to provide a transparent, immutable audit trail and farm-to-fork traceability of farm products (Bhadra et al., 2022) or in order to safeguard Thai rice exporters' quotas (Kittipanya-Ngam & Tan, 2020). Transparency is combined with access and regulation in one case involving the freight booking business (Tan & Sundarakani, 2021).

Safety mechanisms rank second with 7 cases. These relate to a variety of triggers, domains, and stakeholders, and most often to an extended type of intervention, as there is a need for the cooperation and integration of a high number of entities with a multitude of attained and expected outcomes. The safety mechanism is most often combined with access and transparency associated with migrant workers issues, such as unethical migrant recruitment and fraudulent behaviour (Christ & Helliar, 2021), forced labour and substandard income (Gong et al., 2022), handling medication supply (Humayun et al., 2022), or managing prescription data (Garcia et al., 2022). In the complex case of human trafficking operations and forced labour (Naheem, 2021), the safety mechanism works hand-in-hand with the regulation one, and by adding access and traceability in the case of security among the stakeholders in the supply chain (Abdallah & Nizamuddin, 2023). In the pharma industry, the safety mechanism focuses on reducing counterfeit medication and making operational procedures more efficient by providing a transparent chain (Bapatla et al., 2023) by employing three more mechanisms in the form of access, transparency, and traceability.

Traceability is one of the mechanisms at the top of the pyramid, meaning that it is less frequently encountered (5 cases) and associated with complex intervention and various attained and expected outcomes. This mechanism works best in conjunction with access and transparency in the cases of the agrifood industry (Bhadra et al., 2022; Kittipanya-Ngam & Tan, 2020), adding safety in the pharma industry (Bapatla et al., 2023), or regulation in the green composite wind turbine blade supply chain (Yu et al., 2021).

At the top of the pyramid, we encounter the *regulation* mechanism. Due to its complexity, it is present in only four cases and implies a complex intervention and a multitude of outcomes. This mechanism is proposed only in the case of managing air cargo financing by requiring the consensus of all supply chain members to increase supply chain equality and security (Yadav et al., 2022). Regulation is proposed in combination with safety in the case of human trafficking problems (Naheem, 2021) or with access and traceability for the green turbine composite wind machine blade situation (Yu et al., 2021). Finally, the regulation mechanism works in conjunction with access and transparency in the case of an application for a freight booking business (Tan & Sundarakani, 2021).

5 Discussion

5.1 The Mechanisms Supporting Blockchain-Based Supply Chains

The analysis of blockchain technology within supply chains highlights its potential to address various dimensions of inequality in alignment with SDG10. The research identified five primary mechanisms through which blockchain technology influences inequality in supply chains: access, transparency, safety, traceability, and regulation. These mechanisms are rather explanations of how blockchain technology can generate SDG10-related outcomes in supply chains. Considering the presence of these explanations in papers, some mechanisms are broad and are revealed in almost all cases, while some are narrower. Secondly, these mechanisms are pyramidal, meaning that the first layers sustain the next narrower ones. Explained simply, and considering the identified mechanisms, transparency exists only if access to information for all stakeholders exists, but it is not necessarily the case that if access is provided, all supply chains will achieve transparency.

The most frequently revealed mechanism, *access*, is crucial in terms of mitigating inequality by providing all stakeholders with identical information, reducing asymmetries, and ensuring trust. For example, in the fashion industry, blockchain technology enhances consumer access to product origins, promotes ethical consumption, and reduces exploitation (Benstead et al., 2022). In addition, in agriculture, blockchain technology ensures farmers' access to market information and fair pricing, thus supporting economic inclusion (Bhadra et al., 2022). The second-mentioned mechanism, *transparency*, is essential for addressing inequalities related to unfair labour practices and environmental sustainability and diminishing the capability of stakeholders to commit unobservable acts. In the pharmaceutical industry, transparency mechanisms help track medication origins, reducing counterfeit risks and ensuring equitable access to quality drugs (Bapatla et al., 2023). The third mechanism, *safety*, can be considered as one of affordance related to transparency. Ensuring the safety of workers and consumers is a critical outcome facilitated by blockchain technology, by securing supply chains and monitoring working conditions, to prevent modern slavery and human trafficking, and to protect vulnerable populations (Christ & Helliar, 2021; Naheem, 2021). *Traceability* is a more advanced and narrower mechanism that is related to the integrity of products and services, meaning the creation of an advanced audit trail from production to consumption. Traceability is particularly valuable in industries such as agriculture and pharmaceuticals, where it enhances product safety and compliance with regulatory standards (Humayun et al., 2022; Yu et al., 2021). Finally, blockchain technology is used in supply chains for regulation, supporting compliance with international standards and policies, and promoting equitable practices across supply chains. For example, the adoption of blockchain technology in air cargo financing ensures adherence to regulations, thus improving supply chain equality and security (Yadav et al., 2022).

The findings contribute to the theoretical understanding of blockchain technology's role in supply chains by revealing the mechanisms through which it affects inequality. The CIMO framework (Context-Intervention-Mechanism-Outcome) is a robust approach for identifying the mechanisms, highlighting the interplay between contexts where blockchain technology is adopted, the nature of blockchain technology adoption in supply chains, and the subsequent outcomes. This study also extends the literature on the use of blockchain technology in supply chains beyond operational efficiency to its broader social impacts, particularly in promoting SDGs.

5.2 Future Areas of Research

Regarding future research, the road ahead concerning blockchain technology adoption in supply chains and its impact on SDGs in general and SDG10 in particular is wide open. Given the relatively small number of real case studies that reveal these types of outcomes, it could be concluded that this phenomenon is still in its early stages. One research avenue could be to analyse and estimate the value of these SDG10 outcomes because of blockchain technology use in supply chains. As for now, these outcomes are just expected or perceived. Another research avenue could be the long-term exploration of blockchain technology adoption impact in various supply chains. Finally, interdisciplinary approaches integrating technology, economics, and social sciences will be crucial in developing holistic strategies for leveraging blockchain technology to combat inequality.

6 Conclusion

For practitioners, this study explained the different facets that blockchain technology use in supply chains could have. SDG10 outcomes are somewhat secondary outcomes related to blockchain technology use, but they underlie the importance of integrating blockchain technology in supply chains to enhance transparency, traceability, and safety, all of which are crucial for reducing inequality. Blockchain technology is not only used for financial operations or for the operational benefits but also to create economic and social equity and value for many stakeholders. The use of blockchain technology in many industries reveals its potential to be tailored to create value in different contexts. Regulators and policymakers should consider the transformative potential of blockchain technology and should promote equitable supply chains. This equity has been revealed as having many layers, all of which should be considered, starting with access, transparency, safety, traceability, and regulation. Regulatory frameworks could be developed to ensure that blockchain technology adoption in supply chains aligns with broader social objectives.

This study, despite being comprehensive, has several limitations that should be considered: given the relatively small number of cases (17), the results of this paper should not be generalised, and more analysis is needed with regard to other industries and geographical regions; given the secondary data analysis approach of this paper, more real case analysis is needed to understand the wide impact blockchain technology has on supply chains and in generating SDG10-related outcomes. As most cases have focused on positive outcomes, negative outcomes are underrepresented and more research is needed to cover this gap. Finally, as blockchain technology is rapidly evolving, this study is just a picture of the moment, a snapshot, with more changes being put in place, some of them being able to lead to the identification of other mechanisms or to the disappearance of some that have already been identified. In conclusion, while blockchain technology offers promising solutions for creating equity in supply chains, the attainment of the full potential of this enhancement is limited by the increased number of stakeholders involved in this process. While practitioners should widen their focus and adopt social objectives while adopting blockchain technology, more research is needed to understand the real non-economic value these projects create.

References

Abdallah, S., & Nizamuddin, N. (2023). Blockchain-based solution for pharma supply chain industry. *Computers & Industrial Engineering, 177*, 108997.

Adams, R., Kewell, B., & Parry, G. (2018). Blockchain for good? Digital ledger technology and sustainable development goals. In W. Leal Filho, R. W. Marans, & J. Callewaert (Eds.), *Handbook of sustainability and social science research* (pp. 127–140). Springer International Publishing. https://doi.org/10.1007/978-3-319-67122-2_7

Aysan, A. F., Bergigui, F., & Disli, M. (2021). Blockchain-based solutions in achieving SDGs after COVID-19. *Journal of Open Innovation: Technology, Market, and Complexity, 7*(2), 151.

Bapatla, A. K., Mohanty, S. P., Kougianos, E., Puthal, D., & Bapatla, A. (2023). PharmaChain: A blockchain to ensure counterfeit-free pharmaceutical supply chain. *IET Networks, 12*(2), 53–76. https://doi.org/10.1049/ntw2.12041

Benstead, A. V., Mwesiumo, D., Moradlou, H., & Boffelli, A. (2022). Entering the world behind the clothes that we wear: Practical applications of blockchain technology. *Production Planning & Control, 1–18*. https://doi.org/10.1080/09537287.2022.2063173

Bhadra, O., Sahoo, S., Halder, R., & Mohan Kumar, C. (2022). AgroBLF: Blockchain-based framework for smart agriculture. *Innovations in Systems and Software Engineering*. https://doi.org/10.1007/s11334-022-00468-0

Caldarelli, G., Zardini, A., & Rossignoli, C. (2021). Blockchain adoption in the fashion sustainable supply chain: Pragmatically addressing barriers. *Journal of Organizational Change Management, 34*(2), 507–524.

Casino, F., Dasaklis, T. K., & Patsakis, C. (2019). A systematic literature review of blockchain-based applications: Current status, classification and open issues. *Telematics and Informatics, 36*, 55–81.

Chandan, A., John, M., & Potdar, V. (2023). Achieving UN SDGs in the food supply chain using blockchain technology. *Sustainability, 15*(3), 2109.

Choi, T.-M., Guo, S., Liu, N., & Shi, X. (2020). Optimal pricing in on-demand-service-platform-operations with hired agents and risk-sensitive customers in the blockchain era. *European Journal of Operational Research, 284*(3), 1031–1042.

Christ, K. L., & Helliar, C. V. (2021). Blockchain technology and modern slavery: Reducing deceptive recruitment in migrant worker populations. *Journal of Business Research, 131*, 112–120.

Crişan, E. L., & Marincean, A. (2023). The digital transformation of management consulting companies: A review. *Information Systems and e-Business Management, 21*(2), 415–436. https://doi.org/10.1007/s10257-023-00624-4

Crosby, M., Pattanayak, P., Verma, S., & Kalyanaraman, V. (2016). Blockchain technology: Beyond bitcoin. *Applied Innovation, 2*(6–10), 71.

De Villiers, C., Kuruppu, S., & Dissanayake, D. (2021). A (new) role for business–Promoting the United Nations' Sustainable Development Goals through the internet-of-things and blockchain technology. *Journal of Business Research, 131*, 598–609.

Denyer, D., Tranfield, D., & Van Aken, J. E. (2008). Developing design propositions through research synthesis. *Organization Studies, 29*(3), 393–413.

Dutta, P., Choi, T.-M., Somani, S., & Butala, R. (2020). Blockchain technology in supply chain operations: Applications, challenges and research opportunities. *Transportation Research Part E: Logistics and Transportation Review, 142*, 102067.

Dybå, T., & Dingsøyr, T. (2008). Empirical studies of agile software development: A systematic review. *Information and Software Technology, 50*(9–10), 833–859.

Fawcett, S. E., Wallin, C., Allred, C., & Magnan, G. (2009). Supply chain information-sharing: Benchmarking a proven path. *Benchmarking: An International Journal, 16*(2), 222–246.

Garcia, R. D., Ramachandran, G., & Ueyama, J. (2022). Exploiting smart contracts in PBFT-based blockchains: A case study in medical prescription systems. *Computer Networks, 211*, 109003.

Goldstein, D., & Tillemann, T. (2020). *Blockchain impact ledger. 2020*. New America.

Gong, Y., Wang, Y., Frei, R., Wang, B., & Zhao, C. (2022). Blockchain application in circular marine plastic debris management. *Industrial Marketing Management, 102*, 164–176.

Griffiths, P. D. R., & Baudier, P. (2022). Enabling responsible banking through the application of Blockchain. *Journal of Innovation Economics & Management*, I–XXXIII.

Halminen, O., Chen, A., Tenhunen, H., & Lillrank, P. (2021). Demonstrating the value of digital health: Guidance on contextual evidence gathering for companies in different stages of maturity. *Health Services Management Research, 34*(1), 13–20. https://doi.org/10.1177/0951484820971447

Hausberg, P., & Korreck, S. (2018). Analysis-based, systematic literature review. In *Speedboating into the future-how organizations use open foresight and business incubation as strategic means to explore trends and promote innovation* (p. 114).

Humayun, M., Jhanjhi, N. Z., Niazi, M., Amsaad, F., & Masood, I. (2022). Securing drug distribution systems from tampering using blockchain. *Electronics, 11*(8), 1195.

Ivanov, D., Das, A., & Choi, T.-M. (2018). New flexibility drivers for manufacturing, supply chain and service operations. *International Journal of Production Research, 56*(10), 3359–3368. https://doi.org/10.1080/00207543.2018.1457813

Jum'a, L. (2023). The role of blockchain-enabled supply chain applications in improving supply chain performance: The case of Jordanian manufacturing sector. *Management Research Review, 46*(10), 1315–1333.

Kittipanya-Ngam, P., & Tan, K. H. (2020). A framework for food supply chain digitalization: Lessons from Thailand. *Production Planning & Control, 31*(2–3), 158–172. https://doi.org/10.1080/09537287.2019.1631462

Kouhizadeh, M., & Sarkis, J. (2018). Blockchain practices, potentials, and perspectives in greening supply chains. *Sustainability, 10*(10), 3652.

Li, Q.-L., Ma, J.-Y., Chang, Y.-X., Ma, F.-Q., & Yu, H.-B. (2019). Markov processes in blockchain systems. *Computational Social Networks, 6*(1), 5. https://doi.org/10.1186/s40649-019-0066-1

Madhani, P. M. (2021). Enhancing supply chain capabilities with blockchain deployment: An RBV perspective. *IUP Journal of Business Strategy, 18*(4), 7–31.

Min, H. (2019). Blockchain technology for enhancing supply chain resilience. *Business Horizons, 62*(1), 35–45.

Naheem, M. A. (2021). Do cryptocurrencies enable and facilitate modern slavery? *Journal of Money Laundering Control, 24*(3), 491–501.

Nations, U. (2015). *Transforming our world: The 2030 agenda for sustainable development* (Vol. 1, p. 41). United Nations, Department of Economic and Social Affairs.

Parmentola, A., Petrillo, A., Tutore, I., & De Felice, F. (2022). Is blockchain able to enhance environmental sustainability? A systematic review and research agenda from the perspective of Sustainable Development Goals (SDGs). *Business Strategy and the Environment, 31*(1), 194–217. https://doi.org/10.1002/bse.2882

Saberi, S., Kouhizadeh, M., Sarkis, J., & Shen, L. (2019). Blockchain technology and its relationships to sustainable supply chain management. *International Journal of Production Research, 57*(7), 2117–2135. https://doi.org/10.1080/00207543.2018.1533261

Sarpong, S. (2014). Traceability and supply chain complexity: Confronting the issues and concerns. *European Business Review, 26*(3), 271–284.

Tan, W. K. A., & Sundarakani, B. (2021). Assessing Blockchain Technology application for freight booking business: A case study from Technology Acceptance Model perspective. *Journal of Global Operations and Strategic Sourcing, 14*(1), 202–223.

Tranfield, D., Denyer, D., & Smart, P. (2003). Towards a methodology for developing evidence-informed management knowledge by means of systematic review. *British Journal of Management, 14*(3), 207–222. https://doi.org/10.1111/1467-8551.00375

Turnbull James, K., & Denyer, D. (2009). Historical roots and future directions: New challenges for management learning. *Management Learning, 40*(4), 363–370. https://doi.org/10.1177/1350507609335844

Wang, G., Gunasekaran, A., Ngai, E. W., & Papadopoulos, T. (2016). Big data analytics in logistics and supply chain management: Certain investigations for research and applications. *International Journal of Production Economics, 176*, 98–110.

Yadav, P., Bhosale, R., Sahoo, R., Khanzode, V., & Tiwari, M. K. (2022). Advances in air cargo financing using a consortium blockchain. *IFAC-PapersOnLine, 55*(10), 737–742.

Yu, H., Zhu, S., & Yang, J. (2021). The quality control system of green composite wind turbine blade supply chain based on blockchain technology. *Sustainability, 13*(15), 8331.

Ovidiu Niculae Bordean received the Ph.D. degree in management from the Babeş-Bolyai University of Cluj-Napoca, Romania, in 2009 and was awarded a postdoctoral scholarship from the University of Economics in Bucharest, Romania, in 2010. He is currently an associate professor in the Department of Management from the Faculty of Economics and Business Administration of the Babeş-Bolyai University and the Strategy and Value Creation module leader of the EMBA program managed by Transylvania Executive Education in Cluj-Napoca, Romania. He was a Fulbright RAF Scholar hosted by the University of Rochester, Ain Centre for Entrepreneurship. His research interests include issues within the entrepreneurship field and exploring the formation of entrepreneurial intentions. He also has expertise within the areas of business strategies, showing an interest in the digital transformation process.

Ioana Natalia Beleiu is a lecturer at the Faculty of Economics and Business Administration, Babeş-Bolyai University, from Cluj-Napoca, Romania. She finished her PhD studies in 2012, and obtained the habilitation in 2022. Her research interests relate mainly to project management, but she was also involved in research that addresses topics such as governance, entrepreneurship, or quality management. She has conducted research internships at renowned universities such as Wirtschaftsuniversität Wien and UniWien, which have contributed to the development of the research capabilities. She has published in a range of journals, including the *Journal of Technology Transfer*, *Management Learning*, *International Journal of Managing Projects in Business*, and *Annals of Public and Cooperative Economics*.

Irina-Iulia Salanţă is an associate professor at Babeş Bolyai University of Cluj-Napoca, a former visiting professor at Texas A&M University, San Antonio. She has had a PhD in Management since 2011. She is a published author/co-author of a wide number of papers on innovation, digital innovation, accelerators, digital transformation, technology transfer, entrepreneurship, internationalisations of small- and medium-sized enterprises, and many others.

Emil Lucian Crişan is an associate professor at the Department of Management, the Faculty of Economics and Business Administration, Babes-Bolyai University, Romania. His main research areas are digital transformation, entrepreneurship, and education. His work explores how management practices across different fields are established and how they are changing given the digital transformation of our society. He has published in a range of journals, including the *Journal of Technology Transfer*, *Management Learning*, *Journal of International Management*, and *Technological Forecasting and Social Change*.

Exploring the Relationship Between Democracy and the Environment: The Moderating Role of FDI

João Bento and Miguel M. Torres

1 Introduction

This book chapter builds upon Bento and Torres (2024) to extend its empirical findings and theoretical frameworks into a broader analysis of democracy's interaction with FDI in the context of environmental sustainability. While Bento and Torres (2024) focus on quantifying the moderating effects of varieties of democracy on the relationship between inward FDI and carbon intensity, here we advance these findings by incorporating additional variables and theoretical insights, such as international environmental agreements and industry-specific impacts of FDI. Additionally, this chapter critiques the universality of the "halo effect" hypothesis as presented in Bento and Torres (2024), which assumes that FDI reduces carbon intensity under democratic governance. By exploring scenarios where democratic mechanisms may fail due to lobbying or competing economic priorities, this chapter challenges the straightforward causal pathways implied in Bento and Torres (2024). Moreover, it contextualises the empirical results within specific policy environments, offering more recommendations for aligning FDI and democratic governance with the Sustainable Development Goals (SDGs). We validate the quantitative findings of

This work has been financially supported by the Research Unit on Governance, Competitiveness, and Public Policy (GOVCOPP) and national funds through The Foundation for Science and Technology and the Ministry of Science, Technology and Higher Education in Portugal (UIDB/04058/2020) + (UIDP/04058/2020). This chapter is modified from a previous version published as Bento & Torres (2024). Used with permission.

J. Bento (✉)
Lisbon Accounting and Business School, Polytechnical Institute of Lisbon, Lisbon, Portugal
e-mail: jpbento@iscal.ipl.pt

M. M. Torres
Kent Business School, University of Kent, Canterbury, UK
e-mail: m.torres@kent.ac.uk

J. Bento, M. Torres (eds.), *Global Perspectives on Climate Change, Inequality, and Multinational Corporations*, CSR, Sustainability, Ethics & Governance,
https://doi.org/10.1007/978-3-031-80797-8_12

previous work and enrich its theoretical implications by addressing gaps, such as the roles of institutional quality and governance capacity, which were less emphasised previously.

Growing concerns about sustainable development and climate change have prompted international business scholars to emphasise the crucial role that multinational enterprises (MNEs) play in addressing societal issues (Buckley et al., 2022). There is an increasing recognition that climate change presents significant challenges to governments and businesses (Kolk & van Tulder, 2010). MNEs are considered pivotal in addressing global-scale issues due to their substantial influence and involvement in the international transfer of capital and environmentally friendly technological capabilities across national borders (Buckley et al., 2017).

Extensive literature examines how democracy influences the selection of host-country locations for foreign direct investment (FDI) (Li & Resnick, 2003). The literature also shows that the presence of FDI by MNEs, in general, results in environmental improvements within the host country, but findings still need to be exhaustive and conclusive in this area (Cole et al., 2017). There needs to be more governance systems that have mechanisms consisting of checks and balances to oversee the actions of elected officials, which research on the role of MNEs in either aggravating or mitigating climate change in their host markets. Additionally, the impact of different forms of democracy on the effectiveness of FDI in addressing climate change remains underexplored.

We aim to advance empirical knowledge about the environmental effects of FDI and democracy, particularly the interaction between FDI, democracy, and the natural environment as proposed and tested. More specifically, our goal is to contribute to the ongoing debate about whether FDI and democracy are crucial for enabling governments to meet their commitments to achieving the SDGs. Researchers posit that higher levels of democracy are essential for enabling nations to achieve sustainable development (Neumayer, 2002; Pickering et al., 2022).

Democracies facilitate public interest litigation, improve citizen awareness, and promote environmental protection (Farzin & Bond, 2006; Li & Reuveny, 2006). Proponents of democracy's positive impact on the natural environment argue that democracies are more cooperative in climate change mitigation efforts due to their longer history of established democratic institutions (Clulow, 2019; Fredriksson et al., 2005; Fredriksson & Neumayer, 2013). Democratic governance systems include mechanisms of checks and balances that oversee the actions of elected officials, helping to prevent arbitrary government intervention (North & Weingast, 1989; Quan, 2009).

On the other hand, some researchers argue that autocratic governments tend to prioritise short-term gains over long-term benefits, which leads to insufficient implementation and enforcement of climate action (Clarke & Stone, 2008). Hollyer et al. (2011) contend that political leaders in autocracies are less inclined to tolerate mass protests and environmental demonstrations. Unlike a democracy, autocratic regimes could potentially enhance their ability to attract private investors by offering more robust incentive packages and ensuring protection against labour unions (Clegg et al., 2018; Li & Resnick, 2003). Environmental lobbying groups may sometimes prioritise individual gains over collective environmental interests (Carlitz & Povitkina, 2021; Culhane & Roberts, 2021). Thus, the relationship between democracy, FDI, and the environment must be established through empirical analysis (Abdouli & Omri, 2021). Therefore, this chapter answers the following research questions: How do FDI and democracy affect the quality of the environment? How do democratic institutions moderate the relationship between FDI and environmental quality?

To address these research questions, we employ a dataset of 160 countries from 1990 to 2021 and use the Generalised Method of Moments (GMM) estimation technique. Our approach provides a more accurate measure of democracy by considering its various forms based on data from the Varieties of Democracy Project (Coppedge et al., 2016, 2020). Building on the literature that explores the relationship between FDI and the environment, this study adds to the existing body of research on the institutional determinants of FDI (Bénassy-Quéré et al., 2007; Demena & Afesorgbor, 2020) and examines how governance influences FDI (Antonietti & Mondolo, 2023; Globerman & Shapiro, 2002; Kammerlander & Schulze, 2020; Midlarsky, 1998). Research findings reveal that FDI significantly enhances the quality of the environment. Additionally, the estimation results show that democracy moderates the relationship between FDI and environmental quality.

2 Literature Review

The growing significance of MNEs and their FDI activities is often seen as generating positive externalities for the host country (Alfaro et al., 2010; Yao & Wei, 2007). MNEs serve as a channel for the international diffusion of technology (Kokko, 1994; Pfaffermayr, 1999; Xu, 2000). Foreign firms have firm-specific assets that enable them to compete successfully in international markets (Cantwell & Narula, 2001; Dunning, 1980; Markusen, 1995). They have a greater ability to adopt new environmental technologies than their domestic counterparts (Albornoz et al., 2014; Zhu et al., 2012). Environmental management is a key component of a business's environmental engagement strategy (Albornoz et al., 2009).

In the literature on the relationship between FDI and the environment, the pollution halo hypothesis suggests that strict environmental regulations motivate companies to invest more in clean and efficient technologies (Eiadat et al., 2008; Porter & Van Der Linde, 1995). In contrast, the pollution haven hypothesis argues that multinationals relocate their polluting production to countries with weaker environmental standards (Bu et al., 2023; Copeland & Taylor, 2004). Several studies confirm the pollution halo hypothesis (Ben Kheder & Zugravu, 2012; Cole et al., 2008; Kim & Adilov, 2012; Nguyen et al., 2021; Pazienza, 2019; Zhu et al., 2016), while others challenge its validity (Abdouli & Hammami, 2017; Blanco et al., 2013; Demena & Afesorgbor, 2020; Doytch & Uctum, 2016; Omri et al., 2014; Wang et al., 2019; Waqih et al., 2019). Some scholars argue that the pollution haven hypothesis applies primarily to developing countries, while the pollution halo hypothesis is more relevant to developed countries (Benzerrouk et al., 2021).

Several studies have explored the FDI environment in relation to the stringency of environmental regulations (Cole et al., 2006). Eskeland and Harrison (2003) have found no significant relationship between MNEs' location decisions and environmental regulation in host countries. Chung (2014) reports that polluting industries are more likely to invest in countries with weaker environmental regulations. More regulated countries are generally more capital-intensive, while less regulated countries tend to be more labour-intensive (Cole et al., 2017; Zugravu-Soilita, 2017).

The literature on whether democracy benefits the environment has compared the environmental performance of democratic countries with that of authoritarian regimes (Arvin & Lew, 2009; Clulow, 2019; Li & Reuveny, 2006; Policardo, 2016; Povitkina, 2018). A few studies have failed to identify any significant environmental impact of democracy on environmental quality (Adams & Klobodu, 2018; Joshi & Beck, 2018; Povitkina & Jagers, 2022; Selseng et al., 2022; Sulemana et al., 2017). The literature has highlighted specific elements that contribute to environmental protection, including democratic and incorrupt governments, the level of civil society participation, civil liberties, the strength of labour unions, and the involvement of green parties (Bernauer & Koubi, 2009; Fredriksson & Neumayer, 2016; Neumayer, 2003). Democratic institutions have characteristics that constrain corruption, particularly when income inequality is low (Cole, 2007; Jorgenson et al., 2017; Lægreid & Povitkina, 2018; Povitkina, 2018).

Scholars have found that, as a general rule, higher levels of democracy are associated with reductions in emissions (Adams & Acheampong, 2019; Farzanegan & Markwardt, 2018; You et al., 2015). Other studies have yielded inconclusive results (Arvin & Lew, 2009; Clulow, 2019; Selseng et al., 2022). Democracy is found to have a mitigating effect on the environment in developing countries (Acheampong et al., 2022). Democracies implement stricter and more effective environmental policies (Povitkina & Jagers, 2022).

3 Methodology

3.1 Empirical Model

In this study, we use the Stochastic Impacts by Regression on Population, Affluence, and Technology (STRIPAT) model (Dietz & Rosa, 1994), which is given by the following equation:

$$I_{it} = \alpha P_{it}^{b} A_{it}^{c} T_{it}^{d} e_{it} \tag{1}$$

where *I*, *P*, *A*, *T*, and e denote environmental impact, population size, affluence, the level of technology, and random error term in country *i* and year *t*, respectively. The intercept term is α, and letters *b*, *c*, *d* are the exponential terms of *P*, *A*, and *T*, respectively.

The STRIPAT model is an extension of the IPAT identity (Ehrlich & Holdren, 1971), which offers a comprehensive framework for evaluating environmental impact, considering not only the core elements of population size, affluence, and technology but also incorporating the influence of various other drivers. To examine how foreign direct investment and democracy affect environmental quality, we have estimated an extended version of the STIRPAT model.[1] The main empirical model specification becomes:

$$\begin{aligned}\text{Environment}_{it} &= \alpha + \beta_1\text{Population}_{it} + \beta_2\text{Affluence}_{it} + \beta_3\text{Technology}_{it} + \beta_4\text{FDI}_{it} \\ &+\beta_5\text{Democracy}_{it} + \beta_6\left(\text{Democracy}_{it}\ x\,\text{FDI}_{it}\right) + \beta_7\text{Control}_{it} + \tau + \epsilon_{it}\end{aligned} \tag{2}$$

where *Environment* is the dependent variable. The *Population*, *Affluence*, and *Technology* are the explanatory variables of the STRIPAT model. The regression coefficients of foreign direct investment (*FDI*), the democracy variable (*Democracy*), and the interaction term of FDI and democracy (*FDI x Democracy*) are the parameters of interest in this study. In addition to the variables suggested by the STRIPAT model, we add a series of control variables. Control is a vector of other explanatory variables, which include trade openness, industrialisation, financial development, electrification, corruption, and civil liberties. We incorporate a time-specific effect denoted as τ to account for factors common across countries but changing over time, and ε represents the error term.[2]

In what follows, we describe the variables in detail (for an overview, see Table 2). Environmental quality is measured by the level of CO_2 emissions in tonnes of CO_2 by multiplying tonnes of carbon by the conversion factor of 3.664.[3] The primary contributor to global warming, particularly concerning carbon emissions from human activities, has garnered particular scrutiny. Population is measured by the total number of persons based on the de facto definition of the population with end-year estimates. The relationship between population size and environmental quality is often described as a key impact factor (Shuai et al., 2017). We hypothesise that larger populations lead to increased energy demand, contributing to the depletion of natural resources and environmental degradation (Zhu & Peng, 2012). Affluence

[1] Many studies have estimated an inverted U-shape environmental Kuznets curve (EKC) hypothesis. Thus, the outcomes of these studies have been mixed. Instead, we employ the STRIPAT model over the competing EKC to avoid various complications (Kaika & Zervas, 2013). First, the EKC model has faced criticism for its econometric shortcomings (Müller-Fürstenberger & Wagner, 2007; Wagner, 2008). First, using the STRIPAT model mitigates some of these deficiencies by avoiding the quadratic transformation of the per capita income variable (Liddle, 2015). Second, it alleviates the constraint imposed by the EKC model, which mandates that the emissions elasticity of population is unity (Menz & Welsch, 2012). Third, the STRIPAT model is formulated to analyse aggregate emissions rather than pollution/emission on a per capita basis, aiming to capture the ecological system's carrying capacity, which is more pertinent for climate change mitigation efforts. We follow Yao et al. (2020) and select the STRIPAT model as our empirical model.

[2] All variables are expressed in logarithmic form except the variables of democracy, corruption and civil liberties.

[3] We adopt the strategy by Jahanger et al. (2022) and estimate reductions in carbon dioxide emissions, that is the leading greenhouse gas responsible for anthropogenically forced climate change.

measures domestic income, computed per capita, proxied by GDP in PPP constant 2017 international US dollars divided by the total population. We hypothesise a positive association between domestic income and environmental quality since higher domestic incomes imply a higher demand for goods and services, raising emissions with income (Büchs & Schnepf, 2013; Parker & Bhatti, 2020). Technology represents the technological orientation in the energy sector and captures the use of renewable energy sources in its economic operations (Aziz et al., 2023). We hypothesise that a higher percentage of renewable energy in GDP signifies a stronger focus on innovation in clean energy technology and the transition towards sustainable energy practices. On the contrary, a smaller share implies the reliance on fossil fuels rather than cleaner energy sources.

As is standard in the literature, our dependent variable is inward FDI stock divided by GDP to reflect the scale of multinational presence in the host country. This is the most commonly used construct in FDI studies (Asiedu & Lien, 2011; Paul & Feliciano-Cestero, 2021). We discern the implications of democracy and MNEs investments, represented by inward FDI stock, on the dependent variable. We hypothesise that FDI may deter emissions through the positive spillover effect within the environmental domain. FDI is instrumental in the diffusion of environmentally superior innovations and performance since multinationals often incorporate environment-efficient technologies in their foreign operations (Pisani et al., 2019). In contrast to the study conducted by Abdouli and Hammami (2017), our emphasis lies on the interactive effect of democracy and FDI rather than on the direct impact of covariates on environmental quality.

Our democracy variable is derived from the democracy indices from the Varieties of Democracy Project[4]: electoral, liberal democracy, participatory, deliberative, and egalitarian democracy.[5] We compute a composite democracy variable using the

---

[4] We use data from the Varieties of Democracy Project version 12. This dataset outperforms commonly available democracy indices and provides conceptually unique measures of democracy (Boese, 2019). It consists of a large set of indicators that conciliate five separate multidimensional democracy indices (electoral, liberal, participatory, deliberative, egalitarian). Data is obtained from an expert survey of more than 3000 country experts from around the world involved in measuring democracy, institutions and governance. Data collection uses on average five independent experts for every year and country to code each of available indicators. This ensures that the data is not only available but also maintains high standards of quality and reliability, thereby elevating its overall quality when compared to alternative sources.

[5] *Electoral democracy* is a government system wherein citizens possess political rights and engage in the decision-making process through the act of voting (Teorell et al., 2019). In an electoral democracy, multiple political parties and leaders actively compete for votes, offering citizens a range of choices in the political arena. *Liberal democracy* is a form of representative democracy characterised by its endorsement of political pluralism. It places strong emphasis on safeguarding civil rights and liberties, promoting a robust rule of law, establishing an independent judiciary, and implementing effective checks and balances. These measures collectively serve to limit the exercise of executive power (Coppedge et al., 2016). In a *participatory democracy*, citizens actively engage in the decision-making process through both electoral and non-electoral mechanisms to enhance the link between citizens and their government (Coppedge et al., 2011). One key feature of this system of governance is the inclusion of local and regional governments as public collabo-

principal component analysis. All variables are rescaled to range between 0 and 1; a higher number implies more democracy, and a lower number less democracy. The action term *FDI x Democracy* in Eq. (2) captures the interplay between FDI and the composite democracy indicator. The central hypothesis underpinning our study is that democracy positively affects environmental quality. Specifically, we hypothesise that democracy complements FDI to improve environmental quality in host countries, ceteris paribus.

The coefficient $\beta 6$ in Eq. (2), capturing the interaction effect of FDI and democracy on the dependent variable, is the key focus of this study. The coefficient $\beta 4$ in Eq. (2) captures the impact of FDI on environmental quality. It is important to highlight that calculating the marginal effects is essential for assessing the role of democracy on the impact of FDI on environmental quality using Eq. (2). The partial (marginal) effect of FDI on environmental quality can be computed as follows:

$$\text{Environment}_{it} / \text{FDI}_{it} = \beta_4 + \beta_6 \text{Democracy}_{it} \tag{3}$$

Subsequently, we evaluate the effect of FDI on the mean and various percentile values of the democracy variables. We include additional control variables in our regressions. These controls capture the multifaceted nature of the economic and political elements influencing environmental quality. Since the dependent variable does not include the emissions from international trade, we use the combined value of imports and exports as the proportion of GDP as a measure of trade openness. Following the literature, we hypothesise a mixed effect between trade openness and environmental quality (Kim et al., 2019). We employ three measures to capture the level of economic development in host countries: industrial activity measured by industry value added as a percentage of GDP; financial development proxied by the ratio of credit to the private sector to GDP; and electrification, that is, the percentage of the population with access to electricity within a country. All else being equal, we hypothesise that the industrial sector, financial development, and electrification should worsen environmental quality, ceteris paribus (Lin & Li, 2020; Shakouri et al., 2023; Tao et al., 2023).

---

rators in decision-making processes. The *deliberative democracy* emphasises the importance of informed and inclusive discussions as a means of reaching collective decisions (Coppedge et al., 2016). Deliberative mechanisms in the context of environmental issues are defined by respectful dialogue, open and transparent debates that involve the presentation, analysis, and discussion of diverse viewpoints. Elected officials can gather a wide range of perspectives on a policy before its implementation (Dryzek & Pickering, 2017). *Egalitarian democracy* is a political system that grants citizens equal rights, ensures equitable access to resources within society, and provides an equal voice in the political process (Sigman & Lindberg, 2015). In an egalitarian democracy, it is imperative to fulfil basic needs such as access to education and healthcare. This approach aims to encourage citizens to participate in decision-making processes and serves as a safeguard against concentrating political power in the hands of a privileged group (Kashwan, 2017).

Following standard practice in the literature (Acheampong and Opoku (2023); Bernauer and Koubi (2009); Hargrove et al. (2019); Lægreid and Povitkina (2018); Povitkina and Jagers (2022)), we control for civil society liberties and corruption. Non-state actors, such as environmental interest groups, have become increasingly influential in both the formulation and execution of environmental policies (Bernauer & Betzold, 2012). Civil society participation enables the pursuit of shared interests and ideals, providing the capacity to exert pressure on elected officials (Bernauer et al., 2016). In turn, this ability depends on the availability of freedom in civil society (Grauel & Gotthardt, 2017). In this context, we hypothesise that civil liberties harm environmental quality, all being equal. We include perceptions of corruption in our regressions to test how much public power is used for private benefits, including small and large forms of corruption and the "capture" of the state by politicians and personal interests. We recode and normalise values from zero rather than using the original variable to aid interpretation. A higher value of the new variable corresponds to better corruption control (lower level). We hypothesise that corruption worsens environmental quality, ceteris paribus.

3.2 Econometric Methods

The empirical model is estimated using the instrumental variables in the context of a generalised method of moments estimation technique (Baum et al., 2003; Hansen, 1982). This method has several advantages over the conventional counterparts. An essential advantage of this estimation method is its ability to address potential issues of endogeneity and to allow for efficient estimation in the presence of heteroscedasticity and serial correlation (Baum et al., 2007). Endogeneity occurs when the explanatory variables in the regression model are not independent of the error term, resulting in unreliable parameter estimates and inferences, which provide a misleading interpretation of the results.

We use the instrumental variables estimation with the two-step system GMM procedure to derive estimation results from panel data and estimate the parameters of interest (Blundell & Bond, 1998). This approach can simultaneously address endogeneity by dealing with unobserved heterogeneity. The system GMM is more efficient than estimating only the GMM estimator in the first-differenced model because it combines moment conditions for the model in first differences with moment conditions for the model in levels (Arellano & Bond, 1991). However, it imposes orthogonality conditions because the assumptions for the level equation and the first-differenced equation must be satisfied.[6]

[6]The presence of the interaction term poses a challenge in the estimation process. To overcome this problem, we use the difference of the product of the constitutive variables of the interaction term rather than the product of their difference as valid instruments. We opt for utilising lags of the variables deemed endogenous as instruments.

The reliability of the estimates depends on a set of diagnostic tests. First, the AR(1) and AR(2) test statistics for the differenced error terms should be reported, with the AR(2) test showing no second-order serial correlation. In other words, if the null hypothesis of the absence of second-order serial correlation in disturbances is not rejected, the results can be considered reliable. The Hansen J over-identification test is used to test the validity of the instruments. The null hypothesis that the instruments used in the difference equation are exogenous cannot be rejected, indicating that the instruments used are exogenous. In that case, such findings suggest that the estimates are reliable.

3.3 Data

The democracy variables are from the Varieties of Democracy Project by the University of Gothenburg. The remaining variables are obtained from the World Development Indicators (WDI) and World Governance Indicators (WGI) databases of the World Bank, the United Nations Conference on Trade and Development (UNCTAD), and Freedom House. The sample comprises 160 countries and covers the period from 1990 to 2021. The selection of countries and the chosen period relies on the accessibility of data and the objective of including as many countries as possible.

As shown in Table 2, the mean of environmental quality is about 178 tonnes of carbon dioxide. The minimum value of environmental quality is about 0.05 tonnes of CO_2 observed for São Tomé and Princípe in 1990. The maximum value of environmental quality is 11,472 tonnes of CO_2, which will be observed in China in 2021. Regarding the FDI variable, the mean value is about 51%. The mean value of the composite democracy variable is about 0.45, indicating that, on average, the democracy level in our sampled countries is relatively low. All the individual democracy variables except one (electoral democracy) have mean values below 0.5, showing the low democracy portrayed by the composite democracy variable (Tables 1 and 2).

4 Results

The estimation results of the direct and interaction effects of democracy and FDI on environmental quality are reported in Table 3. It contains the estimates of four models: Model 1, which incorporates FDI as the primary explanatory variable; Model 2, where the main explanatory variable is the composite democracy variable; Model 3, which includes both FDI and the composite democracy variables; and Model 4, which incorporates the variables from Model 3 along with the interaction terms of FDI and the composite democracy variables. In Model 1, the coefficient of the FDI variable is positive and statistically significant (at the 10% level). Model 2 reveals a positive and statistically significant coefficient for democracy (at the 1% level). In Model 3, both FDI and democracy variables exhibit positive and statistically significant coefficients (at the 10% level for FDI and 1% for democracy). In the main model of interest, Model 4, the coefficients of both FDI and democracy variables remain positive and statistically significant (at the 1% level). The interaction term between FDI and democracy displays a negative coefficient and is statistically significant (at the 1% level).

The findings generally suggest that FDI is directly associated with higher levels of CO_2 emissions in our sample. Focusing solely on the FDI variable, the results indicate that a 1% increase in FDI is associated with about a 0.10% increase in CO_2 emissions, holding other variables constant (Model 4, Table 3). This indicates that an increase in FDI is associated with potentially worsening environmental quality. This study shows that FDI directly and positively affects environmental quality in our sampled countries. The potential environmental cost of FDI due to increased emissions may undermine the economic benefits linked with the growth of foreign investments in host countries (Demena & Afesorgbor, 2020). It is plausible that FDI may occur due to relocating environmentally harmful production processes in response to stringent environmental regulations (Kearsley & Riddel, 2010; Mani & Wheeler, 1998). The results corroborate those of Doytch and Uctum (2016), who found evidence to support the pollution haven hypothesis.

Regarding the composite democracy variable, the results indicate that democracy directly affects environmental quality. The coefficient of democracy is statistically significant and positive in Models 2–4 (Table 3). A 1% increase in democracy is associated with about a 0.61% increase in emissions, ceteris paribus. These results align with Acheampong et al. (2022), who report that individual democracy variables spur emissions, and contrast with Selseng et al. (2022), who found no direct effect of democracy on environmental quality. There is likely an absence of trust that hampers building a mature democratic society and hinders the development of democratic practices (Jun & Kim, 2002). Besides, inside lobbying prioritising economic competitiveness over climate change mitigation might exert significant political influence (Vesa et al., 2020).

Focusing on the gap in the literature, that is, on the interaction effects between FDI and the composite democracy variable (Model 4, Table 3), we find that the coefficient of the interaction term is negative and statistically significant (at the 1% level). A 1% increase in the interaction term leads to a reduction of about 0.15% in emissions, ceteris paribus. All things being equal, these findings imply that the various forms of democracy serve as a necessary condition by which FDI contributes to reducing CO_2 emissions. This study provides empirical support for the pollution halo hypothesis that FDI can promote cleaner development of host countries. One interpretation of these results is that enhancements in democracy encourage FDI in the adoption and diffusion of cleaner production technologies (Wang et al., 2021).

MNEs internalise the negative environmental impacts linked to industrial production by facilitating the transfer of technology and the dissemination of technical knowledge from their foreign operations (Narula & Driffield, 2012). Democratic countries generally exhibit superior environmental performance compared to

Table 1 List of countries

Afghanistan	Georgia	North Korea
Albania	Germany	North Macedonia
Algeria	Ghana	Norway
Angola	Greece	Oman
Argentina	Guatemala	Pakistan
Armenia	Guinea	Panama
Australia	Guinea-Bissau	Paraguay
Austria	Honduras	Peru
Azerbaijan	Hong Kong	Philippines
Bahrain	Hungary	Poland
Bangladesh	Iceland	Portugal
Belarus	India	Qatar
Belgium	Indonesia	Romania
Benin	Iran	Russia
Bhutan	Iraq	Rwanda
Bolivia	Ireland	Sao Tome and Principe
Bosnia and Herzegovina	Israel	Saudi Arabia
Botswana	Italy	Senegal
Brazil	Jamaica	Serbia
Bulgaria	Japan	Sierra Leone
Burkina Faso	Jordan	Singapore
Burundi	Kazakhstan	Slovakia
Cambodia	Kenya	Slovenia
Cameroon	Kuwait	Somalia
Canada	Kyrgyzstan	South Africa
Cape Verde	Laos	South Korea
Central African Republic	Latvia	Spain
Chad	Lebanon	Sri Lanka
Chile	Lesotho	Sudan
China	Liberia	Sweden
Colombia	Libya	Switzerland
Comoros	Lithuania	Syria
Congo	Luxembourg	Tajikistan
Costa Rica	Madagascar	Tanzania
Côte d'Ivoire	Malawi	Thailand
Croatia	Malaysia	Togo
Cyprus	Mali	Trinidad and Tobago
Czech Republic	Malta	Tunisia
Democratic Republic of Congo	Mauritania	Turkey
Denmark	Mauritius	Turkmenistan
Dominican Republic	Mexico	Uganda
Ecuador	Moldova	Ukraine
Egypt	Mongolia	United Arab Emirates
El Salvador	Montenegro	United Kingdom
Equatorial Guinea	Morocco	United States

(continued)

Table 1 (continued)

Eritrea	Mozambique	Uruguay
Estonia	Myanmar	Uzbekistan
Eswatini	Namibia	Venezuela
Ethiopia	Nepal	Vietnam
Fiji	Netherlands	Yemen
Finland	New Zealand	Zambia
France	Nicaragua	Zimbabwe
Gabon	Niger	
Gambia	Nigeria	

Source: Authors' elaboration
Note: Author's own elaboration

Table 2 Description of variables

Variable	Definition	Mean	Std. Dev.	Min.	Max.	Source
Environment	CO_2 emissions in tonnes of CO_2 using a conversion factor of 3.664	178.3	746.2	0.04	11472.3	GCP
Population	Total population in million persons	40.8	141.7	0.12	1425.8	WDI
Affluence	Per capita GDP in PPP constant 2017 international US dollars	1758.8	1954.5	43.6	12067.7	WDI
Technology	Renewable energy consumption in percentage of total	39.5	31.0	1	98.3	WDI
FDI	Stock of inward FDI in percentage of GDP	51.3	142.4	1	2062.0	UNCTAD
Trade openness	Sum of exports and imports as a percentage of GDP	81.8	52.3	1.9	442.6	WDI
Industrialisation	Industry value added as a percentage of GDP	28.1	11.3	2.7	86.6	WDI
Financial development	Credit to the private sector as a percentage of GDP	47.9	45.6	1	304.5	WDI
Electrification	Percentage of population with access to electricity	78.3	31.1	1	100	WDI
Corruption	Control of corruption estimate recoded	2.1	0.9	0	4.3	WGI
Civil liberties	Civil liberties index on a scale from 1 to 7	3.6	1.8	1	7	Freedom House
Democracy	A composite (PCA) index of the five democracy variables	0.4	0.2	0	1	V-Dem

Source: Authors' elaboration
Note: Global Carbon Project (GCP); World Development Indicators (WDI); United Nations Conference on Trade and Development (UNCTAD); World Governance Indicators (WGI); Varieties of Democracy Project (V-Dem)

Table 3 Direct and interaction effects of democracy and FDI on environmental quality

	Model 1	Model 2	Model 3	Model 4
Lagged dependent variable	0.868***	0.880***	0.898***	0.812***
	(0.031)	(0.030)	(0.029)	(0.032)
Population	0.148***	0.145***	0.121***	0.206***
	(0.032)	(0.029)	(0.030)	(0.033)
Affluence	0.075**	0.076***	0.057**	0.133***
	(0.030)	(0.029)	(0.028)	(0.033)
Technology	−0.020**	−0.021**	−0.017**	−0.031***
	(0.009)	(0.009)	(0.008)	(0.010)
FDI	0.017*		0.016*	0.099***
	(0.009)		(0.009)	(0.017)
Trade openness	−0.045*	−0.003	−0.032	0.018
	(0.025)	(0.023)	(0.023)	(0.027)
Industrialisation	0.111***	0.075**	0.095***	0.081**
	(0.033)	(0.032)	(0.030)	(0.034)
Financial development	0.024**	0.025***	0.016*	0.010
	(0.010)	(0.009)	(0.009)	(0.008)
Electrification	0.054**	0.042*	0.031	0.061**
	(0.022)	(0.022)	(0.020)	(0.025)
Corruption	−0.001	−0.011	−0.011	0.003
	(0.018)	(0.017)	(0.017)	(0.018)
Civil liberties	−0.024***	0.008	−0.003	−0.010
	(0.009)	(0.010)	(0.009)	(0.009)
Democracy		0.152***	0.128***	0.610***
		(0.036)	(0.037)	(0.096)
FDI x democracy				−0.154***
				(0.031)
Intercept	−3.077***	−3.208***	−2.557***	−4.894***
	(0.668)	(0.634)	(0.624)	(0.760)
Number of observations	2563	2594	2563	2557
Number of instruments	78	78	83	84
Year fixed effects	yes	yes	yes	yes
Arellano-Bond test for AR(1) (p-value)	0.000	0.000	0.000	0.000
Arellano-Bond test for AR(2) (p-value)	0.783	0.940	0.781	0.722
Hansen test of overidentification (p-value)	0.324	0.446	0.274	0.469

Source: Authors' elaboration

Note: All models are estimated using the Blundell and Bond (1998) dynamic panel system GMM estimations. The standard errors are reported in parentheses. *, **, and *** indicate significance at 10%, 5%, and 1% levels, respectively. Time dummies are included in the model specification, but the results are not reported to save space. The instruments used in the first-differenced equation are two or more lags of the levels of the endogenous variables

Table 4 Marginal effects of FDI on environmental quality

	Coefficient	Standard error	P-value
Mean	0.029**	0.009	0.002
Percentiles			
1%	0.095***	0.016	0.000
5%	0.091***	0.015	0.000
10%	0.086***	0.014	0.000
25%	0.071***	0.012	0.000
50%	0.367***	0.009	0.000
75%	−0.012	0.012	0.321
90%	−0.039**	0.016	0.019
95%	−0.042**	0.017	0.014
99%	−0.052**	0.018	0.006

Source: Authors' elaboration
Note: * $p < 0.10$, ** $p < 0.05$, *** $p < 0.01$

non-democratic ones, primarily due to their inclination to invest more in environmental technologies and climate governance (Fredriksson & Wollscheid, 2007). Democracies have more resources to develop promotional strategies to attract the so-called "green FDI" that comes into their countries (Narula & Dunning, 2000). In the presence of FDI-attracting factors such as technology, countries benefit from technology absorption and the green innovation capacity of firms (Kayalvizhi & Thenmozhi, 2018; Luo et al., 2020).

Table 4 presents the marginal effect of FDI on environmental quality assessed at both the mean and various percentiles of the composite democracy variable. Evaluated at the mean value of democracy, the estimated coefficient is positive and statistically significant (at the 5% level). Positive and statistically significant marginal effects of FDI on emissions are found for levels of democracy up to the 50th percentile. In contrast, the upper percentiles of democracy show adverse and significant marginal effects. This implies that FDI is detrimental to CO_2 emissions in countries with average levels of democracy. The empirical results suggest that FDI improves the quality of the environment when democracy improves. Otherwise, emissions increase when the level of democracy worsens. Hence, environmental quality declines when the level of democracy decreases.

In the subsequent analyses, we divide the sample into distinct groups to investigate if the results vary and to account for heterogeneity (Table 5). Across the subsamples, these examinations enable us to ascertain whether differences in environmental quality exist based on whether a country either receives lower or higher amounts of FDI or if a country can be categorised as either a "weak democracy" or a "strong democracy". The results show that in the sub-sample of "strong democracy", FDI and democracy have both positive and statistically significant coefficients (at the 5% level). The direct effects of FDI and democracy are also more pronounced in the subsample of countries with "high FDI" (at the 5% level).

Table 5 Direct and moderating effects of democracy and FDI on environmental quality

	Weak democracy	Strong democracy	Low FDI	High FDI
Lagged dependent variable	0.722***	0.884***	0.850***	0.366***
	(0.039)	(0.030)	(0.043)	(0.066)
Population	0.318***	0.124***	0.156***	0.327***
	(0.049)	(0.033)	(0.041)	(0.118)
Affluence	0.254***	0.056**	0.104**	0.282***
	(0.042)	(0.026)	(0.049)	(0.100)
Technology	−0.066***	−0.007	−0.052***	−0.119***
	(0.015)	(0.010)	(0.012)	(0.040)
FDI	0.006	0.108**	0.092***	0.219**
	(0.024)	(0.051)	(0.021)	(0.085)
Trade openness	0.161***	−0.018	0.070*	−0.120
	(0.039)	(0.019)	(0.037)	(0.088)
Industrialisation	−0.085	0.134***	0.039	0.320***
	(0.054)	(0.033)	(0.028)	(0.097)
Financial development	−0.019**	0.014	−0.003	0.277***
	(0.009)	(0.013)	(0.008)	(0.094)
Electrification	0.102***	0.048	0.041	0.283
	(0.036)	(0.037)	(0.028)	(0.217)
Corruption	−0.047	0.012	0.010	−0.125
	(0.040)	(0.015)	(0.024)	(0.090)
Civil liberties	−0.009	−0.006	−0.018	−0.147***
	(0.016)	(0.013)	(0.012)	(0.039)
Democracy	−0.028	0.421*	0.340***	1.194*
	(0.274)	(0.248)	(0.130)	(0.628)
FDI x Democracy	0.043	−0.113*	−0.135**	−0.340***
	(0.093)	(0.064)	(0.055)	(0.119)
Intercept	−7.175***	−3.157***	−3.648***	−8.624***
	(0.957)	(0.742)	(1.018)	(2.456)
Number of observations	1335	1175	1841	617
Number of instruments	72	72	72	66
Year fixed effects	yes	yes	yes	yes
Arellano-Bond test for AR(1) (p-value)	0.000	0.000	0.000	0.044
Arellano-Bond test for AR(2) (p-value)	0.929	0.180	0.841	0.817
Hansen test of overidentification (p-value)	0.603	0.659	0.100	0.353

Source: Authors' elaboration

Note: All models are estimated using the Blundell and Bond (1998) dynamic panel system GMM estimations. The standard errors are reported in parentheses. *, **, and *** indicate significance at 10%, 5%, and 1% levels, respectively. Time dummies are included in the model specification, but the results are not reported to save space. The instruments used in the first-differenced equation are two or more lags of the levels of the endogenous variables

Table 6 Marginal effects of FDI on environmental quality

	Coefficient	Standard error	P-value
Weak democracy			
Mean	0.016**	0.013	0.225
Percentiles			
1%	0.007	0.023	0.755
5%	0.007	0.022	0.728
10%	0.008	0.020	0.668
25%	0.011	0.016	0.490
50%	0.015	0.013	0.243
75%	0.021	0.017	0.233
90%	0.025	0.024	0.300
95%	0.026	0.026	0.319
99%	0.027	0.028	0.340
Strong democracy			
Mean	0.021**	0.008	0.012
Percentiles			
1%	0.050**	0.019	0.010
5%	0.048**	0.018	0.009
10%	0.044**	0.016	0.007
25%	0.035**	0.011	0.004
50%	0.016**	0.008	0.048
75%	0.016***	0.008	0.048
90%	0.003	0.012	0.766
95%	0.001	0.013	0.998
99%	−0.003	0.015	0.825
Low FDI			
Mean	0.033**	0.012	0.011
Percentiles			
1%	0.089***	0.020	0.000
5%	0.086***	0.018	0.000
10%	0.082***	0.017	0.000
25%	0.068***	0.013	0.000
50%	0.039**	0.011	0.001
75%	−0.001	0.023	0.966
90%	−0.027	0.033	0.409
95%	−0.031	0.034	0.363
99%	−0.040	0.038	0.295
High FDI			
Mean	0.051	0.043	0.237
Percentiles			
1%	0.210**	0.082	0.013
5%	0.200**	0.079	0.014
10%	0.188**	0.076	0.016

(continued)

Table 6 (continued)

	Coefficient	Standard error	P-value
25%	0.154**	0.066	0.023
50%	0.057	0.044	0.196
75%	−0.055	0.045	0.223
90%	−0.087*	0.051	0.094
95%	−0.093*	0.052	0.081
99%	−0.116*	0.058	0.051

Source: Authors' elaboration
Note: * $p < 0.10$, ** $p < 0.05$, *** $p < 0.01$

Regarding the interaction term, the coefficients are always negative and statistically significant, except in the subsample of "weak democracy". The strongest interaction effect is registered in the "high FDI" sub-sample. Looking at the mean and various percentiles of the composite democracy variable confirms the insignificant result for "weak democracy" (Table 6). Generally, there are positive and statistically significant coefficients for margins of democracy at lower percentiles. The most apparent finding emerging from the analysis is that the higher percentile values of democracy indicate that FDI improves environmental quality.

The empirical results reveal that FDI does not contribute to reducing emissions when the level of democracy is low. The mitigation effect of FDI on emissions manifests itself in countries with higher levels of democracy. These results suggest that democratic practices can foster transformations towards sustainability (Dryzek & Pickering, 2017; Pickering et al., 2022). In addition, environmental preferences are more relevant for democratic countries that are paying more attention to improving environmental quality (Bättig & Bernauer, 2009). Consolidated democracies can reduce political risk and attract more foreign capital (Lacroix et al., 2021). At the same time, democracy can advance technological changes and generate scope for economic development (Acemoglu et al., 2018). The behaviour of stakeholders and the potential for regulatory challenges are expected to motivate corporations to adopt practices that align with the objectives of broader social movements (Reid & Toffel, 2009).

Most of the remaining variables are statistically significant and show the expected signs (Tables 3, 4, 5 and 6). Aligning with Bernauer and Koubi (2009), our estimates indicate that income significantly affects emissions. Following Dietz et al. (2007), an increase in population size generates higher emissions levels, worsening environmental quality. An improvement in technology leads to an improvement in environmental quality. These results affirm Lindvall and Karlsson (2023), who argue that investments in renewable energy can enhance the capacity of democracies to decrease emissions. Ghosh et al. (2023) argue that using renewable energy combined with democracy and economic growth supports a cleaner environment. The results for the corruption variable are incongruent with Abid (2016), who claims that corruption adversely affects emissions. While civil liberties do not affect emissions, this result contrasts with Guo et al. (2022), who argue that civil liberties enable the media, social movements, and stakeholders to expect companies to be more transparent and disclose relevant corporate environmental information.

5 Conclusion

This chapter investigates the environmental impacts of FDI and democracy through the lens of the Regression on Population, Affluence, and Technology (STRIPAT) model. Employing instrumental variables within the generalised method of moments estimation framework and analysing cross-sectional data from 160 countries spanning 1980 to 2021, the study reveals that FDI significantly enhances environmental quality when accompanied by higher levels of democracy. The findings further highlight democracy's role as a moderating factor in the FDI-environment relationship, reducing environmental emissions and substantiating the pollution halo hypothesis. The results suggest that increased FDI inflows and democratic governance lead to a notable decline in environmental emissions, thereby improving environmental quality in host countries. This effect is particularly pronounced in nations with established democratic systems, which are better positioned to attract FDI. Conversely, in countries with lower levels of democracy, the likelihood of reducing emissions and improving environmental conditions remains limited. The study concludes that fostering democratic institutions, alongside increasing FDI inflows, offers a promising pathway for enhancing environmental quality globally.

With that in mind, there is evidence that FDI and democracy are necessary to achieve the Sustainable Development Goals (SDGs). The findings serve as a reference for policy formulations at the national level to assist governments in reducing environmental emissions. An essential message of this chapter is that incremental improvements in FDI and democracy operating for mutual benefit can effectively augment environmental quality. Therefore, we advocate that countries improve their democratic institutions to protect the environment. At the same time, it is crucial to strengthen democratic governance among political leaders and citizens to produce a more equal society and a better rule of law, enhance political accountability, and facilitate people's participation in the governance process.

The existence of solid democratic institutions creates the need for better environmental standards and reinforces climate action. To this end, leaders and practitioners should promote civic education and raise public awareness to establish strong institutions and reduce environmental emissions. The message to policymakers is that the varieties of democracy seem compatible with climate goals, and governments do not have to become more authoritarian; instead, they have to create incentives to develop democracy to tackle climate change. Moreover, governments should pursue a more attractive investment climate and improve their countries' locations for FDI. For instance, strengthening foreign investment promotion regimes for FDI can go hand in hand with boosting national innovation capacity. For this reason, governments should not hesitate to give away generous tax breaks and guarantee state protection of foreign property. Overall, political leaders should aim to increase technology spillovers through foreign capital. Therefore, MNEs need more environmentally oriented policies to support green innovation and clean technology transfer. In this context, it is crucial that policymakers put in place mechanisms, such as quality indicators, to mobilise FDI that maximises inclusive and sustainable growth and supports climate challenges.

Scholars can extend this line of inquiry by decomposing the democracy varieties into their components to capture different features of democracy in the FDI-environment nexus. This study is limited by the availability of firm-level data that could shed some light on the firm's motivations and interactions with foreign subsidiaries, which are known to affect international technology transfer, diffusion, and emission behaviours within and across countries. Besides these limitations, our research, based on cross-sectional analysis, informs the debate on whether democracy promotes FDI to meet environmental challenges.

References

Abdouli, M., & Hammami, S. (2017). Investigating the causality links between environmental quality, foreign direct investment and economic growth in MENA countries. *International Business Review, 26*(2), 264–278. https://doi.org/10.1016/j.ibusrev.2016.07.004

Abdouli, M., & Omri, A. (2021). Exploring the nexus among FDI inflows, environmental quality, human capital, and economic growth in the Mediterranean region. *Journal of the Knowledge Economy, 12*, 788–810. https://doi.org/10.1007/s13132-020-00641-5

Abid, M. (2016). Impact of economic, financial, and institutional factors on CO2 emissions: Evidence from Sub-Saharan Africa economies. *Utilities Policy, 41*, 85–94. https://doi.org/10.1016/j.jup.2016.06.009

Acemoglu, D., Naidu, S., Restrepo, P., & Robinson, J. (2018). Democracy does cause growth. *Journal of Political Economy, 127*(1), 47–100. https://doi.org/10.1086/700936

Acheampong, A., & Opoku, E. (2023). Energy justice, democracy and deforestation. *Journal of Environmental Management, 341*, 118012. https://doi.org/10.1016/j.jenvman.2023.118012

Acheampong, A., Opoku, E., & Dzator, J. (2022). Does democracy really improve environmental quality? Empirical contribution to the environmental politics debate. *Energy Economics, 109*, 105942. https://doi.org/10.1016/j.eneco.2022.105942

Adams, S., & Acheampong, A. (2019). Reducing carbon emissions: The role of renewable energy and democracy. *Journal of Cleaner Production, 240*, 118245. https://doi.org/10.1016/j.jclepro.2019.118245

Adams, S., & Klobodu, E. (2018). Financial development and environmental degradation: Does political regime matter? *Journal of Cleaner Production, 197*, 1472–1479. https://doi.org/10.1016/j.jclepro.2018.06.252

Albornoz, F., Cole, M., Elliott, R., & Ercolani, M. (2009). In search of environmental spillovers. *The World Economy, 32*(1), 136–163. https://doi.org/10.1111/j.1467-9701.2009.01160.x

Albornoz, F., Cole, M. A., Elliott, R., & Ercolani, M. (2014). The environmental actions of firms: Examining the role of spillovers, networks and absorptive Capacity. *Journal of Environmental Management, 146*, 150–163. https://doi.org/10.1016/j.jenvman.2014.07.011

Alfaro, L., Chanda, A., Kalemli-Ozcan, S., & Sayek, S. (2010). Does foreign direct investment promote growth? Exploring the role of financial markets on linkages. *Journal of Development Economics, 91*(2), 242–256. https://doi.org/10.1016/j.jdeveco.2009.09.004

Antonietti, R., & Mondolo, J. (2023). Inward FDI and the quality of domestic institutions: A cross-country panel VAR analysis. *Economic Systems, 47*(3), 101078. https://doi.org/10.1016/j.ecosys.2023.101078

Arellano, M., & Bond, S. (1991). Some tests of specification for panel data: Monte Carlo evidence and an application to employment equations. *The Review of Economic Studies, 58*(2), 277–297. https://doi.org/10.2307/2297968

Arvin, M., & Lew, B. (2009). Does democracy affect environmental quality in developing countries? *Applied Economics, 43*(9), 1151–1160. https://doi.org/10.1080/00036840802600277

Asiedu, E., & Lien, D. (2011). Democracy, foreign direct investment and natural resources. *Journal of International Economics, 84*(1), 99–111. https://doi.org/10.1016/j.jinteco.2010.12.001

Aziz, G., Sarwar, S., Hussan, M., & Saeed, A. (2023). The importance of extended-STIRPAT in responding to the environmental footprint: Inclusion of environmental technologies and environmental taxation. *Energy Strategy Reviews, 50,* 101216. https://doi.org/10.1016/j.esr.2023.101216

Bättig, M., & Bernauer, T. (2009). National institutions and global public goods: Are democracies more cooperative in climate change policy? *International Organization, 63*(2), 281–308. https://doi.org/10.1017/S0020818309090092

Baum, C., Schaffer, M., & Stillman, S. (2003). Instrumental variables and GMM: Estimation and testing. *The Stata Journal, 3*(1), 1–31. https://doi.org/10.1177/1536867X0300300101

Baum, C., Schaffer, M., & Stillman, S. (2007). Enhanced routines for instrumental variables/generalized method of moments estimation and testing. *The Stata Journal, 7*(4), 465–506. https://doi.org/10.1177/1536867X0800700402

Ben Kheder, S., & Zugravu, N. (2012). Environmental regulation and French firms location abroad: An economic geography model in an international comparative study. *Ecological Economics, 77,* 48–61. https://doi.org/10.1016/j.ecolecon.2011.10.005

Bénassy-Quéré, A., Coupet, M., & Mayer, T. (2007). Institutional determinants of foreign direct investment. *The World Economy, 30*(5), 764–782. https://doi.org/10.1111/j.1467-9701.2007.01022.x

Bento, J., & Torres, M. (2024). Towards environmental impact of inward foreign direct investment: The moderating role of varieties of democracy. *Multinational Business Review, 32*(4), 500–539. https://doi.org/10.1108/MBR-01-2024-0008

Benzerrouk, Z., Abid, M., & Sekrafi, H. (2021). Pollution haven or halo effect? A comparative analysis of developing and developed countries. *Energy Reports, 7,* 4862–4871. https://doi.org/10.1016/j.egyr.2021.07.076

Bernauer, T., & Betzold, C. (2012). Civil society in global environmental governance. *The Journal of Environment & Development, 21*(1), 62–66. https://doi.org/10.1177/1070496511435551

Bernauer, T., & Koubi, V. (2009). Effects of political institutions on air quality. *Ecological Economics, 68*(5), 1355–1365. https://doi.org/10.1016/j.ecolecon.2008.09.003

Bernauer, T., Gampfer, R., Meng, T., & Su, Y.-S. (2016). Could more civil society involvement increase public support for climate policy-making? Evidence from a survey experiment in China. *Global Environmental Change, 40,* 1–12. https://doi.org/10.1016/j.gloenvcha.2016.06.001

Blanco, L., Gonzalez, F., & Ruiz, I. (2013). The impact of FDI on CO2 emissions in Latin America. *Oxford Development Studies, 41*(1), 104–121. https://doi.org/10.1080/13600818.2012.732055

Blundell, R., & Bond, S. (1998). Initial conditions and moment restrictions in dynamic panel data models. *Journal of Econometrics, 87*(1), 115–143. https://doi.org/10.1016/S0304-4076(98)00009-8

Boese, V. (2019). How (not) to measure democracy. *International Area Studies Review, 22*(2), 95–127. https://doi.org/10.1177/2233865918815571

Bu, M., Xu, L., & Tang, R. (2023). MNEs' transfer of socially irresponsible practices: A replication with new extensions. *Journal of World Business, 58*(2), 101384. https://doi.org/10.1016/j.jwb.2022.101384

Büchs, M., & Schnepf, S. (2013). Who emits most? Associations between socio-economic factors and UK households' home energy, transport, indirect and total CO2 emissions. *Ecological Economics, 90,* 114–123. https://doi.org/10.1016/j.ecolecon.2013.03.007

Buckley, P., Doh, J., & Benischke, M. H. (2017). Towards a renaissance in international business research? Big questions, grand challenges, and the future of IB scholarship. *Journal of International Business Studies, 48*(9), 1045–1064. https://doi.org/10.1057/s41267-017-0102-z

Buckley, P., Strange, R., Timmer, M., & de Vries, G. (2022). Rent appropriation in global value chains: The past, present, and future of intangible assets. *Global Strategy Journal, 12*(4), 679–696. https://doi.org/10.1002/gsj.1438

Cantwell, J., & Narula, R. (2001). The eclectic paradigm in the global economy. *International Journal of the Economics of Business, 8*(2), 155–172. https://doi.org/10.1080/13571510110051504

Carlitz, R., & Povitkina, M. (2021). Local interest group activity and environmental degradation in authoritarian regimes. *World Development, 142*, 105425. https://doi.org/10.1016/j.worlddev.2021.105425

Chung, S. (2014). Environmental regulation and foreign direct investment: Evidence from South Korea. *Journal of Development Economics, 108*, 222–236. https://doi.org/10.1016/j.jdeveco.2014.01.003

Clarke, K., & Stone, R. (2008). Democracy and the logic of political survival. *The American Political Science Review, 102*(3), 387–392.

Clegg, L., Voss, H., & Tardios, J. (2018). The autocratic advantage: internationalization of state-owned multinationals. *Journal of World Business, 53*(5), 668–681. https://doi.org/10.1016/j.jwb.2018.03.009

Clulow, Z. (2019). Democracy, electoral systems and emissions: explaining when and why democratization promotes mitigation. *Climate Policy, 19*(2), 244–257. https://doi.org/10.1080/14693062.2018.1497938

Cole, M. (2007). Corruption, income and the environment: An empirical analysis. *Ecological Economics, 62*(3), 637–647. https://doi.org/10.1016/j.ecolecon.2006.08.003

Cole, M., Elliott, R., & Fredriksson, P. (2006). Endogenous pollution havens: Does FDI influence environmental regulations? *The Scandinavian Journal of Economics, 108*(1), 157–178. https://doi.org/10.1111/j.1467-9442.2006.00439.x

Cole, M., Elliott, R., & Strobl, E. (2008). The environmental performance of firms: The role of foreign ownership, training, and experience. *Ecological Economics, 65*(3), 538–546. https://doi.org/10.1016/j.ecolecon.2007.07.025

Cole, M., Elliott, R., & Zhang, L. (2017). Foreign direct investment and the environment. *Annual Review of Environment and Resources, 42*(1), 465–487. https://doi.org/10.1146/annurev-environ-102016-060916

Copeland, B., & Taylor, M. (2004). Trade, growth, and the environment. *Journal of Economic Literature, 42*(1), 7–71. http://www.jstor.org/stable/3217036

Coppedge, M., Gerring, J., Altman, D., Bernhard, M., Fish, S., Hicken, A., Kroenig, M., Lindberg, S., McMann, K., Paxton, P., Semetko, H. A., Skaaning, S., Staton, J., & Teorell, J. (2011). Conceptualizing and measuring democracy: A new approach. *Perspectives on Politics, 9*(2), 247–267. https://doi.org/10.1017/S1537592711000880

Coppedge, M., Lindberg, S., Skaaning, S. E., & Teorell, J. (2016). Measuring high level democratic principles using the V-Dem data. *International Political Science Review, 37*(5), 580–593. https://doi.org/10.1177/0192512115622046

Coppedge, M., Gerring, J., Glynn, A., Knutsen, C. H., Lindberg, S., Pemstein, D., Seim, B., Skaaning, S.-E., & Teorell, J. (2020). Dimensions and components of democracy. In A. Glynn, B. Seim, C. H. Knutsen, D. Pemstein, J. Teorell, J. Gerring, M. Coppedge, S. Lindberg, & S.-E. Skaaning (Eds.), *Varieties of democracy: Measuring two centuries of political change* (pp. 90–129). Cambridge University Press. https://doi.org/10.1017/9781108347860.005

Culhane, T. G., & Roberts, J. (2021). Who delays climate action? Interest groups and coalitions in state legislative struggles in the United States. *Energy Research & Social Science, 79*, 102114. https://doi.org/10.1016/j.erss.2021.102114

Demena, B., & Afesorgbor, S. (2020). The effect of FDI on environmental emissions: Evidence from a meta-analysis. *Energy Policy, 138*, 111192. https://doi.org/10.1016/j.enpol.2019.111192

Dietz, T., & Rosa, E. (1994). Rethinking the environmental impacts of population, affluence and technology. *Human Ecology Review, 1*(2), 277–300. http://www.jstor.org/stable/24706840

Dietz, T., Rosa, E., & York, R. (2007). Driving the human ecological footprint. *Frontiers in Ecology and the Environment, 5*(1), 13–18. https://doi.org/10.1890/1540-9295(2007)5[13:DTHEF]2.0.CO;2

Doytch, N., & Uctum, M. (2016). Globalization and the environmental impact of sectoral FDI. *Economic Systems, 40*(4), 582–594. https://doi.org/10.1016/j.ecosys.2016.02.005

Dryzek, J., & Pickering, J. (2017). Deliberation as a catalyst for reflexive environmental governance. *Ecological Economics, 131*, 353–360. https://doi.org/10.1016/j.ecolecon.2016.09.011

Dunning, J. (1980). Toward an eclectic theory of international production: some empirical tests. *Journal of International Business Studies, 11*(1), 9–31. https://doi.org/10.1057/palgrave.jibs.8490593

Ehrlich, P., & Holdren, J. (1971). Impact of population growth. *Science, 171*(3977), 1212–1217.

Eiadat, Y., Kelly, A., Roche, F., & Eyadat, H. (2008). Green and competitive? An empirical test of the mediating role of environmental innovation strategy. *Journal of World Business, 43*(2), 131–145. https://doi.org/10.1016/j.jwb.2007.11.012

Eskeland, G., & Harrison, A. (2003). Moving to greener pastures? Multinationals and the pollution haven hypothesis. *Journal of Development Economics, 70*(1), 1–23. https://doi.org/10.1016/S0304-3878(02)00084-6

Farzanegan, M., & Markwardt, G. (2018). Development and pollution in the Middle East and North Africa: Democracy matters. *Journal of Policy Modeling, 40*(2), 350–374. https://doi.org/10.1016/j.jpolmod.2018.01.010

Farzin, Y., & Bond, C. (2006). Democracy and environmental quality. *Journal of Development Economics, 81*(1), 213–235. https://doi.org/10.1016/j.jdeveco.2005.04.003

Fredriksson, P., & Neumayer, E. (2013). Democracy and climate change policies: Is history important? *Ecological Economics, 95*, 11–19. https://doi.org/10.1016/j.ecolecon.2013.08.002

Fredriksson, P., & Neumayer, E. (2016). Corruption and climate change policies: Do the bad old days matter? *Environmental and Resource Economics, 63*(2), 451–469. https://doi.org/10.1007/s10640-014-9869-6

Fredriksson, P., & Wollscheid, J. (2007). Democratic institutions versus autocratic regimes: The case of environmental policy. *Public Choice, 130*(3), 381–393. https://doi.org/10.1007/s11127-006-9093-1

Fredriksson, P., Neumayer, E., Damania, R., & Gates, S. (2005). Environmentalism, democracy, and pollution control. *Journal of Environmental Economics and Management, 49*(2), 343–365. https://doi.org/10.1016/j.jeem.2004.04.004

Ghosh, S., Hossain, M., Voumik, L., Raihan, A., Ridzuan, A., & Esquivias, M. (2023). Unveiling the spillover effects of democracy and renewable energy consumption on the environmental quality of BRICS countries: A new insight from different quantile regression approaches. *Renewable Energy Focus, 46*, 222–235. https://doi.org/10.1016/j.ref.2023.06.004

Globerman, S., & Shapiro, D. (2002). Global foreign direct investment flows: The role of governance infrastructure. *World Development, 30*(11), 1899–1919. https://doi.org/10.1016/S0305-750X(02)00110-9

Grauel, J., & Gotthardt, D. (2017). Carbon disclosure, freedom and democracy. *Social Responsibility Journal, 13*(3), 428–456. https://doi.org/10.1108/SRJ-08-2016-0151

Guo, J., Islam, M. A., Jain, A., & van Staden, C. J. (2022). Civil liberties and social and environmental information transparency: A global investigation of financial institutions. *The British Accounting Review, 54*(1), 101018. https://doi.org/10.1016/j.bar.2021.101018

Hansen, L. (1982). Large sample properties of generalized method of moments estimators. *Econometrica, 50*(4), 1029–1054. https://doi.org/10.2307/1912775

Hargrove, A., Qandeel, M., & Sommer, J. (2019). Global governance for climate justice: A cross-national analysis of CO2 emissions. *Global Transitions, 1*, 190–199. https://doi.org/10.1016/j.glt.2019.11.001

Hollyer, J., Rosendorff, P., & Vreeland, J. (2011). Democracy and transparency. *The Journal of Politics, 73*(4), 1191–1205.

Jahanger, A., Usman, M., & Balsalobre-Lorente, D. (2022). Autocracy, democracy, globalization, and environmental pollution in developing world: Fresh evidence from STIRPAT model. *Journal of Public Affairs, 22*(4), e2753. https://doi.org/10.1002/pa.2753

Jorgenson, A., Schor, J., & Huang, X. (2017). Income inequality and carbon emissions in the United States: A state-level analysis, 1997–2012. *Ecological Economics, 134*, 40–48. https://doi.org/10.1016/j.ecolecon.2016.12.016

Joshi, P., & Beck, K. (2018). Democracy and carbon dioxide emissions: Assessing the interactions of political and economic freedom and the environmental Kuznets curve. *Energy Research & Social Science, 39*, 46–54. https://doi.org/10.1016/j.erss.2017.10.020

Jun, J., & Kim, T. (2002). Distrust as a hindrance to democratic governance in South Korea. *Asian Journal of Political Science, 10*(1), 1–16. https://doi.org/10.1080/02185370208434199

Kaika, D., & Zervas, E. (2013). The environmental Kuznets curve (EKC) theory. Part B: Critical issues. *Energy Policy, 62*, 1403–1411. https://doi.org/10.1016/j.enpol.2013.07.130

Kammerlander, A., & Schulze, G. (2020). Are democracies cleaner? *European Journal of Political Economy, 64*, 101920. https://doi.org/10.1016/j.ejpoleco.2020.101920

Kashwan, P. (2017). Inequality, democracy, and the environment: A cross-national analysis. *Ecological Economics, 131*, 139–151. https://doi.org/10.1016/j.ecolecon.2016.08.018

Kayalvizhi, P., & Thenmozhi, M. (2018). Does quality of innovation, culture and governance drive FDI? Evidence from emerging markets. *Emerging Markets Review, 34*, 175–191. https://doi.org/10.1016/j.ememar.2017.11.007

Kearsley, A., & Riddel, M. (2010). A further inquiry into the pollution haven hypothesis and the environmental Kuznets curve. *Ecological Economics, 69*(4), 905–919. https://doi.org/10.1016/j.ecolecon.2009.11.014

Kim, M., & Adilov, N. (2012). The lesser of two evils: An empirical investigation of foreign direct investment-pollution tradeoff. *Applied Economics, 44*(20), 2597–2606. https://doi.org/10.1080/00036846.2011.566187

Kim, D.-H., Suen, Y.-B., & Lin, S.-C. (2019). Carbon dioxide emissions and trade: Evidence from disaggregate trade data. *Energy Economics, 78*, 13–28. https://doi.org/10.1016/j.eneco.2018.08.019

Kokko, A. (1994). Technology, market characteristics, and spillovers. *Journal of Development Economics, 43*(2), 279–293. https://doi.org/10.1016/0304-3878(94)90008-6

Kolk, A., & van Tulder, R. (2010). International business, corporate social responsibility and sustainable development. *International Business Review, 19*(2), 119–125. https://doi.org/10.1016/j.ibusrev.2009.12.003

Lacroix, J., Méon, P.-G., & Sekkat, K. (2021). Democratic transitions can attract foreign direct investment: Effect, trajectories, and the role of political risk. *Journal of Comparative Economics, 49*(2), 340–357. https://doi.org/10.1016/j.jce.2020.09.003

Lægreid, O., & Povitkina, M. (2018). Do political institutions moderate the GDP-CO2 relationship? *Ecological Economics, 145*, 441–450. https://doi.org/10.1016/j.ecolecon.2017.11.014

Li, Q., & Resnick, A. (2003). Reversal of fortunes: democratic institutions and foreign direct investment inflows to developing countries. *International Organization, 57*(1), 175–211. https://doi.org/10.1017/S0020818303571077

Li, Q., & Reuveny, R. (2006). Democracy and environmental degradation. *International Studies Quarterly, 50*(4), 935–956. http://www.jstor.org/stable/4092786

Liddle, B. (2015). What are the carbon emissions elasticities for income and population? Bridging STIRPAT and EKC via robust heterogeneous panel estimates. *Global Environmental Change, 31*, 62–73. https://doi.org/10.1016/j.gloenvcha.2014.10.016

Lin, B., & Li, Z. (2020). Is more use of electricity leading to less carbon emission growth? An analysis with a panel threshold model. *Energy Policy, 137*, 111121. https://doi.org/10.1016/j.enpol.2019.111121

Lindvall, D., & Karlsson, M. (2023). Exploring the democracy-climate nexus: A review of correlations between democracy and climate policy performance. *Climate Policy*, 1–17. https://doi.org/10.1080/14693062.2023.2256697

Luo, Y., Lu, Z., & Long, X. (2020). Heterogeneous effects of endogenous and foreign innovation on CO2 emissions stochastic convergence across China. *Energy Economics, 91*, 104893. https://doi.org/10.1016/j.eneco.2020.104893

Mani, M., & Wheeler, D. (1998). In search of pollution havens? Dirty industry in the world economy, 1960 to 1995. *The Journal of Environment & Development, 7*(3), 215–247. https://doi.org/10.1177/107049659800700302

Markusen, J. (1995). The boundaries of multinational enterprises and the theory of international trade. *The Journal of Economic Perspectives, 9*(2), 169–189. http://www.jstor.org/stable/2138172

Menz, T., & Welsch, H. (2012). Population aging and carbon emissions in OECD countries: Accounting for life-cycle and cohort effects. *Energy Economics, 34*(3), 842–849. https://doi.org/10.1016/j.eneco.2011.07.016

Midlarsky, M. (1998). Democracy and the environment: An empirical assessment. *Journal of Peace Research, 35*(3), 341–361. https://doi.org/10.1177/0022343398035003005

Müller-Fürstenberger, G., & Wagner, M. (2007). Exploring the environmental Kuznets hypothesis: Theoretical and econometric problems. *Ecological Economics, 62*(3), 648–660. https://doi.org/10.1016/j.ecolecon.2006.08.005

Narula, R., & Driffield, N. (2012). Does FDI cause development? The ambiguity of the evidence and why it matters. *The European Journal of Development Research, 24*(1), 1–7. https://doi.org/10.1057/ejdr.2011.51

Narula, R., & Dunning, J. (2000). Industrial development, globalization and multinational enterprises: New realities for developing countries. *Oxford Development Studies, 28*(2), 141–167. https://doi.org/10.1080/713688313

Neumayer, E. (2002). Do democracies exhibit stronger international environmental commitment? A cross-country analysis. *Journal of Peace Research, 39*(2), 139–164. http://www.jstor.org/stable/1555296

Neumayer, E. (2003). Are left-wing party strength and corporatism good for the environment? Evidence from panel analysis of air pollution in OECD countries. *Ecological Economics, 45*(2), 203–220. https://doi.org/10.1016/S0921-8009(03)00012-0

Nguyen, D., Huynh, T., & Nasir, M. (2021). Carbon emissions determinants and forecasting: Evidence from G6 countries. *Journal of Environmental Management, 285*, 111988. https://doi.org/10.1016/j.jenvman.2021.111988

North, D., & Weingast, B. (1989). Constitutions and commitment: the evolution of institutions governing public choice in seventeenth-century England. *The Journal of Economic History, 49*(4), 803–832. http://www.jstor.org/stable/2122739

Omri, A., Nguyen, D., & Rault, C. (2014). Causal interactions between CO2 emissions, FDI, and economic growth: Evidence from dynamic simultaneous-equation models. *Economic Modelling, 42*, 382–389. https://doi.org/10.1016/j.econmod.2014.07.026

Parker, S., & Bhatti, M. (2020). Dynamics and drivers of per capita CO2 emissions in Asia. *Energy Economics, 89*, 104798. https://doi.org/10.1016/j.eneco.2020.104798

Paul, J., & Feliciano-Cestero, M. (2021). Five decades of research on foreign direct investment by MNEs: An overview and research agenda. *Journal of Business Research, 124*, 800–812. https://doi.org/10.1016/j.jbusres.2020.04.017

Pazienza, P. (2019). The impact of FDI in the OECD manufacturing sector on CO2 emission: evidence and policy issues. *Environmental Impact Assessment Review, 77*, 60–68. https://doi.org/10.1016/j.eiar.2019.04.002

Pfaffermayr, M. (1999). Ownership advantages, foreign production and productivity: Evidence from Austrian manufacturing firms. *Review of Industrial Organization, 15*(4), 379–396. https://doi.org/10.1023/A:1007779832246

Pickering, J., Hickmann, T., Bäckstrand, K., Kalfagianni, A., Bloomfield, M., Mert, A., Ransan-Cooper, H., & Lo, A. (2022). Democratising sustainability transformations: Assessing the transformative potential of democratic practices in environmental governance. *Earth System Governance, 11*, 100131. https://doi.org/10.1016/j.esg.2021.100131

Pisani, N., Kolk, A., Ocelík, V., & Wu, G. (2019). Does it pay for cities to be green? An investigation of FDI inflows and environmental sustainability. *Journal of International Business Policy, 2*(1), 62–85. https://doi.org/10.1057/s42214-018-00017-2

Policardo, L. (2016). Is democracy good for the environment? Quasi-experimental evidence from regime transitions. *Environmental and Resource Economics, 64*(2), 275–300. https://doi.org/10.1007/s10640-014-9870-0

Porter, M., & Van der Linde, C. (1995). Toward a new conception of the environment-competitiveness relationship. *The Journal of Economic Perspectives, 9*(4), 97–118. http://www.jstor.org/stable/2138392

Povitkina, M. (2018). The limits of democracy in tackling climate change. *Environmental Politics, 27*(3), 411–432. https://doi.org/10.1080/09644016.2018.1444723

Povitkina, M., & Jagers, S. (2022). Environmental commitments in different types of democracies: The role of liberal, social-liberal, and deliberative politics. *Global Environmental Change, 74*, 102523. https://doi.org/10.1016/j.gloenvcha.2022.102523

Quan, L. (2009). Democracy, autocracy, and expropriation of foreign direct investment. *Comparative Political Studies, 42*(8), 1098–1127. https://doi.org/10.1177/0010414009331723

Reid, E., & Toffel, M. (2009). Responding to public and private politics: Corporate disclosure of climate change strategies. *Strategic Management Journal, 30*(11), 1157–1178. https://doi.org/10.1002/smj.796

Selseng, T., Linnerud, K., & Holden, E. (2022). Unpacking democracy: The effects of different democratic qualities on climate change performance over time. *Environmental Science & Policy, 128*, 326–335. https://doi.org/10.1016/j.envsci.2021.12.009

Shakouri, H., Pandey, S., Rahmatian, F., & Paaso, E. (2023). Does the increased electricity consumption (provided by capacity expansion and/or reliability improvement) cause economic growth? *Energy Policy, 182*, 113704. https://doi.org/10.1016/j.enpol.2023.113704

Shuai, C., Shen, L., Jiao, L., Wu, Y., & Tan, Y. (2017). Identifying key impact factors on carbon emission: Evidences from panel and time-series data of 125 countries from 1990 to 2011. *Applied Energy, 187*, 310–325. https://doi.org/10.1016/j.apenergy.2016.11.029

Sigman, R., & Lindberg, S. (2015). *The index of egalitarian democracy and its components: V-Dem's conceptualization and measurement* (V-Dem Working Paper Series 2015, Issue). http://www.ssrn.com/abstract=2727612

Sulemana, I., James, H., & Rikoon, J. (2017). Environmental Kuznets curves for air pollution in African and developed countries: Exploring turning point incomes and the role of democracy. *Journal of Environmental Economics and Policy, 6*(2), 134–152. https://doi.org/10.1080/21606544.2016.1231635

Tao, M., Sheng, M., & Wen, L. (2023). How does financial development influence carbon emission intensity in the OECD countries: Some insights from the information and communication technology perspective. *Journal of Environmental Management, 335*, 117553. https://doi.org/10.1016/j.jenvman.2023.117553

Teorell, J., Coppedge, M., Lindberg, S., & Skaaning, S.-E. (2019). Measuring polyarchy across the globe, 1900–2017. *Studies in Comparative International Development, 54*(1), 71–95. https://doi.org/10.1007/s12116-018-9268-z

Vesa, J., Gronow, A., & Ylä-Anttila, T. (2020). The quiet opposition: How the pro-economy lobby influences climate policy. *Global Environmental Change, 63*, 102117. https://doi.org/10.1016/j.gloenvcha.2020.102117

Wagner, M. (2008). The carbon Kuznets curve: A cloudy picture emitted by bad econometrics? *Resource and Energy Economics, 30*(3), 388–408. https://doi.org/10.1016/j.reseneeco.2007.11.001

Wang, X., Zhang, C., & Zhang, Z. (2019). Pollution haven or porter? The impact of environmental regulation on location choices of pollution-intensive firms in China. *Journal of Environmental Management, 248*, 109248. https://doi.org/10.1016/j.jenvman.2019.07.019

Wang, M., Zhang, X., & Hu, Y. (2021). The green spillover effect of the inward foreign direct investment: Market versus innovation. *Journal of Cleaner Production, 328*, 129501. https://doi.org/10.1016/j.jclepro.2021.129501

Waqih, M., Bhutto, N., Ghumro, N., Kumar, S., & Salam, M. (2019). Rising environmental degradation and impact of foreign direct investment: An empirical evidence from SAARC region. *Journal of Environmental Management, 243*, 472–480. https://doi.org/10.1016/j.jenvman.2019.05.001

Xu, B. (2000). Multinational enterprises, technology diffusion, and host country productivity growth. *Journal of Development Economics, 62*(2), 477–493. https://doi.org/10.1016/S0304-3878(00)00093-6

Yao, S., & Wei, K. (2007). Economic growth in the presence of FDI: The perspective of newly industrialising economies. *Journal of Comparative Economics, 35*(1), 211–234. https://doi.org/10.1016/j.jce.2006.10.007

Yao, Y., Ivanovski, K., Inekwe, J., & Smyth, R. (2020). Human capital and CO2 emissions in the long run. *Energy Economics, 91*, 104907. https://doi.org/10.1016/j.eneco.2020.104907

You, W.-H., Zhu, H.-M., Yu, K., & Peng, C. (2015). Democracy, financial openness, and global carbon dioxide emissions: heterogeneity across existing emission levels. *World Development, 66*, 189–207. https://doi.org/10.1016/j.worlddev.2014.08.013

Zhu, Q., & Peng, X. (2012). The impacts of population change on carbon emissions in China during 1978–2008. *Environmental Impact Assessment Review, 36*, 1–8. https://doi.org/10.1016/j.eiar.2012.03.003

Zhu, Q., Cordeiro, J., & Sarkis, J. (2012). International and domestic pressures and responses of Chinese firms to greening. *Ecological Economics, 83*, 144–153. https://doi.org/10.1016/j.ecolecon.2012.04.007

Zhu, H., Duan, L., Guo, Y., & Yu, K. (2016). The effects of FDI, economic growth and energy consumption on carbon emissions in ASEAN-5: Evidence from panel quantile regression. *Economic Modelling, 58*, 237–248. https://doi.org/10.1016/j.econmod.2016.05.003

Zugravu-Soilita, N. (2017). How does foreign direct investment affect pollution? Toward a better understanding of the direct and conditional effects. *Environmental and Resource Economics, 66*(2), 293–338. https://doi.org/10.1007/s10640-015-9950-9

João Bento, PhD in Economics, is a senior lecturer in economics at Lisbon Accounting and Business School (ISCAL), Polytechnic Institute of Lisbon (IPL). He is also an external affiliate at the Competitiveness, Innovation, and Sustainability (CIS) research group within the research unit on Governance, Competitiveness, and Public Policies (GOVCOPP) at the University of Aveiro (UA), Portugal. He has more than 20 years of experience in empirical research, academic knowledge, and expertise in the field of economics. He holds a PhD in economics from the University of Reading, UK. He works, and has published, in the areas of international business, tourism, and environmental economics.

Miguel M. Torres, PhD in Economics, is a Lecturer in International Business at Kent Business School (KBS). His career includes consulting services to the European Commission and World Bank, applied research and teaching roles at Leeds, Milano, Aveiro, KU Leuven, and Uppsala universities, with appointments at the University Federal of Goiás, IF Goiano, and the University Catolica of Portugal. He has held multiple visiting professorships and is an INSOL International and Marie Skłodowska-Curie Fellow. He completed executive programs at Harvard, Carnegie Mellon University (USA), and University College London (UK). His research focuses on the intersection of international business, public policy, and strategic management, focusing on multinational management models, corporate political risk, and negotiation strategies, particularly emphasizing the complexities of cross-border restructuring and foreign direct investment drivers using cross-sectional, time-series, and panel data approaches. His work has been published in the *International Business Review, Journal of Business Ethics, Multinational Business Review, Journal of Economic Methodology, Arbitration International, International Insolvency Review, PLOS One, and AIB Insights*. He co-edited the books *Global Perspectives on Climate Change, Inequality, and Multinational Corporations* (Springer, 2025) and *Dead Firms: Causes and Effects of Cross-Border Corporate Insolvency* (Emerald, 2017).

The manufacturer's authorised representative in the EU is Springer Nature Customer Service Centre GmbH, Europaplatz 3, 69115 Heidelberg, Germany. If you have any concerns regarding our products, please contact ProductSafety@springernature.com

Printed and bound by CPI Group (UK) Ltd, Croydon, CR0 4YY
30/09/2026
02225599-0009